D1529063

Temperature Adaptation in a Changing Climate

Nature at Risk

MIX
Paper from responsible sources
FSC® C013604

CABI Climate Change Series

Climate change is a major environmental challenge to the world today, with significant threats to ecosystems, food security, water resources and economic stability overall. In order to understand and research ways to alleviate the effects of climate change, scientists need access to information that not only provides an overview of and background to the field, but also keeps them up to date with the latest research findings.

This series addresses all topics relating to climate change, including strategies to develop sustainable systems that minimize impact on climate and/or mitigate the effects of human activity on climate change. Coverage will therefore encompass all areas of environmental and agricultural sciences as well as human health and tourism. Aimed at researchers, upper level students and policy makers, titles in the series provide international coverage of topics related to climate change, including both a synthesis of facts and discussions of future research perspectives and possible solutions.

Titles Available

1. Climate Change and Crop Production
   Edited by Matthew P. Reynolds
2. Crop Stress Management and Global Climate Change
   Edited by José L. Araus and Gustavo A. Slafer
3. Temperature Adaptation in a Changing Climate: Nature at Risk
   Edited by Kenneth B. Storey and Karen K. Tanino

Temperature Adaptation in a Changing Climate

Nature at Risk

Edited by

Professor Kenneth B. Storey

Carleton University, Canada

and

Professor Karen K. Tanino

University of Saskatchewan, Canada

www.cabi.org

CABI is a trading name of CAB International

CABI
Nosworthy Way
Wallingford
Oxfordshire OX10 8DE
UK

Tel: +44 (0)1491 832111
Fax: +44 (0)1491 833508
E-mail: cabi@cabi.org
Website: www.cabi.org

CABI
875 Massachusetts Avenue
7th Floor
Cambridge, MA 02139
USA

Tel: +1 617 395 4056
Fax: +1 617 354 6875
E-mail: cabi-nao@cabi.org

© CAB International 2012. All rights reserved. No part of this publication may be reproduced in any form or by any means, electronically, mechanically, by photocopying, recording or otherwise, without the prior permission of the copyright owners.

A catalogue record for this book is available from the British Library, London, UK.

Library of Congress Cataloging-in-Publication Data

Temperature adaptation in a changing climate : nature at risk / edited by Kenneth B. Storey and Karen K. Tanino.
p. cm. -- (CABI climate change series ; 3)
Includes bibliographical references and index.
ISBN 978-1-84593-822-2 (alk. paper)
1. Biodiversity--Climatic factors. 2. Adaptation (Biology) I. Storey, K. B. (Kenneth B.) II. Tanino, Karen K. III. Title. IV. Series.

QH541.15.B56T40 2012
577.2'2--dc23

2011026523

ISBN-13: 978 1 84593 822 2

Commissioning editor: Rachel Cutts
Production editor: Simon Hill

Typeset by Columns Design XML Ltd, Reading.
Printed and bound by CPI Group (UK) Ltd, Croydon, CR0 4YY.

Contents

Contributors

Doris Abele, Department of Functional Ecology, Alfred-Wegener Institute of Polar and Marine Research, Am Handelshaven 12, 27570 Bremerhaven, Germany. Email: doris.abele@awi.de

Marilyn C. Ball, Plant Science Division, Research School of Biology, Australian National University, Canberra, ACT 0200, Australia. Email: marilyn.ball@anu.edu.au

Annick Bertrand, Soils and Crops Research and Development Centre, Agriculture and Agri-Food Canada, 2560 Hochelaga Blvd., Québec City, Québec G1V 2J3, Canada. Email: Annick.bertrand@agr.gc.ca

Michael Castellini, School of Fisheries and Ocean Sciences, University of Alaska Fairbanks, Fairbanks, AK 99775-7220, USA. Email: macastellini@alaska.edu

Steven L. Chown, Centre for Invasion Biology, Department of Botany and Zoology, Stellenbosch University, Private Bag X1, Matieland 7602, South Africa. Email: slchown@sun.ac.za

Richard A. Cunjak, Faculty of Forestry & Environmental Management, University of New Brunswick, Fredericton, New Brunswick E3B 5A3, Canada. Email: cunjak@unb.ca

Jenna M. Dolhi, Department of Microbiology, Miami University, Oxford, Ohio 45056, USA. Email: dolhijm@muohio.edu

J.J.G. Egerton, Plant Science Division, Research School of Biology, Australian National University, Canberra, Australian Capital Territory 0200, Australia. Email: jack.egerton@anu.edu.au

Craig L. Frank, Department of Biological Sciences, Fordham University, Louis Calder Center, P.O. Box 887, Armonk, New York 10504, USA. Email: frank@fordham.edu

Denis A. Gaudet, Agriculture and Agri-Food Canada Research Centre, 5403 1st Avenue South, Lethbridge, Alberta T1J 4B1, Canada. Email: denis.gaudet@agr.gc.ca

Daniel Harris-Pascal, Plant Science Division, Research School of Biology, Australian National University, Canberra, Australian Capital Territory 0200, Australia. Email: daniel.harris.pascal@gmail.com

Sarah Kendall, Department of Biology, University of York, Heslington, York Y010 5DD, UK. Email: slk505@york.ac.uk

Klára Kosová, Department of Genetics and Plant Breeding, Crop Research Institute, Drnovská Street 507, 161 06 Prague 6 – Ruzyně, Czech Republic. Email: kosova@vurv.cz

André Laroche, Agriculture and Agri-Food Canada Research Centre, 5403 1st Avenue South, Lethbridge, Alberta T1J 4B1, Canada. Email: andre.laroche@agr.gc.ca

Y.K. Lee, Department of Plant and Environmental Sciences, Norwegian University of Life Sciences, P.O. Box 5003, N-1432 Ås, Norway. Email: yeonkyeong.lee@umb.no

Thomas Lenné, Plant Science Division, Research School of Biology, Australian National University, Canberra, Australian Capital Territory 0200, Australia. Email: thomas.lenne@anu.edu.au

Tommi Linnansaari, Faculty of Forestry & Environmental Management, University of New Brunswick, Fredericton, New Brunswick E3B 5A3, Canada. Email: tommi.linnansaari@unb.ca

Dana MacGregor, Department of Biology, University of York, Heslington, York YO10 5DD, UK. Email: dm561@york.ac.uk

Rachael M. Morgan-Kiss, Department of Microbiology, Miami University, Oxford, Ohio 45056, USA. Email: morganr2@muohio.edu

D. Neilsen, Pacific Agri-Food Research Centre, Agriculture and Agri-Food Canada, Summerland, British Columbia V0H 1Z0, Canada. Email: denise.neilsen@agr.gc.ca

J.E. Olsen, Department of Plant and Environmental Sciences, Norwegian University of Life Sciences, P.O. Box 5003, N-1432 Ås, Norway. Email: jorunn.olsen@umb.no

Steven Penfield, Department of Biology, University of York, Heslington, York YO10 5DD, UK. Email: steven.penfield@york.ac.uk

Ilja Tom Prášil, Department of Genetics and Plant Breeding, Crop Research Institute, Drnovská Street 507, 161 06 Prague 6 – Ruzyně, Czech Republic. Email: prasil@vurv.cz

H.A. Quamme, Pacific Agri-Food Research Centre, Agriculture and Agri-Food Canada, Summerland, British Columbia V0H 1Z0, Canada. Email: hquamme@vip.net

Janet M. Storey, Institute of Biochemistry, Carleton University, 1125 Colonel By Drive, Ottawa, Ontario K1S 5B6, Canada. Email: jan_storey@carleton.ca

Kenneth B. Storey, Institute of Biochemistry, Carleton University, 1125 Colonel By Drive, Ottawa, Ontario K1S 5B6, Canada. Email: kenneth_storey@carleton.ca

Karen K. Tanino, Department of Plant Sciences, University of Saskatchewan, 51 Campus Drive, Saskatoon, Saskatchewan S7N 5A8, Canada. Email: karen.tanino@usask.ca

Anne-Marte Tronsmo, Institute of Plant and Environmental Sciences, Agricultural University of Norway, PO Box 5003, 1432 Ås, Norway. Email: anne-marte.tronsmo@umb.no

1 Introduction: Nature at Risk

Kenneth B. Storey and Karen K. Tanino

All living organisms on Earth are slaves to environmental temperature change. Temperature rules the fundamental biochemistry of cellular metabolism – temperature change alters the rates of all metabolic reactions and changes the conformations of all macromolecules. Temperature also constrains the lives of most organisms – for example, temperature change affects all aquatic organisms by altering the solubility of oxygen in water, tissue freezing at temperatures below 0°C is lethal to the vast majority of organisms, and sustained high temperatures above about 50°C are also lethal to most. Organisms can use biochemical, physiological or behavioural adaptations to help them cope with temperature variation, within limits, and these have been explored in depth by hundreds of researchers (Hochachka and Somero, 1984; Willmer *et al.*, 2005; Rao *et al.*, 2006). Most organisms on Earth are ectotherms, having limited or no control over their temperature. We humans are 'luckier' than most in being among the small percentage of species on Earth that are endotherms, having internal control over body temperature. Furthermore, we have the capacity to regulate our thermal environment closely by building indoor living spaces that are heated or cooled as necessary. However, the consumption of fossil fuels for this purpose, as well as for many other industrial uses by an ever-growing human population, is 'coming back to haunt us' in a growing, potentially disastrous, man-made change to the Earth's climate. The Earth is warming up, bringing with it not just altered thermal conditions but a host of intertwined effects on micro- and macro-environments that will impact the lives of all organisms on the planet.

In this book we are concerned with the impact of global climate change on non-human life on Earth. We already know that spring is arriving earlier, about 2.3–5.2 days/decade, affecting events such as the arrival of migrants, awakening from hibernation, beginning of breeding/nesting or the start of bud burst/first flowering. Multiple negative consequences can occur. For example, in 2007, US$2 billion in frost damage occurred in a single weekend due to advanced spring bloom and growth of crops (Gu *et al.*, 2008). In particular, we focus here on the effects of global warming on species that currently depend on cool or cold conditions for survival. How will they be affected and how will they cope? We offer multiple analyses across microbial, plant, and animal systems to investigate 'Temperature Adaptation in a Changing Climate'.

According to the 2007 Fourth Assessment Report by the Intergovernmental Panel on Climate Change (IPCC, 2007), global surface temperature increased by 0.74 ± 0.18°C during the 20th century. Subsequently, January 2000 to December 2009 was the warmest decade on record, bringing the average rise in global temperature to about 0.8°C since 1880.[1] Fuelling the rise in temperature, greenhouse gas emissions have been on an exponential curve upwards for about the past 150 years since widespread industrialization began to consume massive amounts of fossil fuels (oil, gas, coal, wood). Glacier and sea ice are decreasing at a rapid pace all over the world. Global action by governments and citizens will not stop the current direction of climate change, but concerted action may be able to minimize the magnitude of the change, although this is not yet happening despite much talk. The summary of the Copenhagen accord in December 2009 (15th Conference of Parties

© CAB International 2012. *Temperature Adaptation in a Changing Climate: Nature at Risk* (eds K.B. Storey and K.K. Tanino)

to the United Nations Framework Convention on Climate Change; UNFCCC, 2010) wrote of the need to curb emissions and take other steps to hold the rise in mean global temperature to less than 2°C warmer than in pre-industrial times. However, less than two years later, this 2°C is beginning to be accepted as the minimum change that will occur this century, and various estimates now suggest that a 4°C increase may be unstoppable (New *et al.*, 2011). Few studies have analysed or modelled the effects of a 4°C temperature rise, nor the difficulties or cutbacks that would be needed if countries were to strive to limit warming to just 2°C. The world needs to look harder at how to live with global warming: not just at the consequences for man, but at how all organisms and ecosystems will be affected, and how they will adjust/adapt to the climate change that is now inevitable.

The history of our planet is, in fact, a history of climate change ranging from conditions that were much hotter to much colder than in the present day. For example, a 5°C–7°C greenhouse warming occurred during the Eocene epoch (beginning about 55 million years ago) (Kerr, 2011). This resulted in ice-free north and south poles, major changes in land and sea flora and fauna, and atmospheric CO_2 levels that may have been 8–16 times higher than in the pre-industrial times of the modern era. Fossil evidence shows that temperate forests grew on high Arctic islands that were also populated by animals (turtles, alligators, tapirs, primates) that today occur only at much lower latitudes (Eberle *et al.*, 2010). In contrast, during the last ice age (about 20,000 years ago) Canada's capital city, Ottawa, was under about 2 km of glacier ice.[2] Most of this climate change is due to variation in the amount of solar radiation received by our planet. This in turn is caused by cyclic changes in three characteristics of Earth's orbit – eccentricity, precession and obliquity – that vary over cycles of about 100,000, 26,000, and 41,000 years, respectively. The greatest cooling and warming of the Earth occurs when the troughs and peaks of all three cycles coincide (the Milankovitch theory) (Pidwirny, 2006). Other significant contributors to climate change include variations in solar output, volcanic eruptions, and changes in atmospheric greenhouse gas levels (Pidwirny, 2006). The latter are often linked with the temperature changes caused by Milankovitch oscillations although can also be independent, as in our current global situation.

Throughout continual cycles of global climate change, life has persisted on Earth. It has obviously shifted many times in complexity, diversity, abundance and geographical distribution, including passing through five great mass extinction events (Veron, 2008) and countless smaller ones. Our climate is once again changing after a relatively stable period over the last millennium, but this time the impetus for rapid change is unique, coming primarily from the activities of man. Humans are influencing climate, ecosystems, resources, and even the geography of the Earth as never before. We pump greenhouse gases into the atmosphere (resulting also in acidification of the oceans), destroy forests, convert land to other uses, deplete freshwater from rivers/lakes and aquifers to irrigate crops, and modify shorelines, wetlands, and river flow for our use, among other changes. The consequences of a warming climate – be it natural and/or driven by human action – are huge. They include melting of polar ice caps, thawing of permafrost, a rise in sea level, shifting ocean currents, ocean acidification, altered weather patterns over continents (especially precipitation patterns), changes in the distribution of arable land, and changes in the habitats available for many plant and animal species including man (Myers and Bernstein, 2011).

A huge range of changes can be envisioned for humans as the climate warms. On the positive side, some governments and corporations (mainly of northern countries) are excited about the prospects for commercial shipping through a year-round ice-free northern passage, and for longer seasons for mining operations or offshore oil drilling in the north. If the risk of frost damage can be addressed, agriculture and forestry in some countries will benefit from longer growing seasons, an ability to grow different and/or more profitable crops, or extensions of plant hardiness zones to higher latitudes or altitudes. However, on the negative side, there could be many devastating consequences. For example, in contrast to the potential positives for agriculture/forestry, the generalization that all trees and shrubs will simply extend their growing season into the autumn does not appear to be correct (Tanino *et al.*, 2010). More seriously, as sea level rises, some small island states (for example Tuvalu) may disappear entirely into the sea or become uninhabitable due to saltwater contamination of groundwater. Ocean currents will be disrupted, affecting weather patterns, frequency and severity of storms, displacing fish stocks by many degrees of latitude, and altering environmental conditions of coastal communities for humans, native wildlife, and agriculture.

Changing weather patterns over land will alter the amount of arable land (including enhanced desertification), melt permafrost, continue or enhance the rates of glacier melting, and reduce groundwater reserves. Groundwater will be particularly affected in areas where increased irrigation is needed to counteract desertification. Global warming will also bring major shifts in the patterns of endemic infectious diseases affecting humans, livestock and wildlife (Myers and Bernstein, 2011) and will allow range expansion by various agricultural pest species that are currently held in check by their inability to endure cold winter temperatures.

Furthermore, even if we manage to regulate our human-generated greenhouse gas emissions, global warming may unleash a huge carbon load from other sources that we cannot control (Kerr, 2011). Warming the ocean depths will melt vast reserves of 'frozen' methane hydrates on the sea floor, releasing methane that can be oxidized to CO_2. Melting permafrost will lead to decomposition of huge amounts of organic matter, again releasing CO_2. Indeed, both of these are plausible explanations for the rapid warming that occurred during the Eocene, possibly over as little as 1000 years. Furthermore, Veron (2008) argues that changes in the carbon cycle, particularly those affecting ocean chemistry, are the most plausible explanation for previous mass extinctions on Earth and that the prospect of renewed ocean acidification as a result of increased anthropogenic CO_2 accumulation is potentially the most serious consequence of global warming.

The scope of recent and predicted changes in global climate are so large that some scholars are beginning to call our current era (beginning about 200 years ago) the Anthropocene, an era in which 'changes to the Earth's climate, land, oceans and biosphere are now so great and so rapid that the concept of a new geological epoch defined by the action of humans' is being seriously considered (Zalasiewicz *et al.*, 2011). There is little doubt that life on Earth will persist through this novel era but we will undoubtedly face major changes to the complexity, diversity, abundance and geographical distribution of organisms. Despite being the primary cause of the climate change and environmental problems that we are now experiencing (via greenhouse gas emissions, deforestation, groundwater depletion, pollution and many other factors), humans are heavily invested in wanting to maintain that status quo. We wish to avoid the potential effects of changing climatic patterns such as the displacement of human populations by rising sea levels or dramatic changes to economies when agricultural zones change. However, this is likely to be impossible; all organisms will have to adapt to changing conditions, and humans will not only have to adapt but also accept responsibility for causing dramatic damage to the Earth's biosphere.

Current and upcoming changes to Earth's climate may be particularly hard on organisms, including humans, which are adapted to and depend upon cold climates. Inuit activist, Sheila Watt-Cloutier, long associated with the International Circumpolar Council, has emphasized the human face of the impacts of global climate change in the Arctic and their 'right to be cold' (Gregoire, 2008). She frames the climate change debate around the issue of indigenous self-determination. Traditional hunting cultures of circumpolar peoples may become impossible to maintain due to the loss of sea ice (the platform for seal and whale hunting) and melting of the permafrost that will affect both land-based hunting and the stability of infrastructure in northern communities. Subsistence hunting will also be affected by the effects of climate change on the target animal species. Loss of sea ice, for example, will change breeding grounds for seals, a year-round ice-free northwest passage could allow killer whales to move in and threaten Arctic beluga and narwhal species, and increased ocean-going traffic will drive animals away from shipping lanes.

Do other organisms on Earth also have a 'right to be cold'? Or is the best that we can hope that most species can adapt and endure in the face of accelerating climate change? In the current book, we look at the consequences of climate change from one perspective – that of the effects of a warming climate on organisms that depend on the cold. How do animals and plants that inhabit constant cold or seasonally cold environments survive and prosper? How will they deal with the effects of rising environmental temperatures and changing weather patterns brought on by climate change? The effects of global warming will undoubtedly be dramatic and wide-ranging, affecting organisms in many different ways. For example, plants and animals with distinct seasonality typically register both photoperiod and thermoperiod cues in triggering seasonal cold-hardiness, spring growth, the number of summer generations, and so on. As temperatures increase, 'mixed messages' will be received, leading to a potential loss of synchrony in various life-stage events (Bale and Hayward, 2010). Reduced snow cover in various environments will jeopardize both agricultural and wild species that rely on a thick winter snowpack that provides not only insulation against extreme low temperatures but also buffers freeze/thaw events and reduces the risk of ice encasement (Bale and Hayward, 2010). Disjoints in seasonal timing are developing between low altitudes (where spring warming is occurring sooner) and high altitudes (where melting of the snowpack is unchanged or later), and this will affect survival and breeding success of migratory species that reproduce at high altitudes during the summer (Inouye *et al.*, 2000). Montane species will also be pushed upwards (and potentially into extinction) as temperatures warm, because of competition from lowland species that also move upwards. In polar regions many species that depend on sea ice will be impacted. These include polar bears, receiving much attention as a 'poster child' for the risk from climate change, even though there is conflicting evidence of changes in polar bear populations (MacDonald, 2011). Changes in the timing and extent of sea ice lead to temporal asynchronies and spatial separations between energy requirements and food availability, and these affect multiple trophic levels and alter reproductive success, abundances, and distribution of species (Moline *et al.*, 2008). Furthermore, abrupt climate change will select for species with high vagility and generalism that are able to track their moving habitat, and will select against highly specialized species (Dynesius and Jansson, 2000) meaning that extinction of some species is inevitable.

We hope that readers will be intrigued and stimulated by the range of articles in this volume – covering both plants and animals, terrestrial and aquatic environments – and will come to appreciate, as we have during the editing process, the huge array of interacting consequences of climate change and the challenges for all cold-adapted organisms. We are extremely grateful to the large group of authors that agreed to contribute to this volume and we thank them for their hard work and commitment to producing such interesting and timely chapters. We also thank our Content Editor Sarah Mellor, Associate Editor Rachel Cutts and Editorial Assistant Alexandra Lainsbury at CABI who kept us on track and nearly on time throughout the project. Thanks also to Jan Storey for her 'eagle-eyed' help during editing of the animal chapters.

Notes

[1] http://earthobservatory.nasa.gov/IOTD/view.php?id=48574, accessed 31 May 2011.
[2] http://geoscape.nrcan.gc.ca/ottawa/index_e.php, accessed 31 May 2011.

References

Bale, J.S. and Hayward, S.A. (2010) Insect overwintering in a changing climate. *Journal of Experimental Biology* 213, 980–994.

Dynesius, M. and Jansson, R. (2000) Evolutionary consequences of changes in species' geographical distributions driven by Milankovitch climate oscillations. *Proceedings of the National Academy of Sciences USA* 97, 9115–9120.

Eberle, J.J., Fricke, H.C., Humphrey, J.D., Hackett, L., Newbrey, M.G. and Hutchinson, J.H. (2010) Seasonal variability in Arctic temperatures during early Eocene time. *Earth and Planetary Science Letters* 296, 481–486.

Gregoire, L. (2008) Cold warriors. *Canadian Geographic* 128, 35–50.

Gu, L., Hanson, P.J., Mac Post, W., Kaiser, D.P., Yang, B., Nemani, R., Pallardy, S.G., *et al.* (2008) The 2007 eastern U.S. spring freeze: Increased cold damage in a warming world? *BioScience* 58, 253–262.

Hochachka, P.W. and Somero G.N. (1984) *Biochemical adaptation.* Princeton University Press, Princeton, New Jersey.

Inouye, D.W., Barr, B., Armitage, K.B. and Inouye, B.D. (2000) Climate change is affecting altitudinal migrants and hibernating species. *Proceedings of the National Academy of Sciences USA* 97, 1630–1633.

IPCC (2007) Climate Change 2007. Working Group I: The Physical Science Basis, www.ipcc.ch/publications_and_data/ar4/wg1/en/faq-3-1.html, accessed 31 May 2011.

Kerr, R.A. (2011) What heated up the Eocene? *Science* 331, 142–143.

MacDonald, J. (2011) Polar opposites. *Canadian Wildlife* 16, 19–24.

Moline, M.A., Karnovsky, N.J., Brown, Z., Divoky, G.J., Frazer, T.K., Jacoby, C.A., Torres, J.J., *et al.* (2008) High latitude changes in ice dynamics and their impact on polar marine ecosystems. *Annals of the New York Academy of Science* 1134, 267–319.

Myers, S.S. and Bernstein, A. (2011) The coming health crisis. *The Scientist* 25, 32–37.

New, M., Liverman, D., Schroeder, H. and Anderson, K. (2011) Four degrees and beyond: the potential for a global temperature increase of four degrees and its implications. *Philosophical Transactions of the Royal Society A* 369, 6–19.

Pidwirny, M. (2006) *Fundamentals of Physical Geography.* 2nd ed., www.physicalgeography.net/fundamentals/7y.html, accessed 31 May 2011.

Rao, K.V.M., Raghavendra, A.S. and Reddy, K.J. (2006) *Physiology and Molecular Biology of Stress Tolerance in Plants.* Springer, Dordrecht.

Tanino, K.K., Kalcsits, L., Silim, S., Kendall, E. and Gray, G. (2010) Temperature-driven plasticity in growth cessation and dormancy development in deciduous woody plants: a working hypothesis suggesting how molecular and cellular function is affected by temperature during dormancy induction. *Plant Molecular Biology* 73, 49–65.

UNFCCC (2010) Copenhagen Accord 2010. http://unfccc.int/resource/docs/2009/cop15/eng/l07.pdf, accessed 31 May 2011.

Veron, J.E.N. (2008) Mass extinctions and ocean acidification: biological constraints on geological dilemmas. *Coral Reefs* 27, 459–472.

Willmer, P., Stone, G. and Johnston, I.A. (2005) *Environmental physiology of animals.* 2nd ed., Blackwell Science Ltd, Oxford.

Zalasiewicz, J., Williams, M., Haywood, A. and Ellis, M. (2011) The Anthropocene: a new epoch of geological time? *Philosophical Transactions of the Royal Society A* 369, 835–841.

2 Temperature Perception and Signal Transduction – Mechanisms Across Multiple Organisms

Steven Penfield, Sarah Kendall and Dana MacGregor

2.1 Introduction

Understanding the temperature of the environment gives adaptive clues central to the life histories of organisms at all trophic levels. In temperate regions, plants and animals use temperature to chart the passing of the seasons and to synchronize their lifecycle to the earth's rotation about the sun. For pathogenic organisms, temperature changes can be a clear signal that they have passed to or from a homeothermic host environment. As humans we are familiar with devices that allow us to measure temperature with great precision, and in this chapter we will take a broad view of temperature sensing across organisms, and discover how evolution has adopted the properties of different molecules to construct biological thermometers. A central theme is that organisms have evolved to exploit the biophysical properties of many molecules to create temperature sensors, and have coupled these to common signal transduction pathways to enable them to elicit responses. This chapter reviews the current state of knowledge of how temperature is measured by all manner of life forms. Some mechanisms turn out to be highly specialized in particular phyla, while some, presumably more ancient, are apparently widely conserved across species.

2.2 Temperature Perception in Animals Through the Gating of Channels

Many organisms, whether bacteria, humans, or plants, have evolved the ability to perceive a change in temperature. This ability allows them to physically avoid adverse environments, and to use temperature as a clue to inform future actions. The gating of particular channels, such as Ca^{2+} permeable channels or cation channels, is a common mechanism that is utilized by organisms to sense temperature. In this section we will discuss the role of such channels in temperature perception.

2.2.1 Sensing by thermo-transient receptor potential channels

Thermo-transient receptor potential (thermoTRP) channels have been proposed to act as temperature sensors across many species. These channels are located in the sensory neurons which connect the skin to the central nervous system (Insler, 2006). Although a large number of ion channels are regulated by temperature, thermoTRP channels are distinct in that they can be activated by temperature in the absence of other signals (Patapoutian *et al.*, 2003). These receptors can operate over a specific

© CAB International 2012. *Temperature Adaptation in a Changing Climate: Nature at Risk* (eds K.B. Storey and K.K. Tanino)

temperature range, and thus provide a potential molecular basis for thermosensation (Schepers and Ringkamp, 2010). All thermoTRP channels are putative six-transmembrane (6TM) polypeptide subunits which assemble as tetramers, forming cation-permeable pores (Clapham, 2003). Additionally, these channels are Ca^{2+}-influx channels and are important in contributing to changes in cytosolic Ca^{2+} concentrations (Pedersen *et al.*, 2005). Some thermoTRP channels are activated by heat and others by cold. These signals may be innocuous, with no injury being caused; or noxious, when signals are sensed as being painful and able to cause injury.

2.2.2 Vertebrate thermoTRP channels

In vertebrates, six thermoTRP channels, belonging to three TRP groups, are involved in temperature perception (Fig. 2.1). The first thermoTRP channel to be identified was TRPV1, which is activated by noxious high temperatures and capsaicin, the active component of chilli peppers (Caterina *et al.*, 1997). In TRPV1-null mice noxious heat-induced currents are reduced in cultured sensory neurons (Caterina *et al.*, 2000). TRPV1-null mice display restricted behavioural responses to noxious thermal stimuli, thus proving a role for TRPV1 in normal thermal nociception (Caterina *et al.*, 2000). Three other TRPV channels (TRPV2–4) have been proposed to be involved in sensing heat; however, in contrast to TRPV3 and TRPV4, knock-out mice have not been isolated for TRPV2, and so the role of TRPV2 in noxious heat sensing is yet to be fully evaluated (Moqrich *et al.*, 2005; Todaka *et al.*, 2004; Lee *et al.*, 2005a; Dhaka *et al.*, 2006). In vertebrates the related TRPM8 calcium channel is responsible for sensing innocuous cold and is also activated by chemical agonists such as menthol, the action of which can mimic cool temperatures (McKemy *et al.*, 2002; Peier *et al.*, 2002). Noxious cold is sensed by TRPA1, which is activated at temperatures below 16°C (Story *et al.*, 2003; Zurborg *et al.*, 2007). TRPA1-null mice have demonstrated reduced and normal sensitivity to cold temperatures on different occasions, and so a more thorough characterization is required to determine a clear role for this channel as a thermosensor of noxious cold (Kwan *et al.*, 2006; Bautista *et al.*, 2006; 2007). More recently it has been shown *in vitro* that recombinant TRPA1 is cold sensitive (Karashima *et al.*, 2009).

2.2.3 Invertebrate thermoTRP channels

TRP channels are also involved in *Drosophila* temperature perception, suggesting that there is a high degree of similarity between the molecular mechanisms regulating thermosensation in mammals and insects (McKemy, 2007). Pathways coupled to phospholipase C are involved in activating all thermoTRP channels in both invertebrates and vertebrates (Venkatachalam and

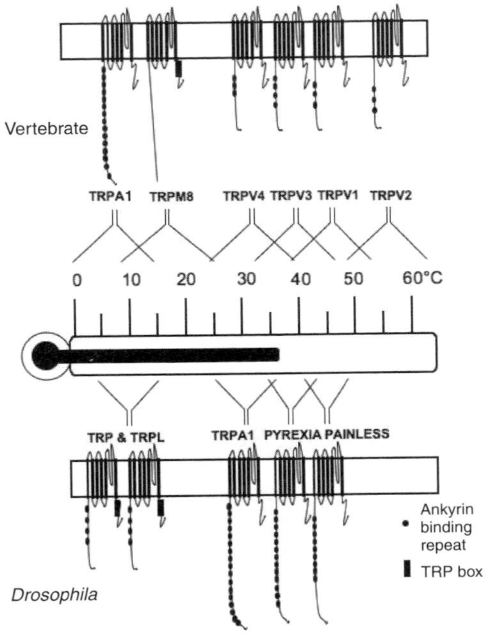

Fig. 2.1. Schematic representation of the thermoTRP channels from vertebrate and *Drosophila*, which are involved in sensing temperatures ranging from noxious cold to noxious heat. The channels include six putative transmembrane units with a pore region between domains five and six. Ankyrin repeats on the amino termini and TRP boxes on the carboxy termini are represented.

Montell, 2007). However, some mechanisms regulating activation of thermoTRP channels do differ between invertebrates and vertebrates. For example, intracellular Ca^{2+} activates mammalian TRPs such as TRPA1 (Zurborg *et al.*, 2007), whereas Ca^{2+} is required by the *Drosophila* TRP PAINLESS as a co-agonist for activation (Sokabe *et al.*, 2008). Until recently, thermoTRP channels involved in *Drosophila* thermosensation were restricted to the TRPA subfamily, but a role for TRPC channels has now been suggested (Rosenzweig *et al.*, 2008). Three members of the TRPA subfamily, PAINLESS, PYREXIA and TRPA1 are involved in perceiving heat, whereas two members of the TRPC subfamily, TRP and TRPL, are involved in perceiving cold (Tracey *et al.*, 2003; Lee *et al.*, 2005b; Viswanath *et al.*, 2003; Rosenzweig *et al.*, 2008) (Fig. 2.1). Whether or not these TRPC channels respond to temperature directly still remains to be determined. Additionally, recent findings have revealed a role for thermoTRP channels in thermosensation by *Caenorhabditis elegans*. TRPA1 is required for cold sensation by the multidendritic PVP neurons, such that the loss of TRPA1 leads to absence of cold-induced calcium transients and escape behaviour (Chatzigeorgiou *et al.*, 2010).

2.2.4 Temperature sensing in plants and the role of calcium channels

Plants are poikilothermic and are exposed to a large range of temperatures, suggesting that they are unlikely to have narrow threshold limits for sensing temperature such as those of the thermoTRP channels (Sung, 2003). Although 19 TRP orthologues have been identified in the green alga *Chlamydomonas reinhardtii*, no obvious orthologues have been found in land plants, suggesting that TRP genes have been lost in land plants during evolution (Wheeler and Brownlee, 2008; Saidi *et al.*, 2009). However, there is some evidence for an important role of Ca^{2+} in temperature responses in plants. Cytosolic Ca^{2+} is affected by increases or decreases in temperature (Knight *et al.*, 1996; Larkindale and Knight, 2002). Tobacco plants overexpressing *Arabidopsis CAX1* {*CALCIUM EXCHANGER 1*} are Ca^{2+} deficient and show increased cold shock sensitivity (Hirschi, 1999). Regulation of the heat shock response (HSR) in the moss *Physcomitrella patens* has been shown to require membrane-regulated entry of extracellular Ca^{2+} ions, and it was found that when external Ca^{2+} was artificially chelated, the heat induction of genes encoding heat shock proteins (HSP) was absent (Saidi *et al.*, 2009). This Ca^{2+} transient is generated by a temperature-sensitive ion channel in the plasma membrane. In lucerne, addition of a calcium channel blocker inhibits the influx of extracellular Ca^{2+} and expression of *CAS* {*COLD ACCLIMATION-SPECIFIC*} genes at 4°C (Monroy and Dhindsa, 1995). Together this evidence suggests that the temperature sensor in plants could be a Ca^{2+}-permeable channel (Monroy and Dhindsa, 1995; Plieth *et al.*, 1999; Saidi *et al.*, 2009). However, genetic evidence for a role for any ion channel in temperature sensing in plants is still absent.

2.3 Perception of Temperature by Microorganisms Uses Diverse Molecular Mechanisms

To initiate a molecular response to a change in temperature, the cell must first perceive that an environmental change has occurred. To function as a thermosensor, it is important that the candidate has a very rapid response to changes in temperature. At the level of a single cell, three obvious candidates with these properties are (i) the lipid bilayer of the membrane; (ii) nucleic acids; and (iii) proteins. Each of these may undergo temperature-dependent physical changes in the phase or conformation of the structure that can dramatically alter their ability to function. As such, it is unsurprising that these three candidates have been extensively studied in their ability to sense a temperature change and respond through various molecular signalling events which result in thermal adaptation.

2.3.1 Thermal adaptation initiated by changes in membrane biophysics

Membranes are exquisitely responsive to temperature. Even relatively small changes of temperature during growth result in changes in phase behaviour and physical properties of the lipids in the membrane (Hazel, 1990,1995). For instance, many bacteria are able to perceive and respond to a 6°C decrease in growth temperature, which in context, corresponds to only a 2% decrease in the molecular motion of the lipids (Murata and Los, 1997; Los and Murata, 1999). Sudden changes in environmental temperature can lead to drastic alterations in membrane fluidity, which can dramatically alter normal membrane functions (Quinn, 1989; Hazel, 1990,1995; Rodgers and Glaser, 1993). Some of the membrane functions altered by temperature include membrane curvature; binding and transport of molecular and ionic species; and the insertion, folding, and function of membrane proteins.

Bacillus subtilis is a single-celled saprophyte that is naturally found in a variety of ecological niches where it is exposed to rapidly fluctuating temperatures. To maintain membrane homeostasis in response to these temperature changes, it uses a two-component regulatory gene pair to control expression of the only desaturase in the genome (Fig. 2.2). DesK is bifunctional histidine kinase-phosphatase and is partnered with a soluble response regulator, DesR, which controls transcription of the *des* gene and therefore production of Δ5-Des (Aguilar *et al.*, 1998; Altabe *et al.*, 2003). The N-terminus of DesK contains a regulatory sunken-buoy (SB) motif, composed of hydrophilic amino acids, that is positioned at the lipid/water interface (Cybulski *et al.*, 2002). The position of the SB motif relative to the membrane bilayer is determined by temperature-dependent alterations to membrane thickness, and its positioning determines whether DesK is stabilized as a phosphatase or as a kinase (Cybulski *et al.*, 2010). During membrane homeostasis, the SB motif is in the aqueous phase, and DesK dephosphorylates DesR, rendering it inactive (Fig. 2.2A). When the membrane is thickened in response to a downward shift in growth temperature (Grau and Mendoza, 1993; Aguilar *et al.*, 1999), or to a restriction in membrane fluidity (Cybulski *et al.*, 2002), the SB motif is buried in the lipid bilayer and DesK changes to a kinase phosphorylating DesR to activate it (Fig. 2.2B) (Cybulski *et al.*, 2010). Only the dephosphorylated form of DesR can activate Des transcription. Together, this two-component regulatory system and the desaturase monitor and maintain membrane homeostasis in response to temperature-dependent changes in membrane thickness.

2.3.2 Temperature-dependent gene expression based on changes in higher-order structures of nucleic acids and proteins

Nucleic and amino acids have the ability to form inter- and intramolecular interactions to create higher-order structures. The formation of these structures is dependent not only on the primary sequence, but also on the strength of the noncovalent bonds that hold the structures together. Because noncovalent bond strength is dependent on temperature, formation of higher-order structures is intrinsically dependent on temperature. Nature has taken advantage of this fact and used the different conformations adopted by DNA, RNA and proteins to coordinate cellular events with environmental temperature (Fig. 2.3). In this section, we will describe three examples of how the formation of secondary structures in DNA, RNA and quaternary structures in proteins taken from pathogenic bacteria are used to regulate the expression in response to environmental temperature.

Pathogenic bacteria experience dramatic differences in their environment when within a host compared to when they are living outside the host. In addition to a myriad of other factors, temperature differs quite dramatically in these two environments. Rather than the typically constant 37°C that pathogens experience in their warm-blooded hosts, non-host temperatures

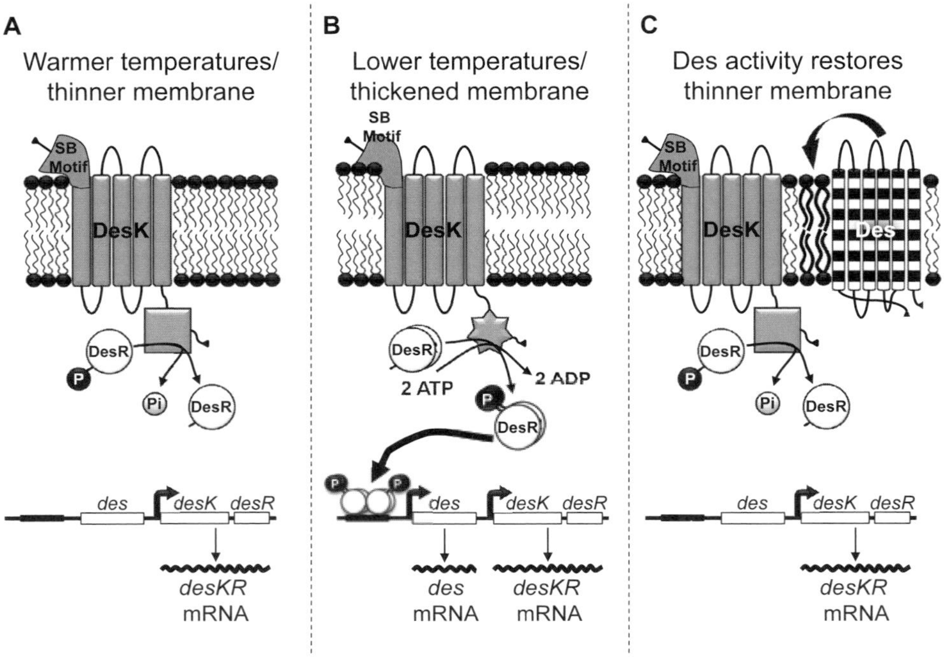

Fig. 2.2. The des pathway in *Bacillus subtilis* monitors and reacts to membrane thickness to control the expression of the sole desaturase. A. DesK (grey structure) is an integral membrane, bifunctional kinase/phosphatase that monitors membrane thickness using the relative position of a Sunken Buoy (SB)-motif to the lipid bilayer (black lipids). When the membrane is relatively thin, the SB-motif is in the aqueous phase and the C-terminus of DesK is stabilized as a phosphatase (grey square). DesK dephosphorylates DesR-P (white circle) rendering it unable to bind to the *des* promoter (black rectangle). B. If there is a thickening of the membrane, the SB-motif becomes buried in the lipid bilayer, destabilizing DesK, and its C-terminus becomes a kinase (grey star). Kinase-active DesK phosphorylates DesR, and homotetramers of DesR-P can promote *des* transcription (white rectangle). C. As Δ5-Des is produced (black and white striped structure), it alters the ratio of unsaturated fatty acids (UFAs, lipids with thick lines) to saturated fatty acids (SFAs, lipids with thin lines), thereby restoring membrane fluidity and homeostasis. The SB-motif is returned to the aqueous phase as the membrane thins, and DesK is stabilized as a DesR-phosphatase (grey square). From Mansilla *et al.* (2004).

fluctuate rapidly but are nevertheless generally lower than 37°C. Pathogenic bacteria have taken advantage of these different thermal signatures to transcribe virulence genes only when at body temperature. Bacterial genomes encode a large number of virulence genes which have evolved specifically to aid the bacteria in colonization and survival within the host; expressing them outside of the host may even confer a selective disadvantage to the bacteria (Maurelli, 1989). Therefore, it is an asset for the bacteria strictly to confine expression of these genes to body temperatures.

2.3.3 DNA topology as a temperature sensor

Prokaryote DNA topology can be affected by temperature, where the unwinding angle changes −0.01 degree per base pair for each °C change (Duguet, 1993). H-NS is a heat-stable nucleoid structuring protein that is conserved among enterobacteria and promotes a DNA conformation that prevents initiation from occurring. H-NS binds to DNA as dimers or higher oligomers at sites that are not conserved at the primary sequence level, but that have a natural curvature (Fig. 2.3A) (Yamada *et al.*,

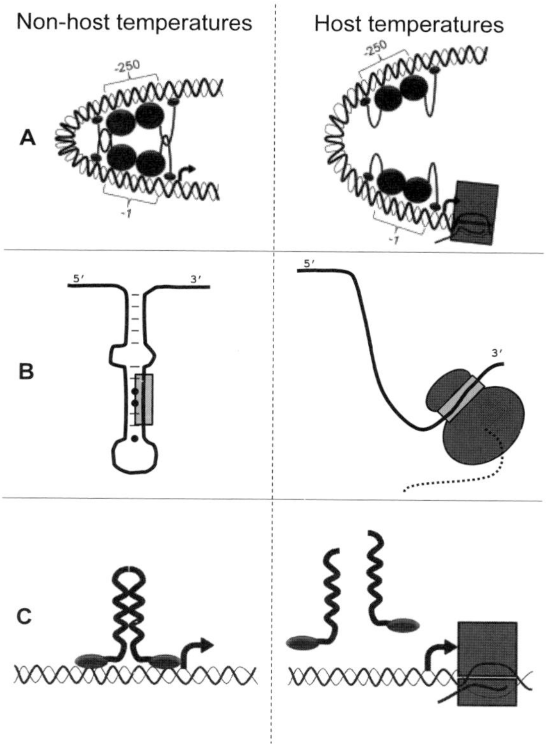

Fig. 2.3. Gene regulation is accomplished by temperature-dependent changes to higher-order structures of DNA, RNA and proteins. A. At non-host temperatures, H-NS, a heat-stable nucleoid structuring protein, stabilizes the topology of its target promoter in a transcriptionally inaccessible form. Both the N-terminal domain (black circle) and C-terminal domain (black oval) of H-NS have nucleic acid-binding activity (Ueguchi, 1997) and are connected by a flexible central linker region (black line). Although the N-terminus of H-NS is needed for dimerization, the central linker region is required for higher-order oligomerization (Bloch *et al*., 2003). At non-host temperatures (T≤28°C) and when both N- and C-termini are bound to DNA (wavy lines), the linker is thought to adopt a more stable structure that allows it to interact with other DNA-bound H-NS dimers to create the transcriptionally silent nucleoprotein structure. At host temperatures, higher-order oligomerization is not sufficient to maintain the closed conformation and RNA polymerase (grey rectangles) can bind and initiate transcription. B. Temperature-dependent RNA secondary structures determine the accessibility of Shine–Dalgarno sequences in *lcrF*. At non-host temperatures, the translation initiation sequence of *lcrF* mRNA (thick black line) forms intramolecular interactions (thin horizontal lines or dots) which stabilize it into a stem loop and sequesters the Shine–Dalgarno sequence (grey box) (Hoe and Goguen, 1993). At higher host temperatures, these intramolecular interactions are unstable and the ribosomes (grey ovals) are free to initiate translation at the Shine–Dalgarno sequence. C. Temperature-dependent dimerization of TlpA is sufficient to regulate its own transcription. TlpA is composed of an N-terminal DNA-binding domain (oval) and a C-terminal α-helical-coiled-coil domain (thick black line). At non-host temperatures, dimeric TlpA binds to its own promoter sequence (wavy lines) to silence transcription. At the host temperature, TlpA exists in a monomeric form that cannot bind DNA, allowing RNA polymerase (grey rectangles) to initiate transcription. A from Dorman *et al*. (2004), B from Hoe and Goguen (1993), C after Eriksson *et al*. (2002).

1990,1991; Koo *et al.*, 1986). The ability of H-NS to modulate transcription – but not its ability to bind DNA – depends on the superhelical state of the DNA, and correlatively, the temperature experienced by the bacteria (Atlung and Ingmer, 1997; Falconi *et al.*, 1998; Prosseda *et al.*, 2004). As such, it is the temperature-dependent DNA topology at these regions, not the binding of H-NS per se, which regulates transcription.

One of the genes H-NS represses encodes for the transcriptional regulator *virF*. Synthesis of *virF* is the primary event occurring after the bacterium *Shigella* shifts from non-host to host temperature, and it promotes the transcription of operons encoding the *Shigella* invasion genes (Tobe *et al.*, 1991,1993; Hale, 1991). *In vitro*, low temperature and H-NS were necessary and sufficient to repress *virF* transcription (Falconi *et al.*, 1998). *virF* repression is initiated by binding of H-NS dimers to two sites centred around –250 and –1 with respect to the transcriptional start site, that are separated by intrinsically curved DNA (Falconi *et al.*, 1998; Prosseda *et al.*, 2004). When the temperature is raised to host body temperatures (>32°C), the local DNA topology at the *virF* promoter undergoes abrupt conformational changes and H-NS bound to promoter regions is insufficient to prevent transcription initiation (Falconi *et al.*, 1998; Prosseda *et al.*, 2004) (Fig. 2.3A). As such, DNA topology, aided and supported by H-NS, is able to limit transcription of the *Shigella* invasion genes until the bacteria have entered the host environment.

2.3.4 RNA as a temperature sensor

Yersinia pestis is historically a very important human pathogen as it causes septicaemic, pneumonic and bubonic plagues. Like other pathogenic bacteria, *Y. pestis* can specifically express plasmid-encoded virulence genes, such as the *yop* genes, at host-body temperature (37°C) but not at non-host temperatures of 26°C (Cornelis *et al.*, 1989b,c; Hoe and Goguen, 1993). Temperature-dependent regulation of the *yop* genes requires the transcription factor encoded for by *lcrF* (Cornelis *et al.*, 1989a; Hoe *et al.*, 1992). Although the transcription of *lcrF* is equal at 25°C and 37°C, the efficiency of translation greatly increases at 37°C (Hoe and Goguen, 1993). This difference in translation is due to the fact that at low temperatures, the translation initiation site of *lcrF* mRNA adopts a secondary structure that sequesters the Shine–Dalgarno sequence, making it inaccessible for translation (Fig. 2.3B) (Hoe and Goguen, 1993). As the cellular temperature increases, the stem-loop structure melts, permitting translation of *lcrF* mRNA (Fig. 2.3B) (Hoe and Goguen, 1993). This mechanism is not specific to *Yersinia* species. The virulence-activating transcription factor *prfA* in *Listeria monocytogenes* also has a thermal regulated stem-loop that is only exposed at host body temperature (Johansson *et al.*, 2002), and bacteriophage λ uses temperature-dependent mRNA conformations to limit the accessibility of the ribosome binding site in the *cIII* gene at low temperatures (Altuvia *et al.*, 1989). Therefore, these mRNAs act as both thermometer and messenger in one molecule to limit the translation of a transcription factor to a particular environment that is recognized by temperature.

2.3.5 Protein as a temperature sensor

Salmonella enterica is a facultative intracellular pathogen that infects human and animal hosts. The 371 amino acid, cytoplasmic protein *TlpA* from *S. enterica* has an N-terminal DNA-binding domain and a C-terminal α-helical coiled-coil domain that can act as a temperature-dependent repressor (Hurme *et al.*, 1996; Gal-Mor *et al.*, 2006). The coiled-coil motif in the C-terminus allows *TlpA* to exist in two concentration- and temperature-dependent states: (i) a non-host temperature (<28°C), oligomeric form that has DNA-binding capacity; and (ii) a high-temperature (>37°C) monomeric form that is unable to bind to and repress expression from the *TlpA* promoter (Fig. 2.3C) (Hurme *et al.*, 1997; Gal-Mor *et al.*, 2006). When the temperature has surpassed the melting midpoint for oligomerization

(43°C), concentration-dependent control is lost because the subunits remain monomeric and inactive, regardless of subunit concentration, and transcription of *TlpA* mRNA continues unchecked (Hurme *et al.*, 1997). To date, the only reported *TlpA*-binding site is in the *TlpA* promoter, and infection of either mouse hosts or tissue culture with a *TlpA* mutant *Salmonella* strain did not attenuate the disease. Although the role that *TlpA* plays in *Salmonella* has yet to be determined, it is well conserved among *S. serovars* and clearly has temperature-dependent protein conformations that are only active over a particular range of temperatures.

2.4 Perception of Temperature by Plants: Chromatin and Transcription

Plants are central to the global carbon cycle with rates of carbon flux through plants dwarfing those from man by a factor of ten. Plant temperature sensing is critical to the understanding of future climates because temperature-dependent changes in plant phenology can modify feedbacks in the carbon cycle and thus affect atmospheric CO_2 levels. A good example of this is a study by Piao *et al.* (2008) who show that warm spring temperatures induce higher annual levels of photosynthesis, while warm autumn temperatures predominantly increase CO_2 release from plants via respiration. These differences arise because warmer temperatures extend the vegetative phase of perennial plants, either by generating an earlier spring, or by delaying senescence during autumn.

In plants membrane fluidity is not generally seen as a good candidate for a temperature-sensing mechanism (Somerville and Browse, 1996). This is despite the fact that many *Arabidopsis* mutants unable to synthesise unsaturated fatty acids have reduced ability to tolerate low temperatures (Miquel *et al.*, 1993; Kunst *et al.*, 1989). Work with these mutants, and with chemical agents that alter membrane fluidity, have failed to find convincing evidence that alteration of membranes has a general effect on the ability of plants to sense temperature. However, in plants, as in lower organisms, temperature can regulate the levels of fatty acid desaturase enzymes, and thus potentially lead to homeostasis in membrane fluidity (Gibson *et al.*, 1994; Dyer, 2001; Matsuda *et al.*, 2005).

The best evidence for a general plant temperature-sensing mechanism comes from a recent study by Kumar and Wigge (2010). These authors show that the regulation of the plant transcriptome by increasing temperatures is dependent on a histone variant known as H2A.Z. Increasing temperatures result in displacement of H2A.Z from the promoters of temperature-responsive genes and this allows their transcriptional activation. As a result, plants deficient in one isoform of H2A.Z known as ARP6 behave as if they were constitutively experiencing warmer temperatures than they really are. This leads to a range of physiological responses consistent with the hypothesis that these plants always behave as if warm adapted, including earlier flowering and increased elongation growth (Kumar and Wigge, 2010). These authors provide evidence that the binding affinity of H2A.Z to DNA is the temperature-sensitive step.

A second hint that chromatin structure is key to temperature sensing in plants comes from analysing the mechanism of vernalization, the process through which flowering is promoted by the experience of winter. Vernalization requires not just that cold is sensed during winter, but that the cumulative experience of cold days is integrated and retained. A large body of evidence shows that the response of MADS box gene expression to low temperatures is central to forming molecular memories of winter. The most important characterized so far is *Arabidopsis* FLOWERING LOCUS C (*FLC*). The initial level of FLC expression is controlled by genetic factors that differ between *Arabidopsis* accessions, but this expression is reduced upon vernalization, with the number of degree days required to silence *FLC* strongly dependent on latitude of origin (Shindo *et al.*, 2006). The mechanism through which temperature controls *FLC* expression is one of the best understood in any eukaryotic organism, and involves the interplay of sense strand transcription with anti-sense transcription, mRNA processing, and chromatin state. The important early step appears to be the low temperature induction of the transcription of a long antisense *FLC* transcript. The processing of the antisense strand appears to induce a chromatin state which silences sense strand expression (Liu *et al.*, 2010). The mechanism through which the *FLC* antisense 'COOLAIR' promoter senses temperature is still unknown, but requires further temperature-sensitive chromatin modifications, and the cold-induced expression of the *VERNALIZATION INDEPENDENT 3* (*VIN3*) gene (Sung and Amasino, 2004), itself an early response to extended cold (Fig. 2.4). There is also evidence that genes related to FLC are important for controlling dormancy in response to temperature and photoperiod. The *evergreen* locus in peach confers an inability to enter winter dormancy, and corresponds to an array of MADS box transcription factors, some of which respond at the level of transcription to prolonged low temperatures in a manner resembling *FLC* (Li *et al.*, 2009; Jiménez *et al.*, 2010).

In *Arabidopsis* one of the most important responses to low temperatures is the induction of *C-REPEAT BINDING FACTOR* (*CBF*) expression. Exposure of *Arabidopsis* to temperatures below around 17°C results in the incremental increases in *CBF* expression around 8 h after dawn, and *CBF* expression increases in proportion to the decrease in temperature. Acute transient exposure to chilling temperatures results in the very rapid but transient induction of high levels of *CBF* expression, and this in turn controls the plant's ability to withstand freezing temperatures. The mechanisms through which cold temperatures induce *CBF* expression have been the subject of much investigation, but are still not entirely clear. In one pathway, the transcriptional activator INDUCER OF CBF EXPRESSION 1 (ICE1) is important for *CBF* transcription, and is controlled at the protein stability level by the cold-responsive E3-ubiquitin ligase HOS1. In another, calmodulin binding transcription activator (CAMTA) factors are activated by cold-induced calcium spikes and bind to the *CBF* promoter (Doherty *et al.*, 2009). The circadian clock is also important for the cold regulation of *CBF* transcription, and defines the time of day at which *CBF* expression is most responsive to temperature (Fowler *et al.*, 2005). *CBF* bind to ABRE-like promoter elements in the promoters of cold-responsive genes, but it has also been shown that evening elements, promoter elements central to the expression of genes in the evening phase of the circadian clock, are also critical for cold-induction of *CBF* target genes (Mikkelsen and Thomashow, 2009). Therefore it is likely that circadian clock proteins which bind evening elements are important for cold-responsive transcription. Interestingly the circadian clock itself is very

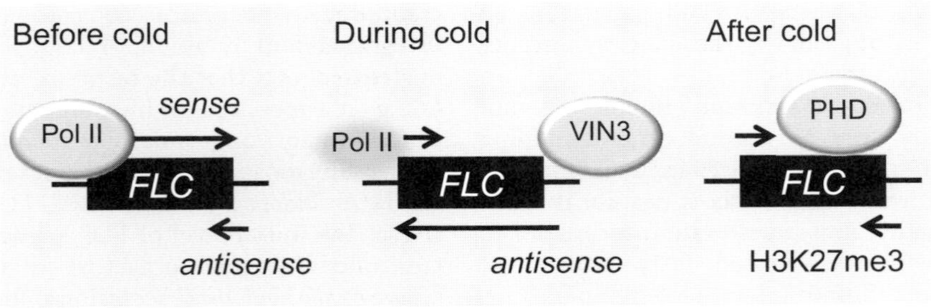

Fig. 2.4. The regulation of *Arabidopsis* FLC expression by cold. After reproduction FLC sense strand is expressed at high levels. Cold induces the expression of the anti-sense strand, the processing of which stalls the polymerase synthesizing the sense strand. Cold-induced VIN3 protein then recruits a polycomb-containing complex which facilitates the stable epigenetic repression of FLC, even after exposure to subsequent warm periods.

responsive to temperature, and it has been widely speculated that the responsiveness of the circadian clock to cold could be used by plants to control seasonal events such as winter dormancy. In poplar, the clock gene *LATE ELONGATED HYPOCOTYL* (*LHY*) is also important for cold hardiness (Ibanez *et al.*, 2010), while disruption of the clock in *Arabidopsis* can create plants that are constitutively cold-adapted, even when they have had no experience of low temperatures (Nakamichi *et al.*, 2009). Thus the action of the circadian clock appears to be intimately associated with low temperature responses in plants.

2.5 Temperature Responses – The Heat Shock Response is Widely Conserved Among Diverse Taxa

The constant exposure of cells and tissues to a variety of acute and chronic stresses has led to the evolution of stress response networks to detect, monitor and respond to environmental changes. In such stressed environments, proteins are adversely affected, and can unfold, misfold, or aggregate (Morimoto, 1998). Through the elevated synthesis of molecular chaperones and proteases that repair protein damage, the HSR is responsible for assisting in the recovery of a cell from such a situation. This response is one of the most evolutionarily-conserved defensive mechanisms against exposure to extreme temperature (Shamovsky and Nudler, 2008). The use of conserved gene families such as the HSP, which are induced by the HSR, allows the investigation of responses to temperature extremes in environmentally important species that are yet to be sequenced. Therefore, understanding the HSR may assist in estimating the effect of climate change, specifically increased temperature, on those species most at risk.

2.5.1 Regulation of the heat shock response by the heat shock factor

The HSR involves the rapid increase in expression of heat shock (HS) genes. These HS genes code for HSP, which act as molecular chaperones involved in protein repair and the prevention of apoptosis. This stress-induced transcription requires the activation of the heat shock factor (HSF). Although HSF from different species share relatively little sequence identity at the amino acid level, the overall domain organization is conserved (Shamovsky and Nudler, 2008). To date, a single HSF protein has been isolated in yeast (Sorger and Pelham, 1988), *Caenorhabditis elegans* (Morley and Morimoto, 2004) and *Drosophila* (Clos *et al.*, 1990). *Xenopus laevis* expresses two HSF (Hilgarth *et al.*, 2004) while mammals express at least three HSF (Rabindran *et al.*, 1991; Nakai *et al.*, 1997). Plants have multiple HSF proteins – the *Arabidopsis thaliana* genome contains 21 open reading frames that encode HSF (Nover *et al.*, 1996, 2001; Miller and Mittler, 2006). HSF1 is proposed to act as the master regulator of heat-inducible HSP gene expression in vertebrates (Shamovsky and Nudler, 2008).

2.5.2 Activation of the heat shock factor

In unstressed conditions, yeast HSF is constitutively bound to the heat shock element (HSE), which are found in the promoters of HSP (Jakobsen and Pelham, 1998). The HSE is characterized by repeating five nucleotide units of nGAAn in head-to-tail orientation (Xiao, 1991). In response to stress, the HSF undergoes heat-shock dependent phosphorylation, which is associated with transcriptional competence (Sorger and Pelham, 1988). In contrast, unstressed vertebrate and *Drosophila* HSF1 exists as a monomer, which is incapable of DNA binding and is distributed mainly, but not exclusively, in the cytoplasm (Sarge *et al.*, 1993; Westwood and Wu, 1993). Upon stress, HSF1 is phosphorylated, converted to a trimer, and relocalized to the nucleus (Sarge *et al.*, 1993). HSF1 is then able to bind to the HSE and induce transcription of the HSPs (Fig. 2.5). This activation-induced trimerization is regulated by three arrays of hydrophobic heptad repeats (HR-A/B). Trimerization is suppressed by another region of hydrophobic heptad repeats (HR-C) (Rabindran *et al.*, 1991). In contrast, yeast HSF is constitutively trimerized, and this may be due to a lack of the HR-C (Jakobsen and Pelham, 1988, 1991).

2.5.3 Negative regulation of the heat shock factor

In vertebrates and *Drosophila*, continuous heat shock leads to the loss of HSF1 DNA binding and a decrease in phosphorylation (Sarge *et al.*, 1993), suggesting the involvement of a feedback mechanism via the expression and accumulation of HSP (Satyal *et al.*, 1998). A substantial body of evidence supports a role for HSP in negative regulation of HSF1. Hsp70 has been shown to interact with HSF1 during the recovery phase of the heat shock response, and this coincides with the loss of HSF1 DNA binding activity and the conversion of the trimer to a monomer state (Satyal *et al.*, 1998). Additionally, a role has been proposed for Hsp90 in regulating the interconversion of HSF1 between the monomeric and trimeric states (Ali *et al.*, 1998).

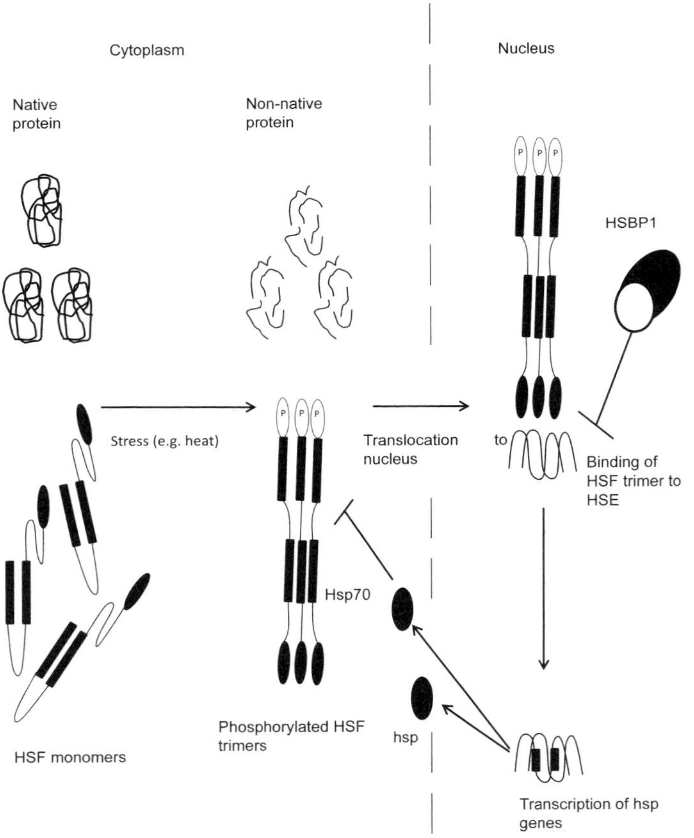

Fig. 2.5. Regulation of transcription of heat shock protein (hsp) genes by the heat shock factor (HSF). HSF is present in the cytoplasm as a monomer, which is unable to bind to DNA. Under stressful conditions, i.e. low or high temperature, the abundance of non-native proteins leads to phosphorylation (P), followed by trimerization of HSF. In the trimeric state, HSF translocates to the nucleus and binds to heat shock element (HSE) in the promoters of hsp genes, where it mediates hsp gene transcription. HSF trimer activity is downregulated by hsps (e.g. Hsp70) and the heat shock binding protein 1 (HSBP1). From Pockley, 2001.

2.5.4 Employment of the heat shock response

The HSR is not only induced by high temperature stress. It is well known that HSP also have roles in protecting against low temperature. The temperature range of HSP induction varies considerably among organisms, although it appears to be associated with the physiological range of supraoptimal temperatures within which active adaptation is observed (Fujita, 1999). For example, Liu *et al.* (1994) showed that in human diploid fibroblasts, the HSR was induced during recovery from cold shock, and that the magnitude and duration of the induction was dependent on the length of cold shock. This induction is transcriptional since an increase in Hsp70 mRNA and activation of HSF are also observed. HS gene expression is required for diapause, an insect overwintering defence strategy (Rinehart *et al.*, 2007). Expression of HS gene is up-regulated at the onset of diapause, and the mRNAs remain until an adult development-initiating signal is received (Rinehart *et al.*, 2007). In older diapausing pupae (20–25 days), RNAi directed against Hsp70 or Hsp23 lead to a significant decrease in cold tolerance.

2.5.5 The heat shock response in a changing environment

The HSR has been described in an array of organisms, however, it appears to be absent in a number of Antarctic organisms. For example, the fish *Trematomus bernacchii*, *Pagothenia borchgrevinki*, *Harapsgifer antarcticus*, and *Lycodichthys dearborni* show constitutive expression of HSP70 along with the absence of HSP70 gene up-regulation in response to an increase in temperature (Hofmann *et al.*, 2000; Clark *et al.*, 2008a; Place and Hofmann, 2005; Place *et al.*, 2004). These stenothermal fish experience an extremely stable environment in which they are exposed to low temperature. In addition, the *hsp* defence system appears to be fully expressed in unstressed flightless midge larvae (*Belgica antarctica*), yet a typical HSR is used by the adults (Rinehart *et al.*, 2006). This difference in ability to generate thermotolerance through the HSR may be due to differences in the thermal stability of the niches occupied by the two stages, whereby the larvae experience an environment which has extreme thermal stability (Rinehart *et al.*, 2006). The constitutive expression of HSP may reflect the requirement of a compensatory mechanism to deal with elevated protein damage at low temperatures (Clark *et al.*, 2008a).

A lack of HSR has also been demonstrated in other antarctic marine species including the ciliate *Euplotes focardii*, a sea star (*Odontaster validus*), and a gammarid (*Paraceradocus gibber*) (La Terza *et al.*, 2001; Clark *et al.*, 2008c). However, this mechanism of utilizing the protective qualities of HSP against the damaging effects of low temperature on proteins is not a global phenomenon of antarctic species. Two Antarctic marine molluscs, *Nacella concinna* and *Laternula elliptica*, are capable of an HSR, but this response occurs at temperatures much higher than would ever be experienced under natural conditions (Clark *et al.*, 2008b). In comparison, a lack of HSR has also been demonstrated in the warm stenothermal coral reef fish, *Pomacentrus moluccensis* (Kassahn *et al.*, 2007), suggesting that this adaptation is not limited to Antarctic species, but can be extended to those inhabiting stable warm environments also.

A recent review by Tomanek (2010) considers the effect of global climate change on an array of organisms inhabiting environments that are separated by absolute temperature and thermal range. Organisms occupying either thermally stable or highly thermally variable environments will be more affected by global climate change due to the absence or maximal employment of the HSR, respectively. Species that inhabit environments in which the temperature varies moderately may be able to utilize the HSR as a means to respond to increasing temperature, although this is not without cost to the organism. This highlights the usefulness of using stress response biomarkers as indicators of species responses to climate change. This type of approach should now be extended to include additional genes and other temperature-response mechanisms.

2.6 Summary

Organisms have evolved multiple mechanisms for sensing temperature, involving many classes of molecules. The properties of DNA, chromatin, RNA, proteins and simple molecules have all been exploited by evolution to construct biological thermometers. The pervasiveness of temperature-sensing mechanisms throughout nature underlines the importance of plastic behaviour among organisms, with experience of the environment underlying developmental modifications with massive adaptive consequences. It is clear that any one organism may have more than one temperature-sensing pathway, perhaps one to control membrane properties, and another to coordinate its lifecycle with the prevailing seasons or host availability. By learning about how organisms respond to temperature we can hope to be able to exploit this knowledge to alter their behaviour. This could involve breeding plants suitable for new environments or to treat diseases, or understanding how ecosystems will respond to a changing climate. The challenge for the scientist is to move beyond describing the negative effects of climate change to providing solutions that will help sustain population in future generations.

References

Aguilar, P., Cronan, J. and De Mendoza, D. (1998) A Bacillus subtilis gene induced by cold shock encodes a membrane phospholipid desaturase. *Journal of Bacteriology* 180, 2194–2200.

Aguilar, P.S., Lopez, P. and De Mendoza, D. (1999) Transcriptional control of the low-temperature-inducible des gene, encoding the delta5 desaturase of Bacillus subtilis. *Journal of Bacteriology* 181, 7028–7033.

Ali, A., Bharadwaj, S., O'Carroll, R. and Ovsenek, N. (1998) HSP90 interacts with and regulates the activity of heat shock factor 1 in Xenopus oocytes. *Molecular and Cellular Biology* 18, 4949–4960.

Altabe, S., Aguilar, P., Caballero, G. and De Mendoza, D. (2003) The Bacillus subtilis acyl lipid desaturase is a {delta}5 desaturase. *Journal of Bacteriology* 185, 3228–3231.

Altuvia, S., Kornitzer, D., Teff, D. and Oppenheim, A.B. (1989) Alternative mRNA structures of the cIII gene of bacteriophage lambda determine the rate of its translation initiation. *Journal of Molecular Biology* 210, 265–280.

Atlung, T. and Ingmer, H. (1997) H-NS: a modulator of environmentally regulated gene expression. *Molecular Microbiology* 24, 7–17.

Bautista, D., Jordt, S.-E., Nikai, T., Tsuruda, P., Read, A., Poblete, J., Yamoah, E., *et al.* (2006) TRPA1 mediates the inflammatory actions of environmental irritants and proalgesic agents. *Cell* 124, 1269–1282.

Bautista, D., Siemens, J., Glazer, J., Tsuruda, P., Basbaum, A., Stucky, C., Jordt, S.-E., *et al.* (2007) The menthol receptor TRPM8 is the principal detector of environmental cold. *Nature* 448, 204–208.

Bloch, V., Yang, Y., Margeat, E., Chavanieu, A., Auge, M.T., Robert, B., Arold, S., *et al.* (2003) The H-NS dimerization domain defines a new fold contributing to DNA recognition. *Nature Structural and Molecular Biology* 10, 212–218.

Caterina, M., Schumacher, M., Tominaga, M., Rosen, T., Levine, J. and Julius, D. (1997) The capsaicin receptor: a heat-activated ion channel in the pain pathway. *Nature* 389, 816–824.

Caterina, M.J., Leffler, A., Malmberg, A.B., Martin, W.J., Trafton, J., Petersen-Zeitz, K.R., Koltzenburg, M., *et al.* (2000) Impaired nociception and pain sensation in mice lacking the capsaicin receptor. *Science* 288, 306–313.

Chatzigeorgiou, M., Yoo, S., Watson, J., Lee, W.-H., Spencer, C., Kindt, K., Hwang, S., *et al.* (2010) Specific roles for DEG/ENaC and TRP channels in touch and thermosensation in C. elegans nociceptors. *Nature Neuroscience* 13, 861–868.

Clapham, D. (2003) TRP channels as cellular sensors. *Nature* 426, 517–524.

Clark, M., Fraser, K., Burns, G. and Peck, L. (2008a) The HSP70 heat shock response in the Antarctic fish Harpagifer antarcticus. *Polar Biology* 31, 171–180.

Clark, M., Fraser, K. and Peck, L. (2008b) Antarctic marine molluscs do have an HSP70 heat shock response. *Cell Stress and Chaperones* 13, 39–49.

Clark, M., Fraser, K. and Peck, L. (2008c) Lack of an HSP70 heat shock response in two Antarctic marine invertebrates. *Polar Biology* 31, 1059–1065.

Clos, J., Westwood, J.T., Becker, P.B., Wilson, S., Lambert, K. and Wu, C. (1990) Molecular cloning and expression of a hexameric Drosophila heat shock factor subject to negative regulation. *Cell* 63, 1085–1097.

Cornelis, G., Sluiters, C., De Rouvroit, C.L. and Michiels, T. (1989a) Homology between virF, the transcriptional activator of the Yersinia virulence regulon, and AraC, the Escherichia coli arabinose operon regulator. *Journal of Bacteriology* 171, 254–262.

Cornelis, G., Sluiters, C., De Rouvroit, C.L. and Michiels, T. (1989b) Homology between virF, the transcriptional activator of the Yersinia virulence regulon, and AraC, the Escherichia coli arabinose operon regulator. *Journal of Bacteriology* 171, 254–262.

Cornelis, G.R., Biot, T., Lambert De Rouvroit, C., Michiels, T., Mulder, B., Sluiters, C., Sory, M.P., *et al.* (1989c) The Yersinia yop regulon. *Molecular Microbiology* 3, 1455–1459.

Cybulski, L., Albanesi, D., Mansilla, M., Altabe, S., Aguilar, P. and De Mendoza, D. (2002) Mechanism of membrane fluidity optimization: isothermal control of the *Bacillus subtilis* acyl-lipid desaturase. *Molecular Microbiology* 45, 1379–1388.

Dhaka, A., Viswanath, V. and Patapoutian, A. (2006) Trp ion channels and temperature sensation. *Annual Review of Neuroscience* 29, 135–161.

Doherty, C., Van Buskirk, H., Myers, S. and Thomashow, M. (2009) Roles for Arabidopsis CAMTA transcription factors in cold-regulated gene expression and freezing tolerance. *Plant Cell* 21, 972–984.

Dorman, C. (2004) H-NS: a universal regulator for a dynamic genome. *Nature Reviews Microbiology* 2, 391–400.

Duguet, M. (1993) The helical repeat of DNA at high temperature. *Nucleic Acids Research* 21, 463–468.

Dyer, J. (2001) Chilling-sensitive, post-transcriptional regulation of a plant fatty acid desaturase expressed in yeast. *Biochemical and Biophysical Research Communications* 282, 1019–1025.

Eriksson, S., Hurme, R. and Rhen, M. (2002) Low temperature sensors in bacteria. *Philosophical Transactions of the Royal Society London B* 357, 887–893.

Falconi, M., Colonna, B., Prosseda, G., Micheli, G. and Gualerzi, C.O. (1998) Thermoregulation of Shigella and Escherichia coli EIEC pathogenicity. A temperature-dependent structural transition of DNA modulates accessibility of virF promoter to transcriptional repressor H-NS. *The EMBO Journal* 17, 7033–7043.

Fowler, S., Cook, D. and Thomashow, M. (2005) Low temperature induction of Arabidopsis CBF1, 2, and 3 is gated by the circadian clock. *Plant Physiology* 137, 961–968.

Fujita, J. (1999) Cold shock response in mammalian cells. *Journal of Molecular Microbiology and Biotechnology* 1, 243–255.

Gal-Mor, O., Valdez, Y. and Finlay, B. (2006) The temperature-sensing protein TlpA is repressed by PhoP and dispensable for virulence of Salmonella enterica serovar Typhimurium in mice. *Microbes and Infection* 8, 2154–2162.

Gibson, S., Arondel, V., Iba, K. and Somerville, C. (1994) Cloning of a temperature-regulated gene encoding a chloroplast [omega]-3 desaturase from Arabidopsis thaliana. *Plant Physiology* 106, 1615–1621.

Grau, R. and Mendoza, D. (1993) Regulation of the synthesis of unsaturated fatty acids by growth temperature in *Bacillus subtilis*. *Molecular Microbiology* 8, 535–542.

Hale, T.L. (1991) Genetic basis of virulence in Shigella species. *Microbiological Reviews* 55, 206–224.

Hazel, J. (1990) The role of alterations in membrane lipid composition in enabling physiological adaptation of organisms to their physical environment. *Progress in Lipid Research* 29, 167–227.

Hazel, J.R. (1995) Thermal adaptation in biological membranes: is homeoviscous adaptation the explanation? *Annual Review of Physiology* 57, 19–42.

Hilgarth, R., Murphy, L., O'Connor, C., Clark, J., Park-Sarge, O.-K. and Sarge, K. (2004) Identification of Xenopus heat shock transcription factor-2: conserved role of sumoylation in regulating deoxyribonucleic acid-binding activity of heat shock transcription factor-2 proteins. *Cell Stress and Chaperones* 9, 214–220.

Hirschi, K. (1999) Expression of Arabidopsis CAX1 in tobacco: altered calcium homeostasis and increased stress sensitivity. *Plant Cell* 11, 2113–2122.

Hoe, N.P. and Goguen, J.D. (1993) Temperature sensing in Yersinia pestis: translation of the LcrF activator protein is thermally regulated. *Journal of Bacteriology* 175, 7901–7909.

Hoe, N.P., Minion, F.C. and Goguen, J.D. (1992)

Temperature sensing in Yersinia pestis: regulation of yopE transcription by lcrF. *Journal of Bacteriology* 174, 4275–4286.

Hofmann, G.E., Buckley, B.A., Airaksinen, S., Keen, J.E. and Somero, G.N. (2000) Heat-shock protein expression is absent in the antarctic fish Trematomus bernacchii (family Nototheniidae). *Journal of Experimental Biology,* 203, 2331–2339.

Hurme, R., Berndt, K., Namork, E. and Rhen, M. (1996) DNA Binding exerted by a bacterial gene regulator with an extensive coiled-coil domain. *Journal of Biological Chemistry* 271, 12626–12631.

Hurme, R., Berndt, K., Normark, S. and Rhen, M. (1997) A proteinaceous gene regulatory thermometer in Salmonella. *Cell* 90, 55–64.

Ibanez, C., Kozarewa, I., Johansson, M., Ogren, E., Rohde, A. and Eriksson, M. (2010) Circadian clock components regulate entry and affect exit of seasonal dormancy as well as winter hardiness in Populus trees. *Plant Physiology* 4, 1823-1833.

Insler, S. (2006) Perioperative thermoregulation and temperature monitoring. *Anesthesiology Clinics of North America* 24, 823–837.

Jakobsen, B.K. and Pelham, H.R. (1998) Constitutive binding of yeast heat shock factor to DNA in vivo. *Molecular and Cellular Biology* 8, 5040–5042.

Jakobsen, B.K. and Pelham, H.R. (1991) A conserved heptapeptide restrains the activity of the yeast heat shock transcription factor. *The EMBO Journal* 10, 369–375.

Jiménez, S., Reighard, G.L. and Bielenberg, D.G. (2010) Gene expression of DAM5 and DAM6 is suppressed by chilling temperatures and inversely correlated with bud break rate. *Plant Molecular Biology* 73, 157–167.

Johansson, J., Mandin, P., Renzoni, A., Chiaruttini, C., Springer, M. and Cossart, P. (2002) An RNA thermosensor controls expression of virulence genes in Listeria monocytogenes. *Cell* 110, 551–561.

Karashima, Y., Talavera, K., Everaerts, W., Janssens, A., Kwan, K., Vennekens, R., Nilius, B., *et al.* (2009) TRPA1 acts as a cold sensor in vitro and in vivo. *Proceedings of the National Academy of Sciences* 106, 1273–1278.

Kassahn, K., Caley, J., Ward, A., Connolly, A., Stone, G. and Crozier, R. (2007) Heterologous microarray experiments used to identify the early gene response to heat stress in a coral reef fish. *Molecular Ecology* 16, 1749–1763.

Knight, H., Trewavas, A. and Knight, M. (1996) Cold calcium signaling in Arabidopsis involves two cellular pools and a change in calcium signature after acclimation. *The Plant Cell* 8, 489–503.

Koo, H.-S., Wu, H.-M. and Crothers, D. (1986) DNA bending at adenine · thymine tracts. *Nature* 320, 501–506.

Kumar, V. and Wigge, P. (2010) H2A.Z-containing nucleosomes mediate the thermosensory response in Arabidopsis. *Cell* 140, 136–147.

Kunst, L., Browse, J. and Somerville, C. (1989) Enhanced thermal tolerance in a mutant of Arabidopsis deficient in palmitic acid unsaturation. *Plant Physiology* 91, 401–408.

Kwan, K., Allchorne, A., Vollrath, M., Christensen, A., Zhang, D., Woolf, C. and Corey, D. (2006) TRPA1 contributes to cold, mechanical, and chemical nociception but is not essential for hair-cell transduction. *Neuron* 50, 277–289.

Larkindale, J. and Knight, M. (2002) Protection against heat stress-induced oxidative damage in Arabidopsis involves calcium, abscisic acid, ethylene, and salicylic acid. *Plant Physiology* 128, 682–695.

La Terza, A., Papa, G., Miceli, C. and Luporini, P. (2001) Divergence between two Antarctic species of the ciliate Euplotes, E. focardii and E. nobilii, in the expression of heat-shock protein 70 genes. *Molecular Ecology* 10, 1061–1067.

Lee, H., Iida, T., Mizuno, A., Suzuki, M. and Caterina, M. (2005a) Altered thermal selection behavior in mice lacking transient receptor potential vanilloid 4. *Journal of Neuroscience* 25, 1304–1310.

Lee, Y., Lee, Y., Lee, J., Bang, S., Hyun, S., Kang, J., Hong, S.-T., *et al.* (2005b) Pyrexia is a new thermal transient receptor potential channel endowing tolerance to high temperatures in Drosophila melanogaster. *Nature Genetics* 37, 305–310.

Li, Z., Reighard, G., Abbott, A. and Bielenberg, D. (2009) Dormancy-associated MADS genes from the EVG locus of peach [Prunus persica (L.) Batsch] have distinct seasonal and photoperiodic expression patterns. *Journal of Experimental Botany* 60, 3521–3530.

Liu, A.Y., Bian, H., Huang, L.E. and Lee, Y.K. (1994) Transient cold shock induces the heat shock response upon recovery at 37 degrees C in human cells. *Journal of Biological Chemistry* 269, 14768–14775.

Liu, F., Marquardt, S., Lister, C., Swiezewski, S. and Dean, C. (2010) Targeted 3' processing of antisense transcripts triggers Arabidopsis FLC chromatin silencing. *Science* 327, 94–97.

Los, D.A. and Murata, N. (1999) Responses to cold shock in cyanobacteria. *Journal of Molecular Microbiology and Biotechnology* 1, 221–230.

Mansilla, M.C., Cybulski, L.E., Albanesi, D. and De Mendoza, D. (2004) Control of membrane lipid fluidity by molecular thermosensors. *Journal of Bacteriology* 186, 6681–6688.

Matsuda, O., Sakamoto, H., Hashimoto, T. and Iba, K. (2005) A temperature-sensitive mechanism that regulates post-translational stability of a plastidial ω-3 fatty acid desaturase (FAD8) in Arabidopsis leaf tissues. *Journal of Biological Chemistry* 280, 3597–3604.

Maurelli, A. (1989) Temperature regulation of virulence genes in pathogenic bacteria: a general strategy for human pathogens? *Microbial Pathogenesis,* 7, 1–10.

McKemy, D. (2007) Temperature sensing across species. *Pflügers Archiv European Journal of Physiology* 454, 777–791.

McKemy, D., Neuhausser, W. and Julius, D. (2002) Identification of a cold receptor reveals a general role for TRP channels in thermo-sensation. *Nature* 416, 52–58.

Mikkelsen, M.D. and Thomashow, M.F. (2009) A role for circadian evening elements in cold-regulated gene expression in Arabidopsis. *The Plant Journal* 60, 328–339.

Miller, G. and Mittler, R. (2006) Could heat shock transcription factors function as hydrogen peroxide sensors in plants? *Annals of Botany* 98, 279–288.

Miquel, M., James, D., Dooner, H. and Browse, J. (1993) Arabidopsis requires polyunsaturated lipids for low-temperature survival. *Proceedings of the National Academy of Sciences USA* 90, 6208–6212.

Monroy, A.F. and Dhindsa, R.S. (1995) Low-temperature signal transduction: induction of cold acclimation-specific genes of alfalfa by calcium at 25°C. *Plant Cell* 7, 321–331.

Moqrich, A., Hwang, S.W., Earley, T., Petrus, M., Murray, A., Spencer, K., Andahazy, M., *et al.* (2005) Impaired thermosensation in mice lacking TRPV3, a heat and camphor sensor in the skin. *Science* 307, 1468–1472.

Morimoto, R. (1998) Regulation of the heat shock transcriptional response: cross talk between a family of heat shock factors, molecular chaperones, and negative regulators. *Genes and Development* 12, 3788–3796.

Morley, J. and Morimoto, R. (2004) Regulation of longevity in Caenorhabditis elegans by heat shock factor and molecular chaperones. *Molecular Biology of the Cell* 15, 657–664.

Murata, N. and Los, D.A. (1997) Membrane fluidity and temperature perception. *Plant Physiology* 115, 875–879.

Nakai, A., Tanabe, M., Kawazoe, Y., Inazawa, J., Morimoto, R.I. and Nagata, K. (1997) HSF4, a new member of the human heat shock factor family which lacks properties of a transcriptional activator. *Molecular and Cellular Biology* 17, 469–481.

Nakamichi, N., Kusano, M., Fukushima, A., Kita, M., Ito, S., Yamashino, T., Saito, K., *et al.* (2009). Transcript profiling of an Arabidopsis PSEUDO RESPONSE REGULATOR arrhythmic triple mutant reveals a role for the circadian clock in cold stress response. *Plant and Cell Physiology* 50, 447–462.

Nover, L., Scharf, K.D., Gagliardi, D., Vergne, P., Czarnecka-Verner, E. and Gurley, W.B. (1996) The Hsf world: classification and properties of plant heat stress transcription factors. *Cell Stress and Chaperones* 1, 215–223.

Nover, L., Bharti, K., Döring, P., Mishra, S.K., Ganguli, A. and Scharf, K.D. (2001) Arabidopsis and the heat stress transcription factor world: how many heat stress transcription factors do we need? *Cell Stress and Chaperones* 6, 177–189.

Patapoutian, A., Peier, A., Story, G. and Viswanath, V. (2003) ThermoTRP channels and beyond: mechanisms of temperature sensation. *Nature Reviews Neuroscience* 4, 529–539.

Pedersen, S., Owsianik, G. and Nilius, B. (2005) TRP channels: An overview. *Cell Calcium* 38, 233–252.

Peier, A., Moqrich, A., Hergarden, A., Reeve, A., Andersson, D., Story, G., Earley, T., *et al.* (2002). A TRP channel that senses cold stimuli and menthol. *Cell* 108, 705–715.

Piao, S., Ciais, P., Friedlingstein, P., Peylin, P., Reichstein, M., Luyssaert, S., Margolis, H., *et al.* (2008) Net carbon dioxide losses of northern ecosystems in response to autumn warming. *Nature* 451, 49–52.

Place, S. and Hofmann, G. (2005) Constitutive expression of a stress-inducible heat shock protein gene, hsp70, in phylogenetically distant Antarctic fish. *Polar Biology,* 28, 261–267.

Place, S., Zippay, M. and Hofmann, G. (2004) Constitutive roles for inducible genes: evidence for the alteration in expression of the inducible hsp70 gene in Antarctic notothenioid fishes. *American Journal of Physiology - Regulatory, Integrative and Comparative Physiology* 287, R429–436.

Plieth, C., Hansen, U.P., Knight, H. and Knight, M.R. (1999) Temperature sensing by plants: the primary characteristics of signal perception and calcium response. *The Plant Journal* 18, 491–497.

Pockley, G. (2001) Heat shock proteins in health

and disease: therapeutic targets or therapeutic agents? *Expert Reviews in Molecular Medicine* 3, 1–21.

Prosseda, G., Falconi, M., Giangrossi, M., Gualerzi, C.O., Micheli, G. and Colonna, B. (2004) The virF promoter in Shigella: more than just a curved DNA stretch. *Molecular Microbiology* 51, 523–537.

Quinn, P. (1989) The role of unsaturated lipids in membrane structure and stability. *Progress in Biophysics and Molecular Biology* 53, 71–103.

Rabindran, S.K., Giorgi, G., Clos, J. and Wu, C. (1991) Molecular cloning and expression of a human heat shock factor, HSF1. *Proceedings of the National Academy of Sciences USA* 88, 6906–6910.

Rinehart, J., Hayward, S., Elnitsky, M., Sandro, L., Lee, R. and Denlinger, D. (2006) Continuous up-regulation of heat shock proteins in larvae, but not adults, of a polar insect. *Proceedings of the National Academy of Sciences USA* 103, 14223–14227.

Rinehart, J., Li, A., Yocum, G., Robich, R., Hayward, S. and Denlinger, D. (2007) Up-regulation of heat shock proteins is essential for cold survival during insect diapause. *Proceedings of the National Academy of Sciences USA* 104, 11130–11137.

Rodgers, W. and Glaser, M. (1993) Distributions of proteins and lipids in the erythrocyte membrane. *Biochemistry* 32, 12591–12598.

Rosenzweig, M., Kang, K. and Garrity, P. (2008) Distinct TRP channels are required for warm and cool avoidance in Drosophila melanogaster. *Proceedings of the National Academy of Sciences USA* 105, 14668–14673.

Saidi, Y., Finka, A., Muriset, M., Bromberg, Z., Weiss, Y., Maathuis, F. and Goloubinoff, P. (2009) The heat shock response in moss plants is regulated by specific calcium-permeable channels in the plasma membrane. *Plant Cell* 21, 2829–2843.

Sarge, K.D., Murphy, S.P. and Morimoto, R.I. (1993) Activation of heat shock gene transcription by heat shock factor 1 involves oligomerization, acquisition of DNA-binding activity, and nuclear localization and can occur in the absence of stress. *Molecular and Cellular Biology* 13, 1392–1407.

Satyal, S., Chen, D., Fox, S., Kramer, J. and Morimoto, R. (1998) Negative regulation of the heat shock transcriptional response by HSBP1. *Genes and Development* 12, 1962–1974.

Schepers, R. and Ringkamp, M. (2010) Thermoreceptors and thermosensitive afferents. *Neuroscience and Biobehavioral Reviews* 34, 177–184.

Shamovsky, I. and Nudler, E. (2008) New insights into the mechanism of heat shock response activation. *Cellular and Molecular Life Sciences* 65, 855–861.

Shindo, C., Lister, C., Crevillen, P., Nordborg, M. and Dean, C. (2006) Variation in the epigenetic silencing of FLC contributes to natural variation in Arabidopsis vernalization response. *Genes and Development* 20, 3079–3083.

Sokabe, T., Tsujiuchi, S., Kadowakl, T. and Tominaga, M. (2008) Drosophila painless is a Ca2+-requiring channel activated by noxious heat. *The Journal of Neuroscience* 28, 9929–9938.

Somerville, C. and Browse, J. (1996) Dissecting desaturation: plants prove advantageous. *Trends in Cell Biology* 6, 148–153.

Sorger, P.K. and Pelham, H.R. (1988) Yeast heat shock factor is an essential DNA-binding protein that exhibits temperature-dependent phosphorylation. *Cell* 54, 855–864.

Story, G., Peier, A., Reeve, A., Eid, S., Mosbacher, J., Hricik, T., Earley, T., *et al.* (2003) ANKTM1, a TRP-like channel expressed in nociceptive neurons, is activated by cold temperatures. *Cell* 112, 819–829.

Sung, D. (2003) Acquired tolerance to temperature extremes. *Trends in Plant Science* 8, 179–187.

Sung, S. and Amasino, R. (2004) Vernalization in Arabidopsis thaliana is mediated by the PHD finger protein VIN3. *Nature* 427, 159–164.

Tobe, T., Nagai, S., Okada, N., Adter, B., Yoshikawa, M. and Sasakawa, C. (1991) Temperature-regulated expression of invasion genes in *Shigella flexneri* is controlled through the transcriptional activation of the *virB* gene on the large plasmid. *Molecular Microbiology* 5, 887–893.

Tobe, T., Yoshikawa, M., Mizuno, T. and Sasakawa, C. (1993) Transcriptional control of the invasion regulatory gene virB of Shigella flexneri: activation by virF and repression by H-NS. *Journal of Bacteriology* 175, 6142–6149.

Todaka, H., Taniguchi, J., Satoh, J.-I., Mizuno, A. and Suzuki, M. (2004) Warm temperature-sensitive transient receptor potential vanilloid 4 (Trpv4) plays an essential role in thermal hyperalgesia. *Journal of Biological Chemistry* 279, 35133–35138.

Tomanek, L. (2010) Variation in the heat shock response and its implication for predicting the effect of global climate change on species' biogeographical distribution ranges and metabolic costs. *Journal of Experimental Biology* 213, 971–979.

Tracey, W.D. Jr, Wilson, R.I., Laurent, G. and Benzer, S. (2003) painless, a Drosophila gene essential for nociception. *Cell,* 113, 261–273.

Ueguchi, C. (1997) Clarification of the dimerization domain and its functional significance for the Escherichia coli nucleoid protein H-NS. *Journal of Molecular Biology* 274, 145–151.

Venkatachalam, K. and Montell, C. (2007) TRP channels. *Annual Review of Biochemistry* 76, 387–417.

Viswanath, V., Story, G., Peier, A., Petrus, M., Lee, V., Hwang, S.W., Patapoutian, A. *et al.* (2003) Opposite thermosensor in fruitfly and mouse. *Nature* 423, 822–823.

Westwood, J.T. and Wu, C. (1993) Activation of Drosophila heat shock factor: conformational change associated with a monomer-to-trimer transition. *Molecular and Cellular Biology* 13, 3481–3486.

Wheeler, G. and Brownlee, C. (2008) Ca2+ signalling in plants and green algae – changing channels. *Trends in Plant Science* 13, 506–514.

Xiao, H. (1991) Cooperative binding of drosophila heat shock factor to arrays of a conserved 5 bp unit. *Cell* 64, 585–593.

Yamada, H., Muramatsu, S. and Mizuno, T. (1990) An Escherichia coli protein that preferentially binds to sharply curved DNA. *Journal of Biochemistry* 108, 420–425.

Yamada, H., Yoshida, T., Tanaka, K., Sasakawa, C. and Mizuno, T. (1991) Molecular analysis of the Escherichia coli hns gene encoding a DNA-binding protein, which preferentially recognizes curved DNA sequences. *Molecular and General Genetics* 230, 332–336.

Zurborg, S., Yurgionas, B., Jira, J., Caspani, O. and Heppenstall, P. (2007) Direct activation of the ion channel TRPA1 by Ca2+. *Nature Neuroscience* 10, 277–279.

3 Microorganisms and Plants: a Photosynthetic Perspective

Rachael M. Morgan-Kiss and Jenna M. Dolhi

3.1 Introduction

Permanently low temperature habitats (ecosystems existing at yearly stable temperatures below or close to the freezing point of water) are one of the most prevalent biomes on earth. Cold habitats that are driven by light-dependent primary production include Antarctic and Arctic marine environments, sea ice, glacial ice, polar soils, montane environments, and lithic communities, as well as high alpine and ice-covered polar lakes (Morgan-Kiss *et al.*, 2006; Cowan *et al.*, 2007). In common with all ecosystems, low temperature food webs are dependent upon energy and carbon derived from autotrophic organisms, the vast majority of which rely on light-driven photosynthetic reactions. Photosynthetic organisms residing in permanently cold habitats (including plants, algae, lichens and photosynthetic bacteria) have evolved a variety of adaptive strategies not only to thrive under the thermodynamic challenges associated with cold temperatures, but also to deal with the unique problem of coupling cellular processes which exhibit wide degrees of temperature sensitivity.

Many cold environments are dominated by microorganisms, including Archaea, yeasts, gram-negative and gram-positive bacteria and fungi, as well as a wide diversity of photosynthetic microorganisms (such as cyanobacteria, anoxygenic phototrophic bacteria, micro- and macro-algae and mixotrophic protists). In addition, a large number of plants (up to 1700 species) live in the Arctic tundra ecosystem, and include flowering plants, dwarf shrubs, herbs, grasses, mosses and lichens. In contrast, only two vascular plants have colonized coastal Antarctic habitats, the Antarctic hair grass and Antarctic pearlwort (Alberdi *et al.*, 2002), although there are approximately 300 species of bryophytes residing on the Antarctic peninsula. However, primary production – which forms the base of inorganic carbon fixation and autotrophic energy production in the vast majority of low temperature ecosystems – is driven largely by photosynthetic microorganisms. Many studies have been devoted to the impact of low temperatures on plants (see, for example, recent reviews in Chinnusamy and Zhu, 2009; Franklin, 2009; Hua, 2009; Yamori *et al.*, 2009). However, much of this work has focused on cold-sensitive plants or cold-tolerant crop species exposed to seasonal low temperatures. Since the purpose of this chapter is to discuss the potential impact of climate change on photosynthetic organisms residing in year-round cold environments, there is an emphasis on studies regarding microbial communities in cold habitats as well as research on microbial isolates of psychrophilic photoautotrophs, or the 'photopsychrophiles'.

Unlike temperate ecosystems, many of the pristine low temperature ecosystems have experienced very little human impact. Global climate change is altering low temperature aquatic habitats in profound ways, and the impact of climate change on sensitive high alpine and polar ecosystems will not be gradual or linear (Williamson *et al.*, 2009). In the past 50 years, temperatures and precipitation have increased in Antarctica, with a rise in mean annual temperatures of about 2.6°C. The effect is

© CAB International 2012. *Temperature Adaptation in a Changing Climate: Nature at Risk* (eds K.B. Storey and K.K. Tanino)

most evident along the west coast where there has been a dramatic retreat of the ice shelves (Sympson, 2000). Perturbation of the primary producer communities will directly affect functioning of the food chain as a whole. It is therefore critical to understand the impact of climate change on the photoautotrophic communities in low temperature ecosystems. Last, a greater appreciation for the contribution of mixotrophic microorganisms (those capable of both photosynthetic carbon fixation and phagotrophic breakdown of organic carbon) has been only recognized recently (Sanders *et al.*, 2000; Moorthi *et al.*, 2009). Thus, the impact of climate change on mixotrophy is unknown, but could have a significant impact on carbon cycling in low temperature aquatic systems such as ice-covered lakes, where microbial eukaryotes play essential roles at all trophic levels in the food web (Priscu *et al.*, 1999; Morgan-Kiss *et al.*, 2006).

3.2 Climate Change and Low Temperature Habitats for Phototrophic Organisms

Permanently low temperature environments provide a myriad of habitats for cold-adapted phototrophs and can vary widely in their physical and chemical characteristics, as well as anthropogenic influences (Overland *et al.*, 2008). While a comprehensive understanding of how climate change is affecting the full range of permanently low temperature environments is still lacking (Hoegh-Guldberg and Bruno, 2010), it is well known that polar habitats are changing rapidly and profoundly in response to anthropogenic climate change. One of the most spectacular alterations is the recent and rapid decline in the extent and seasonality of Arctic sea ice (Symon *et al.*, 2005; Stroeve *et al.*, 2007; Stroeve and Serreze, 2008). For primary producers, sea ice plays an important role as a substrate which limits light availability for photosynthesis and prevents thermal or freshwater stratification during the freezing season, while enhancing it during the thaw season. The decline in sea ice may enhance primary production in some algal groups due to increased nutrient availability from wind mixing and ice-shelf break upwelling (Arrigo *et al.*, 2008). However, reduced sea-ice cover has also been predicted to induce a shift from an algae–benthos- to a phytoplankton–zooplankton-dominated ecosystem (Piepenburg, 2005) and to reduce the supply of labile organic matter and nutrients to bacteria (Kirchman *et al.*, 2009), resulting in reduced export of organic carbon and a decline in pelagic–benthic coupling in the Arctic ocean. In contrast, there is some encouraging evidence that Arctic plants may exhibit resistance to climate change by shifting over long distances via wind, floating sea ice or birds, and moving to lower temperature environments (Walther *et al.*, 2002).

Rapid climate change is occurring in marine ecosystems along several coasts in Antarctica. In the western continental shelf of the Antarctic Peninsula, winter air temperatures are on average ~5 times higher than the yearly average in the last 50 years. This has had the stunning impact of converting the cold-dry polar system climate to one that is more of a warm-humid maritime sub-Antarctic climate, consequently forcing sea ice-dependent metazoa such as Adélie penguins and krill to be displaced further poleward, with an overall effect of reducing the habitat area for polar marine communities. A recent study showed that oceanic phytoplankton productivity has been significantly altered along the Western Antarctic Peninsula (WAP) ice shelf over the last 30 years (Montes-Hugo *et al.*, 2009). Specifically, phytoplankton biomass (estimated as chlorophyll a) was reduced by 12% overall, and trends in chlorophyll a maxima moved southwards in the last three decades. These alterations in the distribution of primary producers may influence the redistribution of higher trophic levels. For example, loss of primary production will enhance the decline in abundance of Antarctic krill in more northerly WAP locations, which is likely to negatively impact the survival of larger marine species such as silverfish and birds.

Lakes and reservoirs extend over a wide range of the globe and form a network of environmental sensors that can act as sentinels of climate change (Williamson *et al.*, 2009). Polar and alpine lakes are experiencing particularly rapid effects of climate change (Vieillette *et al.*, 2008; Mueller *et al.*, 2009). Low elevation reservoirs generally have large watersheds that are heavily impacted by local land use patterns, while high elevation watersheds such as alpine lakes are highly susceptible to episodic climate events including anthropogenic deposition of fixed nitrogen (Saros *et al.*, 2005). Numerous studies on sensitive high alpine lakes have established clear links between climate change and altered alpine lake ecology (Thompson *et al.*, 2005; Parker *et al.*, 2008). Permanently ice-covered lakes are some of the low temperature aquatic habitats that are most sensitive to environmental change. The McMurdo Dry Valleys in southern Victoria Land, Antarctica (77° 00′ S, 162° 52′ E), are some of the coldest, driest locations on earth, with precipitation of <10 cm/year and a yearly temperature range of −55°C to 5°C. Despite these inhospitable conditions, permanently ice-covered lakes provide the sole source of liquid water year round, and are oases for aquatic communities dominated by the microbial loop. The ice cover prevents yearly mixing and allochthonous inputs. Polar lake systems impart extreme stresses on phototrophic life due to year-round low temperatures combined with extreme seasonal light availability and permanent shade conditions in the water columns below the ice covers. These pristine aquatic systems have experienced very little human impact and since ice cover provides year-round constant environmental conditions, the biological components are adapted to one of the most stable aquatic environments in the world.

The Antarctic continent is predicted to warm significantly in the next 50–100 years (Chapman and Walsh, 2007; Walsh, 2009), which will impact the McMurdo Dry Valley ecosystems by increased hydrological activity through increased occurrence in pulse flood events and glacial melting. Episodic events causing short duration variations in major abiotic drivers are now recognized to have profound effects on aquatic ecosystem functioning, and are expected to intensify at the level of frequency and magnitude as a consequence of global climate change (Alley *et al.*, 2003; Meehl *et al.*, 2007; Jentsch *et al.*, 2007). While the dry valley lakes receive little direct anthropogenic contact, increases in episodic events associated with high summer flows will increase nutrient loading to the lakes (Doran *et al.*, 2008). The long-term impact of increased flood events on the dry valley lake primary producers is currently unknown. However, it is predicted that photosynthetic communities residing in ice-covered Antarctic lakes as well as in high latitude alpine aquatic environments, will be more responsive to both climate warming and the effects of increased episodic events compared with lower latitude watersheds (Fig. 3.1). This prediction is based on the general characteristics of the alpine and polar aquatic environments, which are often nutrient limited (i.e. oligotrophic), of low temperatures, and relatively pristine compared with most lower latitude lakes (Fig. 3.1). Physiological adaptation of cold-adapted phototrophic organisms is tailored to their natural environments, thus, predicting the impact of climate change on phototrophic communities within and across cold habitats is complex.

3.3 Photostasis, Photoacclimation and Environmental Adaptation

All photosynthetic organisms capture light energy and transduce it into energy carrier molecules via the photosynthetic electron transport chain. Light energy is absorbed by membrane-embedded pigment-binding protein complexes (named photosystem I, PS I; and photosystem II, PS II). Both photosystems comprise two multi-subunit complexes, a light harvesting apparatus which absorbs light energy, and reaction centres which transduce absorbed light energy into chemical energy (Fig. 3.2). Electron transport is coupled with proton

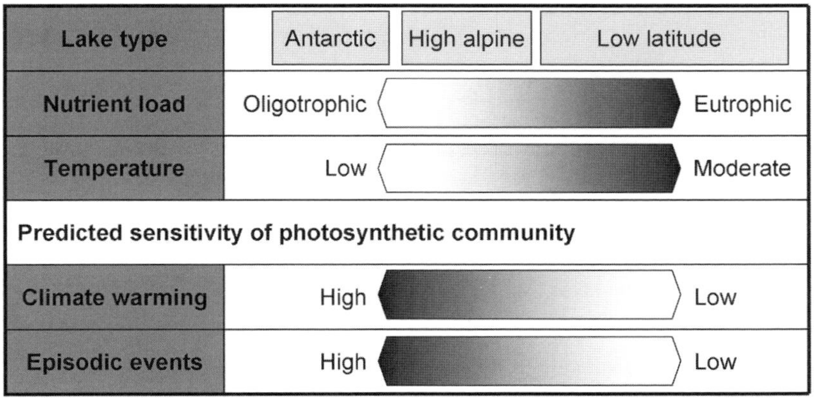

Fig. 3.1. Predicted sensitivity of photosynthetic communities residing in a range of watershed ecosystems to climate change. Relatively pristine ice-covered Antarctic and high alpine lakes receive significantly less allochthonous carbon/nutrient input as well as anthropogenic interactions compared with most lower latitude lakes. As a result, this model predicts that photosynthetic organisms residing in relatively pristine high alpine and Antarctic lakes will exhibit high sensitivity to both global warming and increased frequency of episodic storm or flood events, compared with lakes located in lower latitude environments.

translocation during transfer of electrons between PS II and PS I and the resulting energy potentials are used to produce NADPH and ATP. These energy carriers are regenerated when they are consumed by downstream metabolic reactions, the bulk of which are reactions associated with fixation of carbon, such as those associated with the Calvin–Benson–Bassham (CBB) cycle (Fig. 3.2). Photosynthetic organisms possess a myriad of acclimative to adaptive mechanisms to perform energy transduction at low temperatures and to tightly coordinate energy produced with energy consumed, through a process called photostasis. In this chapter we will define photosynthetic adaptation as long-term changes over multiple generations that are manifested as genetic and epigenetic changes, while photoacclimation refers to short-term adjustments during an organism's lifetime which allow a photoautotroph to respond to short-duration environmental perturbations. Both adaptive and acclimative processes play separate yet equally important roles in the ability of a photosynthetic organism to colonize and thrive in its natural habitat.

Temperature specifically impacts a photosynthetic organism's ability to maintain photostasis by perturbing the balance between temperature-insensitive photochemical reactions which capture energy, and temperature-sensitive biochemical reactions which consume energy. Photostasis can be described by the simple equation:

$$\sigma_{PSII} \times E_k = \tau^{-1} \quad (3.1)$$

where σ_{PSII} is the effective cross section of the PS II, E_k is the irradiance at which the maximum photosynthetic quantum yield balances the photosynthetic capacity estimated from a photosynthetic light response curve, and τ^{-1} is the rate at which photosynthetic electrons are consumed by a terminal electron acceptor, such as CO_2 or NO_3^- (Falkowski and Chen, 2003). The left side of the equation ($\sigma_{PSII} \times E_k$) is governed largely by temperature-insensitive photochemical reactions, while the right side (τ^{-1}) is typically very sensitive to temperature changes, as much of the downstream reactions are enzyme-catalysed (Fig. 3.2). Photosynthetic organisms growing at low temperatures are unbalanced at the level of photostasis due to a reduction in τ^{-1}. In addition to temperature, a wide number of environmental stresses can cause a

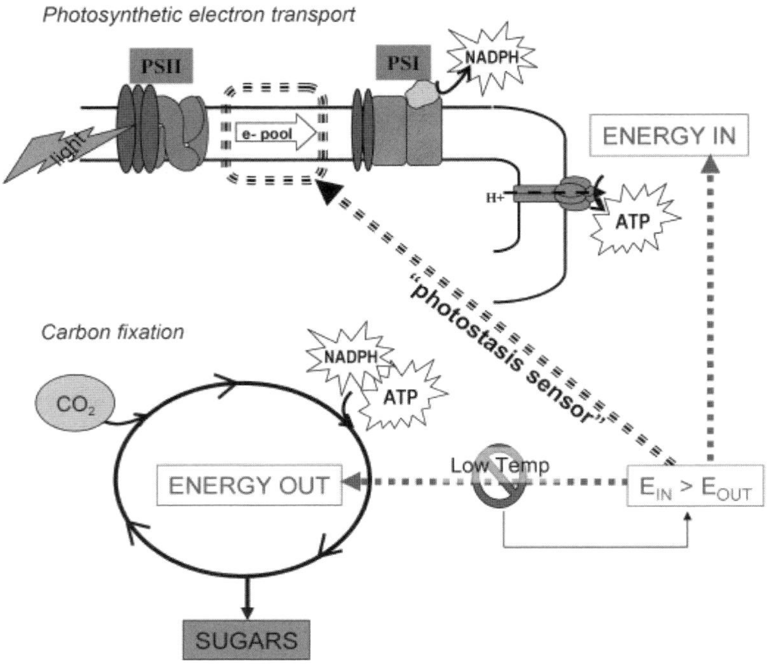

Fig. 3.2. The energy balance between photosynthetic electron transport and fixation of inorganic carbon. In oxygenic photosynthetic organisms, light absorption and energy transduction are catalysed by two multi-protein pigment-binding complexes, photosystem I (PS I) and photosystem II (PS II). Electron transport between PS II and PS I is linked by lipid-soluble electron carriers. The reduction state of the electron carrier pool (plastoquinone, PQ) is a major sensor of imbalances between energy produced in the form of NADPH and ATP, and consumption of these energy carriers via carbon fixation. Low temperatures disrupt this energy balance by preferentially reducing rates of carbon fixation.

disruption in the balance between these processes. Perturbations in photostasis are manifested as over-reduction of intersystem electron transport, and the redox state of the electron pool is thought to be a major sensor of energy imbalances (Fig. 3.2). The redox state of intersystem electron transport (also termed 'excitation pressure') can be estimated by *in vivo* pulse amplitude-modulated fluorescence as either $1 - qP$ (where qP is photochemical quenching) or $1 - qL$ (where qL is the fraction of oxidized PS II reaction centres). Photosynthetic organisms possess varying degrees of physiological plasticity to respond to conditions leading to high excitation pressure.

In a temperate plant or algal species, short-term low temperature stress impacts photostasis by reducing the rate of turnover of electron sinks (Fig. 3.3A). In response to this stress, photosynthetic organisms rely on a number of acclimatory mechanisms to either reduce light energy absorbed/transduced or to increase the number of electron sinks. Reduction of σ_{PSII} involves decreasing the size of the light-harvesting apparatus in combination with reducing the effective size of PS II by increasing processes which dissipate excess absorbed light energy harmlessly as heat (Maxwell *et al.*, 1995; Falkowski and Chen, 2003; Wilson *et al.*, 2006). Dissipation of excitation pressure and maintenance of photostasis can also be attained by increasing electron sink capacity (τ^{-1}). However, many temperate organisms are unable to adjust τ^{-1} when exposed to low temperatures, due to the thermodynamic effects of low temperatures on enzyme reaction rates.

A

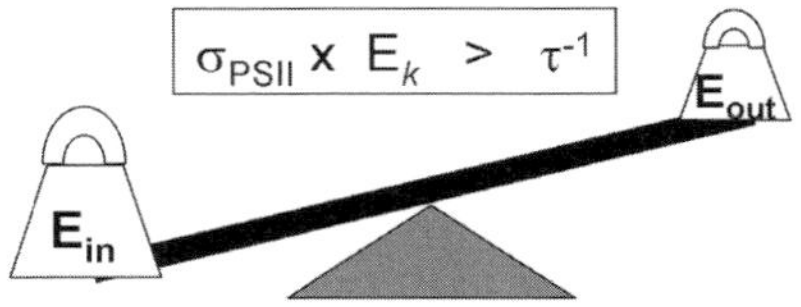

B

Organism	*Chlorella vulgaris*		*Chlamydomonas raudensis UW0241*	
Irradiance level	LL	HL	LL	HL
Culture appearance				
Redox state (1 - *qP*)	0.20	0.60	0.10	0.35
Growth rate (day^{-1})	0.24	0.24	0.24	0.48

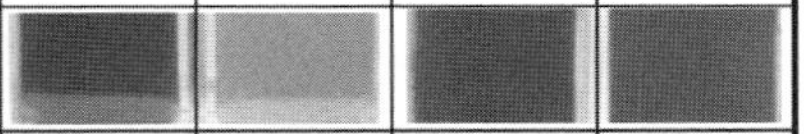

Fig. 3.3 (a) Impact of low temperatures on photostasis. Low temperatures reduce rates of downstream metabolic reactions which consume photochemically derived energy carriers, causing an imbalance between energy produced (E_{in}) and energy consumed (E_{out}). (b) Mesophilic (*Chlorella vulgaris*) and psychrophilic (*Chlamydomonas raudensis*) algae exhibit differential abilities to maintain energy balance under low temperatures. σ_{PSII}, effective cross section of the PS II; E_k, irradiance at which the maximum photosynthetic quantum yield balances the photosynthetic capacity estimated from a photosynthetic light response curve; τ^{-1}, rate at which photosynthetic electrons are consumed by a terminal electron acceptor, such as CO_2. LL, low light; HL, high light.

3.4 Psychrophilic Phototrophs: 'Photopsychrophiles'

Psychrophilic photoautotrophs have the unique ability not only to thrive and reproduce in permanently low temperature environments, but they also possess the ability to transduce light energy into stored chemical energy, and utilize the stored energy to fix inorganic carbon into simple sugar molecules. The combination of low temperatures in the presence of light is potentially one of the most damaging environmental extremes; however, this sub-group of psychrophiles have presumably adapted to maintain photostasis in these conditions. Despite this remarkable ability, studies on the photopsychrophiles have lagged behind research on non-photoautotrophic psychrophiles, and are often overlooked in reviews focused on psychrophiles (D'amico *et al.*, 2006). In this section, we summarize what is known about the few members of this sub-group of psychrophiles that have been studied, and how these studies can inform us of the way climate change may impact the growth, productivity, and survival of these extremophilic life forms.

3.4.1 Permanent low temperatures and membrane composition

Psychrophilic microorganisms have successfully colonized all known permanently cold environments on earth, chiefly through their ability to overcome two major problems associated with low temperatures: temperature effects on biochemical reactions and the increase in viscosity of aqueous environments (Thomas and Deickmann, 2002; D'amico *et al.*, 2006). The effect of membrane rigidification is particularly problematic for photosynthetic organisms, since the process of light energy transduction via the photosynthetic electron transport chain is dependent upon membrane–protein interactions and membrane-soluble electron transporters. Maintenance of membrane fluidity is a key low temperature adaptation in photosynthetic psychrophiles, and involves modifications in the fatty acid tails which include incorporation of polyunsaturated and short-chain fatty acids.

Numerous studies have implicated the role of lipids in protection against low temperature stress in temperate plants and photosynthetic microorganisms (including Gombos *et al.*, 1992; Kodama *et al.*, 1995; Moon *et al.*, 1995; Allakhverdiev *et al.*, 2001; Ariizumi *et al.*, 2002). The role of unsaturation in lipid membranes is one of the most thoroughly studied mediators of cold adaptation in psychrophilic microorganisms (Margesin *et al.*, 2008). Polyunsaturated fatty acids (PUFAs, i.e. fatty acids harbouring >1 unsaturated double bond) have been widely reported in low temperature-adapted sea ice diatoms, dinoflagellates, green algae and macroalgae (Nichols *et al.*, 1989, 1993; Henderson *et al.*, 1998; Mock and Kroon, 2002a,b; Thomson *et al.*, 2004; Ortiz *et al.*, 2006; Fogliano *et al.*, 2009). A number of Antarctic cyanobacterial isolates also exhibit relatively high levels of PUFAs, including very high levels of arachidonic acid (Pushparaj *et al.*, 2008), an essential fatty acid in human diets. High PUFA content aids tolerance to extreme water stress in Antarctic moss species (Wasley *et al.*, 2006). Mock and Kroon reported that maintenance of photosynthetic function in three sea ice diatoms under low temperatures during acclimation to either nitrogen limitation or low irradiance is dependent upon membrane lipid composition (Mock and Kroon, 2002a,b). High levels of unsaturated fatty acids in photosynthetic membranes are important for assembly of major photochemical apparatus (Murata and Wada, 1995). In the Antarctic marine red alga, *Palmaria decipiens*, the ratio of saturated to unsaturated fatty acids in the photosynthetic membranes was proposed to play a role in maintaining an active repair cycle for D1, a major PS II reaction centre protein, under high light stress (Becker *et al.*, 2010). In contrast with these reports on psychrophilic photosynthetic microorganisms, the highly frost-tolerant Antarctic vascular plant, *Deschampsia antarctica*, exhibited a typical degree of membrane fatty acid unsaturation compared with temperate species (Zuniga *et al.*, 1994).

The Antarctic lake alga, *Chlamydomonas raudensis* UWO241, is one of the most well-characterized photopsychrophiles to date. Adaptation to a wide range of environmental extremes has resulted in a number of unique adaptive strategies, including the fatty acid composition of both the cell and photosynthetic membranes (Table 3.1). While distribution of lipid species in cells of *C. raudensis* UWO241 is typical of green algal species, the fatty acid profile of the psychrophilic chlorophyte differs significantly from the model mesophilic green algal species, *C. reinhardtii*. These include novel unsaturated fatty acids with double bonds located close to the polar lipid head group. Most notable, fatty acid moieties associated with plastid-specific galactolipids possess very high levels of PUFAs (Morgan-Kiss *et al.*, 2002), indicating that low-temperature adaptation of the photosynthetic membranes is critical for maintaining functional membrane-associated photosynthetic processes at permanent low temperatures. However, it appears that highly unsaturated membranes are not an absolute requirement for psychrophily in photoautotrophs. The psychrophilic Antarctic mat alga, *Chlorella* BI sp., possesses relatively reduced trienoic and tetraenoic membrane fatty acids compared with *Chlamydomonas raudensis* UWO241 (Morgan-Kiss *et al.*, 2008). While both Antarctic species possess comparable optimal growth temperatures of ~10°C, their natural Antarctic habitats are quite distinct from each other. *C. raudensis* UWO241 is adapted to an extremely stable year-round environmental growth regime (due to a permanent ice-cap), while *Chlorella* BI sp. is exposed to high seasonal variability in multiple environmental factors, in particular its light environment (Morgan-Kiss *et al.*, 2006, 2008). One major difference between *Chlorella* BI sp. and *Chlamydomonas raudensis* UWO241 physiological capabilities is that the former is able to supplement photoautotrophy with acquisition of a variety of organic carbon sources and grows optimally under mixotrophic (a combination of phototrophy with heterotrophy) growth conditions, while *C. raudensis* is a strict photoautotroph (Morgan-Kiss *et al.*, 2006, 2008). Thus, perhaps highly unsaturated fatty acids are more critical in photosynthetic membranes than other cell membranes, as the former allow for maintenance of electron transport function in obligate photosynthetic psychrophiles such as *C. raudensis* UWO241.

Unsaturated fatty acids are synthesized via a desaturation pathway which is catalysed by a family of enzymes called desaturases. Desaturases are specific for their fatty acid substrates and the bond position. Given that psychrophilic microalgae possess multiple forms of PUFAs, it seems likely that they harbour a suite of desaturase enzymes with varying specificity. Recent studies have identified genes that encode fatty acid desaturases, which are essential for the conversion of saturated to unsaturated bonds in membrane lipids. *Chlorella vulgaris* NJ-7, a psychrotrophic Antarctic alga collected from a transitory pond, exhibits

Table 3.1. *Chlamydomonas raudensis* UWO241 is a polyextremophile. More than a decade of research on this enigmatic green alga has produced data on the most well-understood photopsychrophile to date. PSI and PSII, photosystems I and II, respectively; LHCI and LHCII, light harvesting I and II, respectively.

Extreme condition	Physical description	Growth limitations	Physiological consequence	References
Constant physical conditions	No environmental variability	Many	Loss of many acclimatory mechanisms	Morgan-Kiss *et al.*, 2002b Morgan-Kiss *et al.*, 2005 Szyszka *et al.*, 2007
Salinity	Hypersaline	min ~ 9 PSU opt ~ 40 PSU max ~ 150 PSU	Halotolerant	Pocock *et al.*, 2011
Light quality	Blue-green	Blue light requirement	High PSII:PSI ratio; reduced PSI and LHCI; inability to grow in red light	Morgan *et al.*, 1998 Morgan-Kiss *et al.*, 2005
Oxygen tension	High (300%)	None known	Unknown	
Irradiance	Shade	min ~ 1 μmol opt ~ 20 μmol max ~ 250 μmol	Large LHCII antenna; efficient energy transfer between LHCII and PSII; sensitivity to photoinhibition; reduced eyespot	Morgan *et al.*, 1998
Temperature	Low	min = 0°C opt = 10°C max =16°C	Psychrophilic; high levels of polyunsaturated fatty acids; thermolabile photosynthetic apparatus; cold adaptation of enzymes?	Morgan-Kiss *et al.*, 2002a
Winter darkness	5 months' dark	None known	Functional down-regulation of PSOII	Morgan-Kiss *et al.*, 2006

high levels of Δ12-UFAs (16:2, 16:3, 18:2, 18:3) compared with other psychrophilic microalgae (Lu *et al.*, 2009). The latter researchers identified two novel Δ^{12} fatty acid desaturases, CvFAD2 and CvFAD6, encoding microsomal and plastid enzymes, respectively. While low temperatures were associated with increases in transcript levels of both genes, mRNAs of the plastid enzyme accumulated only at temperatures close to freezing, while expression of CvFAD2 was induced under moderate (15°C) or extreme (2°C) low temperatures (Lu et al., 2009, 2010). This suggests that the degree of unsaturation may be regulated differentially in chloroplast versus cell membranes. Maintenance of membrane fluidity is essential for optimal photosynthetic function, including folding of multi-subunit photochemical complexes, electron transport (Margesin *et al.*, 2008), and the photosystem II repair cycle (Kanervo *et al.*, 1997). The Antarctic *Chlamydomonas* sp. ICE-L, which was isolated from the bottom of Antarctic fast ice, requires growth temperatures below 15°C, and exhibits substantially higher PUFA content compared with a mesophilic *Chlamydomonas* sp. (Liu *et al.*, 2006). An omega-3 fatty acid desaturase (CiFAD3) that is important in catalysing the desaturation of n-6 fatty acids to n-3 fatty acids, an essential step in PUFA production, was identified in this organism. Expression of CiFAD3 was induced in response to cold stress, freeze–thawing and salinity stress (Zhang *et al.*, 2010).

Some cold-adapted microalgae exhibit the ability to dramatically modulate their unsaturated fatty acid content in response to variations in their thermal environment. In addition, acclimation to lower temperatures is associated with significant increases in specific PUFAs. For example, a cold-adapted prymnesiophyte responded to a reduction in temperature from 5°C to 2°C by inducing an increase in the levels of octadecapentaenoic acid (18:5w3) and octadecatetraenoic acid (18:4w3) (Okuyama *et al.*, 1992). Two psychrotrophic *Chlorella* sp. from Antarctica (*C. vulgaris* NJ-7 and NJ-18) responded to an increase in growth temperature from 4°C to 30°C by increasing the abundance of C18:2 and decreasing levels of C18:3 (Hu *et al.*, 2008). On the other hand, high levels of trienoic fatty acids in combination with an inability to modulate the degree of membrane fatty acid unsaturation may affect rates of photosynthesis at elevated temperatures in *Chlamydomonas raudensis* UWO241 (Morgan-Kiss *et al.*, 2006; Pocock *et al.*, 2007). Photopsychrophiles that do not possess the ability to modulate fatty acid composition to changes in their environment will probably be more at risk of loss of membrane function during climate warming, including an enhanced risk of lipid peroxidation (Caldwell *et al.*, 2007).

3.4.2 Photochemistry

Photochemical apparatus structure and function

Regardless of the environmental stress, all photoautotrophic organisms must maintain a balance between energy produced and energy utilized to avoid the detrimental effects of photo-oxidative stress caused by over-excitation of the photochemical apparatus (Hüner *et al.*, 1995). As discussed in section 3.3, low temperatures impact the energy budget by causing an imbalance between the photochemical events and downstream metabolic pathways that consume photochemically derived energy products. The redox state of the photochemical electron pool is an important early signal of excessive excitation energy, or high excitation pressure (Fig. 3.2). In temperate phototrophic organisms, excitation pressure is relatively high ($1\text{-}qP \geq 0.7$) in low-temperature grown temperate plants (Gray *et al.*, 1996, 1997; Savitch *et al.*, 2001) and microalgae (Maxwell *et al.*, 1994, 1995; Savitch *et al.*, 1996; Wilson *et al.*, 2003). The low temperature-induced high excitation pressure is largely due to an inability to utilize much of the light energy at growth temperatures suboptimal for temperate plants and algae, resulting in a reduced τ^{-1} (Fig. 3.3A). In contrast with low temperature-grown temperate algae, *C. raudensis* UWO241 cultures grown at their optimal low growth temperature of 8°C exhibited a remarkable capacity to maintain an oxidized intersystem electron transport pool compared with the mesophilic type strain *C. raudensis* SAG 49.72 (Szyszka *et al.*, 2007). In addition, trends in excitation pressure in response to increasing growth irradiance were comparable between 8°C-grown *C. raudensis* UWO241 cultures versus 29°C-grown *C. raudensis* SAG 49.72. Similarly, 8°C-grown cultures of the Antarctic pond alga, *Chlorella* BI sp., exhibit the ability to maintain low excitation pressure under a variety of trophic growth states (i.e. growth in the presence/absence of light and/or an organic carbon source), while low temperature-grown cultures of the mesophilic alga, *Chlorella vulgaris* exhibited variable $1 - qL$ levels, depending on the trophic condition (Jaffri, 2011). These results indicate that psychrophilic algae have adapted photosynthetic processes to maintain a comparable energy balance under low temperatures as that of the mesophilic species grown at moderate temperatures.

The underlying structural and functional mechanisms contributing to this enhanced ability in *C. raudensis* UWO 241 to maintain an oxidized electron transport pool appear to be distinct. Under an optimal growth temperature regime, *C. raudensis* UWO241 possesses an unbalanced ratio PS II:PS I due to reduced levels of PS I reaction centre proteins in addition to all associated light-harvesting complex I (LHC I) polypeptides (Table 3.1; Morgan *et al.*, 1998). Novel PS I proteins are also sites of phosphorylation in the psychrophilic lake alga (Morgan-Kiss *et al.*, 2005; Szyszka *et al.*, 2007). In addition, light-harvesting complex II (LHC II) appears to be significantly larger in the psychrophile photochemical apparatus, as shown by extremely low chlorophyll a/b ratios and high levels of the oligomeric form of LHC II (Morgan *et al.*, 1998). These structural changes in the photochemical apparatus in *C. raudensis* UWO241 are likely to play a role in novel functional mechanisms utilized by this organism to maintain photostasis at permanent low temperatures and extreme shade. For example, the psychrophile possesses a relatively high capacity for energy dissipation which appears to rely on constitutive rather than induced processes to dissipate excess excitation energy (Szyszka *et al.*, 2007). Specifically, a major energy dissipation mechanism in the majority of studied temperate microalgae is heat loss through specific xanthophylls whose production is controlled by the trans-thylakoid pH. The psychrophilic *C. raudensis* UWO241 has some capacity for the xanthophyll cycle-mediated energy-quenching pathway, but also relies on additional alternative dissipative mechanisms, including a high capacity for PS II reaction centre-dependent quenching (Szyszka *et al.*, 2007).

Psychrophilic algae respond to low temperature-induced over-excitation of the photosynthetic apparatus differentially. As mentioned above, growth at low temperatures induces a higher redox state in a temperate alga compared with a psychrophilic alga. Moreover, low temperatures combined with higher light levels exacerbates this response which is manifested as an increase in $1 - qP$ in a mesophilic species (Fig. 3.3B; Maxwell *et al.*, 1994). The rise in PS II excitation pressure is due to an imbalance in energy absorbed and energy utilized, and is manifested in temperate algal species as functional and structural down-regulation of the photochemical apparatus (Maxwell *et al.*, 1994, 1995). This energy imbalance is due to low temperature-associated restrictions in photosynthetic carbon metabolism (Savitch *et al.*, 1996). Similar responses to low temperature have been observed in other mesophilic green algae (Escoubas *et al.*, 1995; Sukenik *et al.*, 1988; Masuda *et al.*, 2003) and cyanobacteria (Miskiewicz *et al.*, 2000, 2002), indicating that maintaining redox poise by down-regulation of the photochemical apparatus is a global mechanism across diverse mesophilic photoautotrophs to deal with low temperature-associated limitations in downstream energy utilization pathways. On the other hand, cultures of the low temperature-grown *C. raudensis* UWO241 exhibit minimal changes in culture pigmentation when shifted from low to high light, but maintain a low PS II excitation pressure (Fig. 3.3B). Biochemical analyses of the thylakoid apparatus showed that unlike the mesophilic *C. vulgaris*, *C. raudensis* UWO241 exhibits minimal changes in either abundance of major light harvesting proteins or distribution of photosynthetic pigments (Morgan-Kiss *et al.*, 2006). Furthermore, acclimation to high light is accompanied by an increase in growth rate (Fig. 3.3B) in conjunction with maintenance of high rates of gross photosynthesis and an exceptionally high quantum yield of O_2 evolution (Pocock *et al.*, 2007). It is likely that photopsychrophiles possess the ability to maintain redox poise at low temperatures via a higher capacity for photosynthetic energy utilization, most likely at the level of carbon fixation rates (Morgan-Kiss *et al.*, 2006).

Photochemistry and temperature

The photochemical apparatus is one of the most thermally sensitive complexes (Schreiber and Berry, 1977; Armond *et al.*, 1980; Berry and Bjorkman, 1980). While the temperature optimum for photosynthesis in temperate organisms can be broad due to daily variations in the daytime thermal environment, inhibition of photosynthesis and losses in primary productivity can occur at temperatures above the optimum. It is likely that global warming will result in thermal environments above the optimum for photosynthesis, and that the impact of thermally induced declines in primary production will be exacerbated in cold-adapted photosynthetic organisms. The manifestations of thermal stress in photosynthetic organisms include a loss of oxygen evolving complex activity (Yamane *et al.*, 1998), dissociation of LHC II from PS II, alterations in antenna oligomerization state (Armond *et al.*, 1980), and an impairment in electron transport activity (Murakami *et al.*, 2000). Recovery from heat-induced photodamage of PS II and the efficiency of the PS II photosynthetic repair machinery is also impaired at high temperatures (Takahashi *et al.*, 2004). A direct effect of impaired photosynthetic electron transport is light-induced production of reactive oxygen species (ROS). Thus, photoinhibition due to thermal stress will also increase the potential for ROS-associated damage in a wide range of cellular targets, including peroxidation of lipids harbouring PUFAs (Lesser, 2006). Global climate change, specifically thermally-induced oxidative damage, is thought to be the primary cause of coral bleaching events around the world (Hoegh-Guldberg, 1999; Lesser, 2004). The bleaching events are a direct effect of photo-oxidative damage of the symbiotic photosynthetic partners of the corals, and the extent of the bleaching response is affected by the thermal tolerance of the algal partner (Abrego *et al.*, 2008). Last, the effect of thermally induced damage to photosynthetic processes will be exacerbated in polar environments where the hole in the ozone layer is maximal, and photosynthetic communities are exposed to high levels of UV radiation.

The heat lability of the photochemical apparatus of the psychrophilic *C. raudensis* UWO241 was explored (Morgan-Kiss *et al.*, 2002). Thermally induced increases in the room temperature fluorescence parameter F_0 (F_0, minimal dark fluorescence, originating from light-harvesting apparatus) have been attributed to dissociation of LHC II from PS II and concomitant alterations in antenna organization. The maximum for temperature-induced F_0 yield in the psychrophile was 10°C lower than that of the mesophilic species *C. reinhardtii*. This was accompanied by higher thermolability of both core complexes of PS I and PS II in the psychrophile. Heat treatment of photosynthetic membranes was also accompanied by alterations in the energy distribution between PS II and PS I, manifested as a shift in light energy to favour PS I photochemistry. This response can be observed at the level of shifts in energy distribution using low temperature (77K) chlorophyll-a fluorescence (Sane *et al.*, 1984) or a stimulation in rates of P700 (PS I reaction centre) photo-oxidation (Gounaris *et al.*, 1983). This response in temperate organisms is an effect of heat-induced disruption of the photosynthetic membranes, but may also be an important mechanism for protection of PS I when PS II is preferentially inhibited (Morgan-Kiss *et al.*, 2006). However, heat-treated cells of *C. raudensis* 241 did not exhibit either an energy redistribution or a stimulation in PS I activity (Morgan-Kiss *et al.*, 2002). Thus, not only is the photochemical apparatus of the psychrophilic *C. raudensis* 241 more thermally sensitive, the effects of heat stress are differential compared with mechanisms common across many temperate plants and algae.

Thermal effects on photoinhibition and oxidative stress

Photoinhibitory stress occurs under any conditions of over-excitation of the photosynthetic apparatus which results in inhibition of PS II activity. In natural environments, a large number of stressful environmental conditions exacerbate photoinhibitory stress, including high salinity, low temperature, high temperature, drought and nutrient deficiency (Berry and Bjorkman, 1980; Allakhverdiev and Murata, 2004; Murata *et al.*, 2007). The extent of photoinhibition is a balance between avoidance of the toxic effect of excess light absorption and the repair of photo-damaged PS II (Aro *et al.*, 1993). ROS play an important role in the photoinhibitory process. In early studies, it was proposed that ROS plays a role in damage to PS II: first by the direct production of ROS either by excessive reduction of Q_A, the primary acceptor of PS II, or from triplet chlorophyll species; and second by directly attacking and damaging PS II reaction centres (Aro *et al.*, 1993). However, an alternate model for the role of ROS in photoinhibition is ROS-mediated inhibition in the repair pathway of PS II, in particular inhibition of D1 synthesis in the presence of either H_2O_2 or $^1O_2$ (Nishiyama *et al.*, 2001, 2004).

Chronic photoinhibition results under any stressful environment where rates of PS II damage exceed rates of repair (Melis, 1999). Since ROS is a major player in photoinhibitory damage, it is also thought that inhibition of PS II repair is one of the key sites for exacerbation of photoinhibition under a variety of environmental stresses (Murata *et al.*, 2007). Low temperatures exacerbate photoinhibitory damage (Hüner *et al.*, 1993) and impact repair of PS II in temperate organisms by inhibiting de novo synthesis of the PS II reaction centre protein, D1 (Aro *et al.*, 1990). Many cold-adapted species exhibit constitutive tolerance to photoinhibitory stress at low temperatures. Psychrotolerant alpine plant species possess higher resistance than their temperate counterparts to light-induced photoinhibition (Shang and Feierabend, 1998; Streb *et al.*, 2003). The photochemical apparatus of the psychrophilic diatom *Fragilariopsis cylindrus* possesses the remarkable ability to tolerate light-induced photoinhibition under freezing conditions. A shift from +5°C to −1.8°C (a temperature below the freezing point of seawater) elicited only a modest decline in photosynthetic quantum yield which fully recovered within one day (Thomson *et al.*, 2004).

Pocock *et al.* (2007) studied the effect of temperatures on the extent of photoinhibition and capacity for repair in the psychrophilic *C. raudensis* UWO241. Not surprisingly, the psychrophile exhibited low sensitivity to low temperature-induced photoinhibition compared with the temperate alga, *C. reinhardtii*. However, the extent of photoinhibition was temperature independent in the psychrophile: cultures exposed to 29°C exhibited comparable tolerance to those exposed to 8°C, despite the fact that the upper temperature was far above the permissive temperature for growth in the psychrophile. In contrast, full recovery from photoinhibitory damage was inhibited in cultures of the psychrophile exposed to 29°C. This impairment was accompanied by a significant reduction in the ability to replace the reaction centre protein D1, indicating the D1 repair cycle is very thermally sensitive in the psychrophilic alga. Thus, in agreement with others (Murata *et al.*, 2007), recovery from photoinhibition rather than greater resistance to photoinhibition is a critical adaptation to low temperatures in the psychrophilic *C. raudensis* UWO241. Since ROS has been implicated to play a key role in an organism's capacity to recover from photoinhibition

(Nishiyama *et al.*, 2001, 2004), it is interesting to speculate that the repair cycle of *C. raudensis* may be either less sensitive to ROS, or that the psychrophile possesses enhanced ability to restrict ROS production at low growth temperatures.

Temperature and circadian cycles

Many photosynthetic processes are controlled by the daily photoperiod in plants and algae (Harmer, 2009), and circadian clocks have been shown to confer a significant advantage in plants at the level of higher growth and survival rates (Dodd *et al.*, 2005). Polar algal species are adapted to naturally occurring seasonal changes from continuous irradiance in the summer to complete darkness in the polar winter. This extreme range in irradiance levels combined with the rapid change in the photoperiod during the transition between seasons makes polar algae ideal candidates for studying the influence of photoperiod on the periodicity of photosynthetic processes. Physiological changes in *C. raudensis* UWO241 exposed to seasonal periodicity (i.e. the transition between polar summer and winter) were investigated. It appears that this Antarctic lake alga maintains a functional photosynthetic apparatus during the polar winter which is similar to over-wintering evergreens that can quickly shift to a functional apparatus during the transition from polar winter to summer (Morgan-Kiss *et al.*, 2006). In addition to the seasonal circadian response, *C. raudensis* exhibited rhythmic changes in PS II maximum photochemical efficiency in response to a daily photoperiod, with maximal levels occurring during mid-day (Fig. 3.4). A diel pattern in photochemical activity has also been reported in natural sea ice diatoms communities (Aikawa *et al.*, 2009), indicating that diel control over photosynthetic activity is important in psychrophilic photoautotrophs for maintaining efficient photosynthetic rates. Furthermore, the diel pattern of PS II photochemical efficiency was disrupted when cultures of *C. raudensis* UWO241 were grown at temperatures above the optimum (Fig. 3.4). In temperate plants, circadian clocks have evolved to avoid the effect of temperature disruption to the timing of daily rhythms via temperature compensation mechanisms (Eckardt, 2006). However, the results shown in Fig. 3.4 suggest that elevated temperatures could also impact the ability of a cold-adapted photoautotroph to regulate photochemical function under diurnally changing irradiance levels, and interfere with daily maintenance of photosynthetic efficiency.

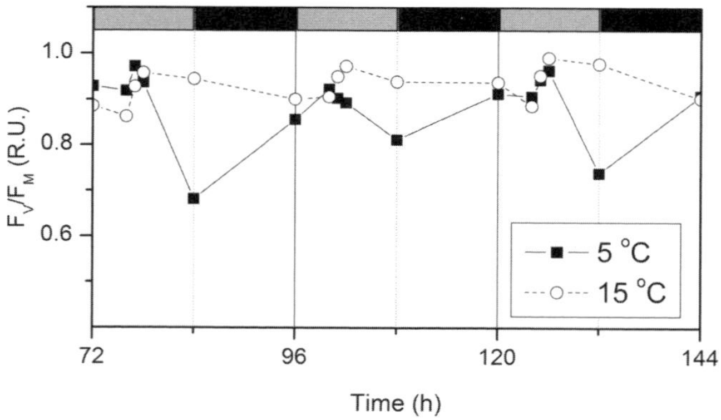

Fig. 3.4. Daily variation in the maximum PS II photochemical efficiency (F_V/F_M, normalized to daily maximum values) in the Antarctic lake alga *Chlamydomonas raudensis* UWO241 during a 12 h:12 h day/night growth cycle. Cultures were grown at either the optimal growth temperature (8°C) or high temperature (15°C). Grey bars, day; black bars, night.

3.5 Enzymes

An important criterion for adaptation to permanent low temperatures is molecular adaptation of enzymes to maintain appropriate reaction rates of enzyme-catalysed reactions. This can be accomplished by either producing higher levels of enzyme concentrations or adapting enzymes to function better at low temperatures. A number of cold-active enzymes have been characterized from psychrophilic photo-autotrophs, including a thermally labile nitrate reductase and agininosuccinase lyase from an Antarctic *Chloromonas* sp. (Loppes *et al.*, 1996). An NADH-nitrate reductase displaying both higher activity at colder temperatures as well as higher thermolability compared with a mesophilic enzyme has also been isolated from the Antarctic alga *Koliella antarctica* (Di Martino *et al.*, 2006). Here we focus on what is known regarding low temperature adaptation of a critical enzyme for fixation of inorganic carbon.

3.5.1 RubisCO

Ribulose-1,5-bisphosphate carboxylase/oxygenase (RubisCO; EC 4.1.1.39) catalyses the key step in the CBB cycle, making it one of the primary determinants of plant and algal primary productivity. This step involves carboxylation of ribulose-1,5-bisphosphate (RuBP) to produce two molecules of phosphoglyceric acid (PGA) that can be used to synthesize cellular biomass. RubisCO also displays oxygenase activity, the degradation of RuBP to PGA and phosphoglycolate. The decline of RubisCO activity at low temperatures has been associated with decreases in catalytic turnover (Bruggemann *et al.*, 1992; Hirotsu *et al.*, 2005) and activation state (Byrd *et al.*, 1995). In plants, regeneration of the substrate for RubisCO (RuBP) via the CBB cycle is also affected by environmental conditions, leading to RuBP-limited rates of photosynthesis (Sage *et al.*, 1990; Zhu *et al.*, 2004). The activation state of RubisCO, which is under the control of the enzyme RubisCO activase (Rca), is influenced by environmental conditions including light, temperature and CO_2 concentration, as well as stromal ADP/ATP ratios (Byrd *et al.*, 1995; Portis Jr., 2003). While RubisCO catalytic activity is sensitive to low temperatures, it appears that activation of the Rca enzyme itself is sensitive to high temperature in temperate and Antarctic plant species (Salvucci and Crafts-Brandner, 2004; Yamori *et al.*, 2006). Investigation into carbon metabolism in cold-adapted algae has been limited to a few studies (Devos *et al.*, 1998; Haslam *et al.*, 2005), and despite its importance in carbon acquisition, a cold-adapted RubisCO has yet to be discovered in a psychrophilic photoautotroph. Devos *et al.* (1998) reported that RubisCO isolated from two psychrophilic *Chloromonas* species exhibited an increase in thermolability compared with the mesophilic *Chlamydomonas reinhardtii* enzyme; however, neither of the psychrophilic RubisCO exhibited changes in catalytic activity at low incubation temperatures.

The temperature-dependence of RubisCO carboxylase activity in the psychrophilic *C. raudensis* UWO241 was recently investigated. Carboxylase activity of the psychrophilic RubisCO exhibited maximum carboxylase turnover rates at 25°C while the temperature for maximum activity in a mesophilic species (*C. raudensis* SAG 49.72) was 40°C (Fig. 3.5A). In addition, 8°C-grown cultures of *C. raudensis* UWO241 exhibited a linear relationship between growth irradiance and *in vivo* RubisCO activity (Fig. 3.5B). These results indicated that *C. raudensis* UWO241 possesses a cold-active RubisCO and that unlike temperate algal species (Savitch *et al.*, 1996), it possesses the ability to modulate RubisCO activity in response to higher irradiance environments. Thus the cold-adapted RubisCO is a major mechanism for maintaining an oxidized electron transport pool and avoiding over-excitation of the photochemical apparatus at low temperatures in the psychrophilic alga (Fig. 3.3A). However, higher environmental temperatures could impact rates of CO_2 fixation in psychrophilic photoautotrophs with cold-adapted and/or thermally labile RubisCO.

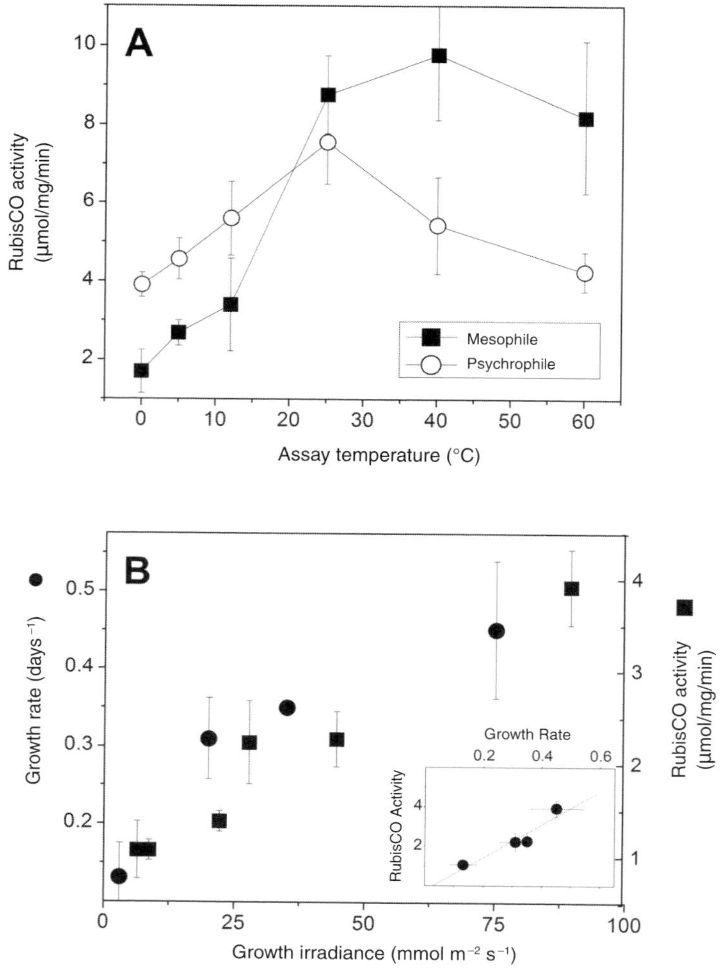

Fig. 3.5. (a) Temperature dependence of maximum RubisCO carboxylase activity (expressed as µmol PGA produced/mg protein/min) in soluble lysates extracted from *Chlamydomonas raudensis* SAG 49.72 (mesophile) versus *Chlamydomonas raudensis* UWO241 (psychrophile). (b) Comparison of specific growth rates and RubisCO carboxylase activity in response to variable growth irradiance in low temperature-grown (8°C) cultures of the Antarctic lake alga, *Chlamydomonas raudensis* UWO241.

3.6 Conclusion and Future Directions

Photosynthetic organisms play critical roles in low temperature ecosystems as the major producers of fixed carbon. Global climate change is rapidly changing polar and alpine aquatic habitats for photosynthetic life forms. Focused laboratory studies on isolated photopsychrophiles indicate that photopsychrophiles exhibiting limited physiological plasticity are likely to be more sensitive to the effects of climate change. For example, studies on cold-adapted microalgal species exhibiting thermosensitivity of key photochemical processes as well as enzymes critical for carbon fixation suggest that climate change is likely to have a detrimental effect in thermally sensitive photoautotrophic populations. On the other hand, it is clear that some psychrophilic photoautotrophs possess the ability to respond to short-term perturbations in their environment. Differential sensitivities to climate change could be manifested as reductions in rates of primary productivity key photosynthetic species and/or shifts in trophic function in the protist populations. Anticipating the ultimate impacts of climate change on the photosynthetic life forms residing in these sensitive ecosystems will require expanding our understanding of the physiology of photopsychrophiles, and improving integration of field work on natural communities with laboratory studies on psychrophilic isolates.

References

Abrego, D., Ulstrup, K., Willis, B.L. and van Oppen, J.H. (2008) Species-specific interactions between algal endosymbionts and coral hosts define their bleaching response to heat and light stress. *Proceedings of the Royal Society B* 275, 2273–2282.

Aikawa, S., Hattori, H., Gomi, Y., Watanabe, K., Kudoh, S., Kashino, Y. and Satoh, K. (2009) Diel tuning of photosynthetic systems in ice algae at Saroma-ko Lagoon, Hokkaido, Japan. *Polar Science* 3, 57–72.

Alberdi, M., Bravo, L.A., Gutiérrez, A., Gidekel, M. and Corcuera, L.J. (2002) Ecophysiology of Antarctic vascular plants. *Physiologia Plantarum* 115, 479–486.

Allakhverdiev, S.I. and Murata, N. (2004) Environmental stress inhibits the synthesis de novo of proteins involved in the photodamage-repair cycle of Photosystem II in *Synechocystis* sp. PCC 6803. *Biochimica et Biophysica Acta – Bioenergetics* 1657, 23–32.

Allakhverdiev, S.I., Kinoshita, M., Inaba, M., Suzuki, I. and Murata, N. (2001) Unsaturated fatty acids in membrane lipids protect the photosynthetic machinery against salt-induced damage in *Synechococcus*. *Plant Physiology* 125, 1842–1853.

Alley, R.B., Marotzke, J., Nordhaus, W.D., Overpeck, J.T., Peteet, D.M., Pielke, Jr R.A., Pierrehumbert, R.T., *et al.* (2003) Abrupt climate change. *Science* 299, 2005–2010.

Ariizumi, T., Kishitani, S., Inatsugi, R., Nishida, I., Murata, N. and Toriyama, K. (2002) An increase in unsaturation of fatty acids in phosphatidylglycerol from leaves improves the rates of photosynthesis and growth at low temperatures in transgenic rice seedlings. *Plant and Cell Physiology* 43, 751–758.

Armond, P.A., Björkman, O. and Staehelin, L.A. (1980) Dissociation of supramolecular complexes in chloroplast membranes. A. Manifestation of heat damage to the photosynthetic apparatus. *Biochimica et Biophysica Acta* 601, 433–442.

Aro, E.M., Hundal, T., Carlberg, I. and Andersson, B. (1990) In vitro studies on light-induced inhibition of photosystem II and D1 protein degradation at low temperatures. *Biochemica et Biophysica Acta* 1019, 269–275.

Aro, E., Virgin, I. and Andersson, B. (1993) Photoinhibition of photosystem II. Inactivation, protein damage and turnover. *Biochimica et Biophysica Acta* 1143, 113–134.

Arrigo, K.R., van Dijken, G. and Pabi, S. (2008) Impact of a shrinking Arctic ice cover on marine primary production. *Geophysical Research Letters* 35, L19603.

Becker, S., Graeve, M. and Bischof, K. (2010) Photosynthesis and lipid composition of the Antarctic endemic rhodophyte Palmaria decipiens: effects of changing light and temperature levels. *Polar Biology* 33, 945–955.

Berry, J. and Bjorkman, O. (1980) Photosynthetic response and adaptation to temperature in higher plants. *Annual Review of Plant Physiology* 31, 491–543.

Bruggemann, W., van der Kooji, T.A.W. and van Hasselt, P.R. (1992) Long-term chilling of young tomato plants under low light and subsequent recovery. II. Chlorophyll fluorescence, carbon metabolism and activity of ribulose-1,5-bisphophate carboxylase/oxygenase. *Planta* 186, 179–187.

Byrd, G.T., Ort, D.R. and Ogren, W.L. (1995) The effect of chilling in the light on ribulose-1,5-bisphosphate carboxylase/oxygenase activation in tomato (Lycopersicon esculentum Mill.). *Plant Physiology* 107, 585–591.

Caldwell, M.M., Bornman, J.F., Ballare, C.L., Flint, S.D. and Kulandaivelu, G. (2007) Terrestrial ecosystems, increased solar ultraviolet radiation, and interactions with other climate change factors. *Photochemical & Photobiological Sciences* 6, 252–266.

Chapman, W.L. and Walsh, J.E. (2007) A synthesis of Antarctic temperatures. *Journal of Climate* 20, 4096–4117.

Chinnusamy, V. and Zhu, J.-K. (2009) Epigenetic regulation of stress responses in plants. *Current Opinion in Plant Biology* 12, 133–139.

Cowan, D.A., Casaneuva, A. and Stafford, W. (2007) Ecology and biodiversity of cold-adapted microorganisms. In: Gerday, C. and Glasnsdorff, N. (eds) *Physiology and Biochemistry of Extremophiles*. ASM Press, Herndon, pp. 119–132.

D'amico, S., Collins, T., Marx, J.C., Feller, G. and Gerday, C. (2006) Psychrophilic microorganisms: challenges for life. *EMBO Report* 7, 385–389.

Devos, N., Ingouff, M., Loppes, R. and Matagne, R. (1998) RUBISCO adaptation to low temperatures: a comparative study in psychrophilic and mesophilic unicellular algae. *Journal of Phycology* 34, 665–660.

Di Martino, R., Vona, V., Carfagna, S., Esposito, S., Carillo, P. and Rigano, C. (2006) Temperature dependence of nitrate reductase in the psychrophilic unicellular alga, Koliella antarctica and the mesophilic Chlorella sorokiniana. *Plant, Cell & Environment* 29, 1300–1409.

Dodd, A.N., Salathia, N., Hall, A., Kavei, E., That, R., Nagy, F., Hibberd, J.M., *et al.* (2005) Plant circadian clocks increase photosynthesis, growth, survival, and competitive advantage. *Science* 309, 630–633.

Doran, P.T., McKay, C.P., Fountain, A.G., Nylen, T., McKnight, D.M., Jaros, C. and Barrett, J.E. (2008) Hydrologic response to extreme warm and cold summers in the McMurdo Dry Valleys, East Antarctica. *Antarctic Science* 20, 499–509.

Eckardt, N.A. (2006) A wheel within a wheel: temperature compensation of the circadian clock. *The Plant Cell* 18, 1105–1108.

Escoubas, J., Lomas, M., LaRoche, J. and Falkowski, P.G. (1995) Light intensity regulation of cab gene transcription is signalled by the redox state of the plastoquinone pool. *Proceedings of the National Academy of Sciences USA* 92, 10237–10241.

Falkowski, P.G. and Chen, Y.-B. (2003) Photo-acclimation of light harvesting systems in eukaryotic algae. In: Green, B.R. and Parson, W.W. (eds) *Advances in Photosynthesis and Respiration. Light Harvesting Systems in Photosynthesis.* Kluwer Academic Publishers, Dordrecht, pp. 423–447.

Fogliano, V., Andreoli, C., Martello, A., Caiazzo, M., Lobosco, O., Formisano, F., Carlino, P.A., *et al.* (2009) Functional ingredients produced by culture of Koliella antarctica. *Aquaculture* 299, 115–120.

Franklin, K.A. (2009) Light and temperature signal crosstalk in plant development. *Current Opinion in Plant Biology* 12, 63–68.

Gombos, Z., Wada, H. and Murata, N. (1992) Unsaturation of fatty acids in membrane lipids enhances tolerance of the cyanobacterium *Synechocystis* PCC6803 to low-temperature photoinhibition. *Proceedings of the National Academy of Sciences USA* 89, 9959–9963.

Gounaris, K., Brian, A.P.R., Quinn, P.J. and Williams, W.P. (1983) Structural reorganization and functional changes associated with heat-induced phase separations of non-bilayer lipids in chloroplast thylakoid membranes. *FEBS Letters* 153, 47–52.

Gray, G.R., Savitch, L.V., Ivanov, A.G. and Huner, N. (1996) Photosystem II excitation pressure and development of resistance to photoinhibition (II. Adjustment of photosynthetic capacity in winter wheat and winter rye). *Plant Physiology* 110, 61–71.

Gray, G.R., Chauvin, L.P., Sarhan, F. and Huner, N. (1997) Cold acclimation and freezing tolerance (a complex interaction of light and temperature). *Plant Physiology* 114, 467–474.

Harmer, S.L. (2009) The circadian system in higher plants. *Annual Review Plant Biology* 60, 357–377.

Haslam, R.P., Keys, A., Andralojc, P.J., Madgwick, P.J., Andersson, I., Grimsrud, A., Eilertsen, H.C., *et al.* (2005) Specificity of diatom Rubisco. In: Omasa, K., Nouchi, I. and De Kok, L.J. (eds) *Plant Responses to Air Pollution and Global Change.* Springer-Verlag, Tokyo.

Henderson, R.J., Hegseth, E.N. and Park, M.T. (1998) Seasonal variation in lipid and fatty acid composition of ice algae from the Barents Sea. *Polar Biology* 20, 48–55.

Hirotsu, N., Makino, A., Yokota, S. and Tadahiko, M. (2005) The photosynthetic properties of rice leaves treated with low temperature and high irradiance. *Plant and Cell Physiology* 46, 1377–1383.

Hoegh-Guldberg, O. (1999) Climate change, coral bleaching and the future of the world's coral reefs. *Marine and Freshwater Research* 50, 839–866.

Hoegh-Guldberg, O. and Bruno, J.F. (2010) The impact of climate change on the world's marine ecosystems. *Science*, 1523–1528.

Hu, H., Li, H. and Xu, X. (2008) Alternative cold response modes in *Chlorella* (Chlorophyta, Trebouxiophyceae) from Antarctica. *Phycologia* 47, 28–34.

Hua, J. (2009) From freezing to scorching, transcriptional responses to temperature variations in plants. *Current Opinion in Plant Biology* 12, 568–573.

Hüner, N.P.A., Hurry, V.M., Kröl, M., Falk, S. and Griffith, M. (1993) Photosynthesis, photo-inhibition and low temperature acclimation in cold tolerant plants. *Photosynthesis Research* 37, 19–39.

Hüner, N.P.A., Maxwell, D.P., Gray, G.R., Savitch, L.V., Laudenbach, D.E. and Falk, S. (1995) Photosynthetic response to light and temperature: PSII excitation pressure and redox signalling. *Acta Physiologia Plantarum* 17, 167–176.

Jaffri, S. (2011) Characterization of the photosynthetic apparatus of *Chlorella* BI sp., an Antarctic mat alga under varying trophic growth states. MSc thesis, Miami University, Oxford, Ohio.

Jentsch, A., Kreyling, J. and Beierkuhnlein, C. (2007) A new generation of climate-change experiments: events, not trends. *Frontiers in Ecology & Environment* 5, 365–374.

Kanervo, E., Tasaka, Y., Murata, N. and Aro, E.M. (1997) Membrane lipid unsaturation modulates processing of the photosystem II reaction-center protein D1 at low temperatures. *Plant Physiology* 114, 841–849.

Kirchman, D.L., Moran, X.A.G. and Ducklow, H. (2009) Microbial growth in the polar oceans - role of temperature and potential impact of climate change. *Nature Reviews in Microbiology* 7, 451–459.

Kodama, H., Horiguchi, G., Nishiuchi, T., Nishimura, M. and Iba, K. (1995) Fatty acid desaturation during chilling acclimation is one of the factors involved in conferring low-temperature tolerance to young tobacco leaves. *Plant Physiology* 107, 1177–1185.

Lesser, M.P. (2004) Experimental biology of coral reef ecosystems. *Journal of Experimental Marine Biology and Ecology* 300, 217–252.

Lesser, M.P. (2006) Oxidative stress in marine environments: Biochemistry and physiological ecology. *Annual Review of Physiology* 68, 253–278.

Liu, C.L., Huang, X.H., Wang, X.L., Zhang, X.C. and Li, G.Y. (2006) Phylogenetic studies on two strains of Antarctic ice algae based on morphological and molecular characteristics. *Phycologia* 45, 190–198.

Loppes, R., Devos, N., Willem, S., Barthelemy, P. and Matagne, R.F. (1996) Effect of temperature on two enzymes from a psychrophilic *Chloromonas* (Chlorophyta). *Journal of Phycology* 32, 276–278.

Lu, Y., Chi, X., Yang, Q., Li, Z., Liu, S., Gan, Q. and Qin, S. (2009) Molecular cloning and stress-dependent expression of a gene encoding Δ12-fatty acid desaturase in the Antarctic microalga *Chlorella vulgaris* NJ-7. *Extremophiles* 13, 875–884.

Lu, Y., Chi, X., Li, Z., Yang, Q., Li, F., Liu, S., Gan, Q. and Qin, S. (2010) Isolation and characterization of a stress–dependent plastidial δ12 fatty acid desaturase from the antarctic microalga *Chlorella vulgaris* NJ-7. *Lipids* 45, 179–187.

Margesin, R., Schinner, F., Marx, J.-C., Gerday, C. and Russell, N.J. (2008) Membrane components and cold sensing. In: Margesin, R., Schinner, F., Marx, J.C. and Gerday, C. (eds) *Psychrophiles: from Biodiversity to Biotechnology*. Springer, Berlin/Heidelberg, pp. 177–190.

Masuda, T., Tanaka, A. and Melis, A. (2003) Chlorophyll antenna size adjustments by irradiance in *Dunaliella salina* involve coordinate regulation of chlorophyll a oxygenase (CAO) and Lhcb gene expression. *Plant Molecular Biology* 51, 757–771.

Maxwell, D.P., Falk, S., Trick, C.G. and Huner, N.P.A. (1994) Growth at low temperature mimics high-light acclimation in *Chlorella vulgaris. Plant Physiology* 105, 535–543.

Maxwell, D.P., Laudenbach, D.E. and Huner, N. (1995) Redox regulation of light-harvesting Complex II and cab mRNA abundance in *Dunaliella salina. Plant Physiology* 109, 787–795.

Meehl, G.A., Stocker, T.F., Collins, W.D., Friedlingstein, P., Gaye, A.T., Gregory, J.M., Kitoh, A. *et al.* (2007) Global climate projections. In: Solomon, S., Qin, D., Manning, M., Chen, Z., Marquis, M., Averyt, K.B., Tignor, M., *et al.* (eds) *Climate Change 2007: The Physical Science Basis*. Contribution of Working Group I to the Fourth Assessment Report of the Inter-governmental Panel on Climate Change, Cambridge University Press, Cambridge.

Melis, A. (1999) Photosystem-II damage and repair cycle in chloroplasts: what modulates the rate of photodamage in vivo? *Trends in Plant Science* 4, 130–135.

Miskiewicz, E., Ivanov, A.G., Williams, J.P., Khan, M.U., Falk, S. and Huner, N.P. (2000) Photosynthetic acclimation of the filamentous cyanobacterium, *Plectonema boryanum* UTEX 485, to temperature and light. *Plant and Cell Physiology* 41, 767–775.

Miskiewicz, E., Ivanov, A.G. and Huner, N.P. (2002) Stoichiometry of the photosynthetic apparatus and phycobilisome structure of the cyanobacterium *Plectonema boryanum* UTEX 485 are regulated by both light and temperature. *Plant Physiology* 130, 1414–1425.

Mock, T. and Kroon, B.M. (2002a) Photosynthetic energy conversion under extreme conditions--I: important role of lipids as structural modulators and energy sink under N-limited growth in Antarctic sea ice diatoms. *Phytochemistry* 61, 41–51.

Mock, T. and Kroon, B.M. (2002b) Photosynthetic energy conversion under extreme conditions. II: the significance of lipids under light limited growth in Antarctic sea ice diatoms. *Phytochemistry* 61, 53–60.

Montes-Hugo, M., Doney, S.C., Ducklow, H.W., Fraser, W., Martinson, D., Stammerjohn, S.E. and Schofield, O. (2009) Recent changes in phytoplankton communities associated with rapid regional climate change along the western Antarctic Peninsula. *Science* 323, 1470–1473.

Moon, B.Y., Higashi, S., Gombos, Z. and Murata, N. (1995) Unsaturation of the membrane lipids of chloroplasts stabilizes the photosynthetic machinery against low-temperature photoinhibition in transgenic tobacco plants. *Proceedings of the National Academy of Sciences USA* 92, 6219–6223.

Moorthi, S., Caron, D.A., Gast, R.J. and Sanders, R.W. (2009) Mixotrophy: a widespread and important ecological strategy for planktonic and sea-ice nanoflagellates in the Ross Sea, Antarctica. *Aquatic Microbial Ecology* 54, 269–277.

Morgan, R.M., Ivanov, A.G., Priscu, J.C., Maxwell, D.P. and Hüner, N.P.A. (1998) Structure and composition of the photochemical apparatus of the Antarctic green alga, *Chlamydomonas subcaudata*. *Photosynthesis Research* 56, 303–314.

Morgan-Kiss, R., Ivanov, A.G., Williams, J., Mobashsher, K. and Hüner, N.P. (2002a) Differential thermal effects on the energy distribution between photosystem II and photosystem I in thylakoid membranes of a psychrophilic and a mesophilic alga. *Biochemica et Biophysica Acta* 1561, 251–265.

Morgan-Kiss, R.M., Ivanov, A.G. and Hüner, N.P.A. (2002b) The Antarctic psychrophile, *Chlamydomonas subcaudata*, is deficient in state I-state II transitions. *Planta* 214, 435–445.

Morgan-Kiss, R.M., Ivanov, A.G., Pocock, T., Król, M., Gudynaite-Savitch, L. and Hüner, N.P.A. (2005) The Antarctic psychrophile, *Chlamydomonas raudensis* Ettl (UWO241) (CHLOROPHYCEAE, CHLOROPHYTA) exhibits a limited capacity to photoacclimate to red light. *Journal of Phycology* 41, 791–800.

Morgan-Kiss, R.M., Priscu, J.P., Pocock, T., Gudynaite-Savitch, L. and Hüner, N.P.A. (2006) Adaptation and acclimation of photosynthetic microorganisms to permanently cold environments. *Microbiology & Molecular Biology Reviews* 70, 222–252.

Morgan-Kiss, R.M., Ivanov, A., Modla, S., Cyzmmek, K., Huner, N.P.A., Priscu, J.C. and Hanson, T.E. (2008) Identity and phylogeny of a new psychrophillic eukaryotic green alga, *Chlorella* sp. strain BI isolated from a transitory pond near Bratina Island, Antarctica. *Extremophiles* 12, 701–711.

Mueller, D.R.L., Van Hove, D., Antoniades, D., Jeffries, M.O. and Vincent, W.F. (2009) High Arctic lakes as sentinel ecosystems: cascading regime shifts in climate, ice-cover, and mixing. *Limnology and Oceanography* 54, 2371–2385.

Murakami, Y., Tsuyama, M., Kobayashi, Y., Kodama, H. and Iba, K. (2000) Trienoic fatty acids and plant tolerance of high temperature. *Science* 287, 476–479.

Murata, N. and Wada, H. (1995) Acyl-lipid desaturases and their importance in the tolerance and acclimatization to cold of cyanobacteria. *Biochemistry Journal* 308, 1–8.

Murata, N., Takahashi, S., Nishiyama, Y. and Allakhverdiev, S.I. (2007) Photoinhibition of photosystem II under environmental stress. *Biochimica et Biophysica Acta - Bioenergetics* 1767, 414–421.

Nichols, D.S., Palmisano, A.C., Rayner, A.C., Smith, G.A. and White, D.C. (1989) Changes in the lipid composition of Antarctic sea-ice diatom communities during a spring bloom: an indication of community physiological status. *Antarctic Science* 1, 133–144.

Nichols, D.S., Nichols, P.D. and Sullivan, C.W. (1993) Fatty acid, sterol and hydrocarbon composition of Antarctic sea ice diatom communities during the spring bloom in McMurdo Sound. *Antarctic Science* 5, 217–278.

Nishiyama, Y., Yamamoto, H., Allakhverdiev, S.I., Inaba, M., Yokota, A. and Murata, N. (2001) Oxidative stress inhibits the repair of photodamage to the photosynthetic machinery. *EMBO Journal* 20, 5587–5594.

Nishiyama, Y., Allakhverdiev, S.I., Yamamoto, H., Hayashi, H. and Murata, N. (2004) Singlet oxygen inhibits the repair of Photosystem II by suppressing the translation elongation of the D1 protein in Synechocystis sp. PCC 6803. *Biochemistry* 43, 11321–11330.

Okuyama, H., Morita, N. and Kogame, K. (1992) Occurrence of octadecapentaenoic acid in lipids of a cold stenothermic alga, Prymnesiophyte Strain B. *Journal of Phycology* 28, 465–472.

Ortiz, J., Romero, N., Robert, P., Araya, J., Lopez-Hernández, J., Bozzo, C., Navarrete, E., *et al.* (2006) Dietary fiber, amino acid, fatty acid and tocopherol contents of the edible seaweeds Ulva lactuca and Durvillaea antarctica. *Food Chemistry* 99, 98–104.

Overland, J., Turner, J., Francis, J., Gillett, N., Marshall, G. and Tjernstrom, M. (2008) The Arctic and Antarctic: two faces of climate change. *EOS* 89, 169–170.

Parker, B.R., Vinebrooke, R.D. and Schindler, D.W. (2008) Recent climate extremes alter alpine lake ecosystems. *Proceedings of the National Academy of Sciences USA* 105, 12927–12931.

Piepenburg, D. (2005) Recent research on Arctic benthos: common notions need to be revised. *Polar Biology* 28, 733–755.

Pocock, T., Koziak, A., Rosso, D., Falk, S. and

Huner, H.P.A. (2007) Chlamydomonas raudensis Ettl (UWO241) exhibits the capacity for rapid D1 repair in response to chronic photoinhibition at low temperature. *Journal of Phycology* 43, 924–936.

Pocock, T., Vetterli, A. and Falk, S. (2011) Evidence for phenotypic plasticity in the Antarctic extremophile *Chlamydomonas raudensis* Ettl UWO 241. *Journal of Experimental Botany* 62, 1169–1177.

Portis Jr, A.R. (2003) Rubisco activase - Rubisco's catalytic chaperone. *Photosynthesis Research* 75, 11–27.

Priscu, J.C., Wolf, C.F., Takacs, C.D., Fritsen, C.H., Laybourn-Parry, J., Roberts, J.K.M. and Berry-Lyons, W. (1999) Carbon transformations in the water column of a perennially ice-covered Antarctic Lake. *Bioscience* 49, 997–1008.

Pushparaj, B., Buccioni, A., Paperi, R., Piccardi, R., Ena, A., Carlozzi, P. and Sili, C. (2008) Fatty acid composition of Antarctic cyanobacteria. *Phycologia* 47, 430–434.

Sage, R.F., Sharkey, T.D. and Pearcy, R.W. (1990) The effect of leaf nitrogen and temperature on the CO2 response of photosynthesis in the C3 dicot Chenopodium album L. *Australian Journal of Plant Physiology* 17, 135–148.

Salvucci, M.E. and Crafts-Brandner, S.J. (2004) Relationship between the heat tolerance of photosynthesis and the thermal stability of rubisco activase in plants from contrasting thermal environments. *Plant Physiology* 134, 1460–1470.

Sanders, R.W., Berninger, U.-G., Lim, E.L., Kemp, P.F. and Caron, D.A. (2000) Heterotrophic and mixotrophic nanoplankton predation on pico-plankton in the Sargasso Sea and on Georges Bank. *Marine Ecological Progress Series* 192, 103–118.

Sane, P.V., Desai, T.S., Tatake, V.G. and Govindjee (1984) Heat induced reversible increase in Photosystem 1 emission in algae, leaves and chloroplasts: spectra, activities, and relation to state changes. *Photosynthetica* 18, 439–444.

Saros, J.E., Interlandi, S.J., Doyle, S., Michel, T.J. and Williamson, C.E. (2005) Are the deep chlorophyll maxima in alpine lakes primarily induced by nutrient availability, not UV avoidance? *Arctic, Antarctic and Alpine Research* 37, 557–563.

Savitch, L.V., Maxwell, D.P. and Huner, N. (1996) Photosystem II excitation pressure and photosynthetic carbon metabolism in *Chlorella vulgaris*. *Plant Physiology* 111, 127–136.

Savitch, L.V., Barker-Astrom, J., Ivanov, A.G., Hurry, V., Öquist, G., Huner, N.P. and Gardestrom, P. (2001) Cold acclimation of *Arabidopsis thaliana* results in incomplete recovery of photosynthetic capacity, associated with an increased reduction of the chloroplast stroma. *Planta* 214, 295–303.

Schreiber, U. and Berry, J. (1977) Heat-induced changes of chlorophyll-fluorescence in intact leaves correlated with damage of the photosynthetic apparatus. *Planta* 136, 233–238.

Shang, W. and Feierabend, J. (1998) Slow turnover of the D1 reaction center protein of photosystem II in leaves of high mountain plants. *FEBS Letters* 425, 97–100.

Streb, P., Aubert, S. and Bligny, R. (2003) High temperature effects on light sensitivity in the two high mountain plant species Soldanella alpina (L.) and Ranunculus glacialis (L.). *Plant Biology* 5, 432–440.

Stroeve, J. and Serreze, M. (2008) Arctic sea ice extent plummets in 2007. *EOS* 89, 13–20.

Stroeve, J., Holland, M.M., Meier, W., Scambos, T. and Serreze, M. (2007) Arctic sea ice decline: Faster than forecast. *Geophysical Research Letters* 34, L09501.

Sukenik, A., Bennett, J. and Falkowski, P.G. (1988) Changes in the abundance of individual apoproteins of light-harvesting chlorophyll a/b complexes of photosystem I and II with growth irradiance in the marine chlorophyte Dunaliella tertiolecta. *Biochemica et Biophysica Acta* 932, 206–215.

Symon, C., Arris, L. and Heal, B. (eds) (2005) *Arctic Climate Impact Assessment.* Cambridge University Press, Cambridge.

Sympson, S. (2000) In focus: melting away. *Scientific American* 281, 14–15.

Szyszka, B., Ivanov, A.G. and Huner, N.P.A. (2007) Psychrophily induces differential energy partitioning, photosystem stoichiometry and polypeptide phosphorylation in Chlamydomonas raudensis. *Biochemica et Biophysica Acta* 1767, 789–800.

Takahashi, S., Nakamura, T., Sakamizu, M., van Woesik, R. and Yamasaki, H. (2004) Repair machinery of symbiotic photosynthesis as the primary target of heat stress for reef-building corals. *Plant and Cell Physiology* 45, 251–255.

Thomas, D.J. and Deickmann, G.S. (2002) Antarctic sea ice: a habitat for extremophiles. *Science* 295, 641–644.

Thompson, R., Kamenik, C. and Schmidt, R. (2005) Ultra-sensitive alpine lakes and climate change. *Journal of Limnology* 64, 139–152.

Thomson, P.G., Wright, S.W., Bolch, C.J.S., Nichols, P.D., Skerratt, J.H. and McMinn, A. (2004) Antarctic distribution, pigment and lipid composition, and molecular identification of the brine dinoflagellate Polarella glacialis (Dinophyceae). *Journal of Phycology* 40, 867–873.

Vieillette, J., Mueller, D.R.L., Antoniades, D. and Vincent, W.F. (2008) Arctic epishelf lakes as sentinel ecosystems: past, present and future. *Journal of Geophysical Research & Biogeosciences* 113, G04014.

Walsh, J.E. (2009) A comparison of Arctic and Antarctic climate change, present and future. *Antarctic Science* 21, 179–188.

Walther, G.-R., Post, E., Convey, P., Menzel, A., Parmesan, C., Beebee, T.J.C., Fromentin, J.-M., *et al.* (2002) Ecological responses to recent climate change. *Nature* 416, 389–395.

Wasley, J., Robinson, S.A., Lovelock, C.E. and Popp, M. (2006) Some like it wet – biological characteristics underpinning tolerance of extreme water stress events in Antarctic bryophytes. *Functional Plant Biology* 33, 443–455.

Williamson, C.E., Saros, J.E., Vincent, W.F. and Smol, J.P. (2009) Lakes and reservoirs as sentinels, integrators, and regulators of climate change. *Limnology and Oceanography* 54, 2273–2282.

Wilson, K.E., Krol, M. and Huner, N.P. (2003) Temperature-induced greening of Chlorella vulgaris. The role of the cellular energy balance and zeaxanthin-dependent nonphotochemical quenching. *Planta* 217, 616–627.

Wilson, K.E., Ivanov, A.G., Oquist, G., Grodzinski, B., Sarhan F. and Huner, N.P.A. (2006) Energy balance, organellar redox status and acclimation to environmental stress. *Journal of Botany* 84, 1355–1370.

Yamane, Y., Kashino, Y., Koike, H. and Satoh, K. (1998) Effects of high temperatures on the photosynthetic systems in spinach: Oxygen-evolving activities, fluorescence characteristics and the denaturation process. *Photosynthesis Research* 57, 51–59.

Yamori, W., Suzuki, K., Noguchi, K., Nakai, M. and Terashima, I. (2006) Effects of Rubisco kinetics and Rubisco activation state on the temperature dependence of the photosynthetic rate in spinach leaves from contrasting growth temperatures. *Plant, Cell & Environment* 29, 1659–1670.

Yamori, W., Noguchi, K., Hikosaka, K. and Terashima, I. (2009) Cold-tolerant crop species have greater temperature homeostasis of leaf respiration and photosynthesis than cold-sensitive species. *Plant and Cell Physiology* 50, 203–215.

Zhang, P., Liu, S., Cong, B., Wu, G., Liu, C., Lin, X., Shen, J. and Huang, X. (2010) A novel omega-3 fatty acid desaturase involved in acclimation processes of polar condition from Antarctic ice algae *Chlamydomonas* sp. ICE-L. *Marine Biotechnology* 13, 393–401.

Zhu, X.-G., Portis Jr, A.R. and Long, S.P. (2004) Would transformation of C3 crop plants with foreign Rubisco increase productivity? A computational analysis extrapolating from kinetic properties to canopy photosynthesis. *Plant, Cell & Environment* 27, 155–165.

Zuniga, G.E., Alberdi, M., Fernandez, J., Montiel, P. and Corcuera, L.J. (1994) Lipid content in leaves of Deschampsia antarctica Desv. from Maritime Antarctic. *Phytochemistry* 37, 669–672.

4 Insects

Steven L. Chown

4.1 Introduction

Insects are virtually ubiquitous. They are absent only from the continental Antarctic, the very highest mountain peaks, and largely from the open oceans. Among the animals they are one of the most species rich and abundant groups, with diversity peaking in the tropics in most groups, but with several fascinating exceptions, such as the sawflies and aphids (Gaston, 1996; Hillebrand, 2004). Given their abundance and diversity, the insects play important functional roles in above- and below-ground terrestrial systems (Wardle *et al.*, 2004; Larsen *et al.*, 2005; Nichols *et al.*, 2008), and in lakes and rivers (Wallace and Webster, 1996), but are less functionally significant, though not absent, from the oceans (Cheng, 1976). Moreover, they are also common and important, both ecologically and economically, in ecosystems modified or largely constructed by humans, whether in the form of disease vectors in cities (Gratz, 1999), biological control agents in agricultural landscapes (Samways, 1994), or invasive species in what might otherwise have been considered lightly disturbed or even pristine landscapes (Frenot *et al.*, 2005). In consequence, any consideration of the impacts of changing climates must, by necessity, include a substantial focus on insects. Not only are changing climates likely to affect the very different members of this class directly, but insects will also be affected by and in turn affect other organisms and systems substantially. For example, changing host plant distributions and the nutritional value of different hosts under varying temperature conditions will affect many herbivorous species (Scriber, 2002; Pelini *et al.*, 2010; Hill *et al.*, 2011), while differential temperature effects on insect prey and their predators will have significant effects on the latter (Jepsen *et al.*, 2008; Both *et al.*, 2009). Climate change impacts on insect vectors of human and domestic animal diseases will also be substantial, but difficult to disentangle from changing human spatial demographics and control interventions (Lafferty, 2009).

The very diversity of the insects means that a wide variety of responses to change can be expected that are contingent on local factors and the population or species in question. None the less, just as there are not 10 million kinds of population dynamics, but a range of reasonably identifiable strategies across the continuum (Lawton, 1992), so too are general principles emerging concerning insect responses to the thermal landscape (Chown, 2001; Chown and Sinclair, 2010) and changing climates (Bale *et al.*, 2002; Berggren *et al.*, 2009). This is much like Mahlman's (1998) pinball game scenario – predicting the trajectory of the ball is almost impossible, but if the machine is angled more steeply, the odds of the ball ending in the gutter increase sharply. These general principles will be explored here in the context of changing temperature regimes. At the same time it must be recognized that changing temperatures will also be accompanied by changing moisture conditions (Sanderson *et al.*, 2011), and often simultaneously affect the dryness of the air in terrestrial systems, in the absence of changes in moisture, because of changing saturation vapour pressure. Terrestrial insects, and especially their eggs and early juvenile stages, are highly susceptible to desiccation (Chown, 2002; Woods, 2010). Likewise, temperature affects the solubility of gases in water. Changing oxygen

© CAB International 2012. *Temperature Adaptation in a Changing Climate: Nature at Risk* (eds K.B. Storey and K.K. Tanino)

concentrations have, in the past, had a substantial influence on insects (Harrison *et al.*, 2010), and may well also affect insect thermal limits, especially in aquatic species or stages, via oxygen limitation of thermal tolerance (see Pörtner, 2001; Klok *et al.*, 2004; Lighton, 2007; Stevens *et al.*, 2010).

Before proceeding with a discussion of how insects are likely to respond to changing thermal regimes, the palaeobiological context of modern change should be considered, if only briefly. While thermal conditions are changing very rapidly now (New *et al.*, 2011), and in the context of a landscape that has been substantially fragmented by humans, making responses more complicated than in the past (Warren *et al.*, 2001; Hill *et al.*, 2011), the changing thermal conditions associated with anthropogenic climate change will not be the first encountered by the insects, though they may be among the fastest. Thus, insects have been present at least since the Devonian (*c.* 400 million years ago (mya)), and in a wide variety of forms since the late Palaeozoic (*c.* 300 mya) (Grimaldi and Engel, 2005). Since then, the earth's climate has changed markedly, having been through periods that are both substantially warmer (e.g. the Palaeocene–Eocene thermal maximum) and substantially cooler (glacial periods recorded over the past 600,000 years) than the present (Archer and Rahmstorf, 2010). These changes have had pronounced effects on the distribution and composition of insect diversity (Labandeira and Sepkoski, 1993; Ashworth and Kuschel, 2003; Wilf *et al.*, 2006), as well as on the diversity of life forms within particular groups (e.g. Kukalová-Peck, 1985; Harrison *et al.*, 2010). However, of the *c.* 29 extant Orders, the majority can be traced back to the Jurassic. This does not mean that the insects will not be affected substantially by current environmental change or that the likely impacts of human activities on the group have not been underestimated (Dunn, 2005; Fonseca, 2009; Hill *et al.*, 2011). Nor does it mean that conservation actions that are aimed at insects (Samways, 2007), should not be undertaken. Rather it suggests that investigations of some periods in the past may provide further insight into what may be expected in a future world. It also suggests that the climate change now under way (as much as a 4°C warming before the end of the century, New *et al.*, 2011) may have more profound consequences for the palaeontologically young primate that is responsible for it, than for the insects as a whole.

4.2 Thermal Adaptation

4.2.1 Temperatures – means, extremes and predictability

With the exception of a relatively small number of heterotherms (see Heinrich, 1993), insects are ectotherms. While they are not wholly subject to the thermal whims of the environment, being capable of behavioural regulation and habitat choice (Chown and Nicolson, 2004), and in some cases habitat manipulation (Turner, 2000), they are affected substantially by ambient conditions. Thus, insect body temperature is determined largely by the abiotic environment through the usual biophysical processes (reviewed in Casey, 1988; Campbell and Norman, 1998). Body temperature determines whether a given individual finds itself close to the optimum for a given trait or suite of traits, or at a limit that might impair functioning or reproduction, or cause death (Vannier, 1994; Chown and Terblanche, 2007; Hoffmann, 2010). The latter translates to effects at the population level, influencing demographic rates, and ultimately the abundance and distribution of a given species via its component populations. Concatenation of variety in occurrence and abundance eventually determines assemblage membership and species richness across the landscape (Gaston *et al.*, 2008; Ricklefs, 2008). Body temperature also determines the rates of given physiological functions within the limits of their functioning. The differential thermal response of various physiological characteristics can have profound impacts on various higher-level characteristics (see, for example, de Jong and Van der Have, 2008) as well as on thermal regulation (Coggan *et al.*, 2011). Insects have evolved a variety of physiological strategies (reviewed in Chown and Nicolson, 2004) to counter what Clarke and Fraser (2004) called the tyranny of Boltzmann, but within a given population temperature has profound effects on vital rates.

Variability in thermal conditions is also of considerable significance. Over the short term, temperature variation can lead to increases in growth rates (see Chown and Nicolson, 2004), and if warmer temperatures interrupt periods of cold, can improve survival, though sometimes with a fitness cost (Marshall and Sinclair, 2010). Repeated low temperature stresses can also increase mortality (Brown *et al.*, 2004; Sinclair and Chown, 2005a), as can warm periods during mid-winter by compromising metabolic stores through an elevation in metabolic rates of diapausing species (Irwin and Lee, 2003; Williams *et al.*, 2003). More broadly, it appears that the extent of temporal temperature variation may also have influenced the geographic distribution of thermal tolerance strategies. For example, among cold-hardy species, moderate freeze tolerance is most common in unpredictable cool environments, strong freeze tolerance in predictably very cold areas (e.g. the High Arctic), and freeze avoidance is typical in predictably cold areas (Sinclair *et al.*, 2003). The predictability of environmental variation also plays important roles in determining the likelihood that plasticity will evolve, with cue unreliability favouring fixed over plastic phenotypes (reviewed in Chown and Terblanche, 2007). Through its effects on mortality and fecundity, environmental variability also has higher-level effects, such as on the geographic range size and body size of insects (Gaston and Chown, 1999; Stillwell *et al.*, 2007). Indeed, the influences of environmental variability on such integrated traits are also regularly found in other taxa. For example, at global scales, environmental variation is an important correlate of variation in clutch size and cooperative breeding in birds (Jetz *et al.*, 2008; Jetz and Rubenstein, 2011), metabolic rates in mammals (Lovegrove, 2000) and thermal limits in reptiles (Clusella-Trullas *et al.*, 2011). Thermal extremes can also have profound effects on insects, including the extinction of local populations (Tenow and Nilssen, 1990; Parmesan *et al.*, 2000), and longer-term changes in distributions (Battisti *et al.*, 2006), and have significant influences on population dynamics (Crozier, 2004). Similarly, precipitation extremes can have continent-wide impacts on insect populations (Hawkins and Holyoak, 1998).

Because average thermal conditions, temporal variation in temperature, and the predictability of this variation have such significance for insects, understanding the geographic and spatial variation of temperature, and likely future changes therein, are important prerequisites for determining the risks from climate change faced by insects generally, or by a given insect population. Indeed, this is true for all organisms (Helmuth *et al.*, 2010). Owing to their small size, microclimatic conditions are especially significant for insects. The exploitation of small-scale variation in conditions associated with, for example, the boundary layers of leaves (Pincebourde and Casas, 2006a,b; Potter *et al.*, 2009), or refuges below the snow or underneath litter or bark (Leather *et al.*, 1993), can substantially alter the thermal conditions encountered by individuals, so influencing their survival and performance. Importantly, it is not only the existence of thermal variation that plays an important role in influencing thermal physiology and therefore local demographics. The spatial arrangement of this variation, and its accessibility to individuals given both intraspecific (such as competition for limited sites) and interspecific (e.g. predation in exposed areas between sites) interactions, influence thermal physiology and the extent to which individuals can realize optimal conditions (Angilletta, 2009). However, not all species are capable of detecting, predicting, or exploiting environmental variation, which can lead to substantial mortality and therefore spatial variation in abundance, such as can be found in *Drosophila* (Feder *et al.*, 1997; Warren *et al.*,

2006), or alter the extent to which they show acclimation ability, even among life stages which differ in mobility (Marais and Chown, 2008).

The importance of documenting microclimates has long been acknowledged (Willmer, 1982; Kingsolver, 1989), and methods for so doing, and for analysing the data, including variation, its predictability, and changes in the nature thereof, have been widely discussed (Unwin, 1991; Kingsolver, 1989; Gaines and Denny, 1993; Chown and Terblanche, 2007). Many studies now document at least some component of the microclimate of the population of interest, although often for periods that are not as extended as they should be. The increasing availability of small dataloggers and global system for mobile communications (GSM)-based telemetry means that these problems are likely to recede in the future. More recently, the significance of microclimate data has been re-emphasized based on demonstrations that the rates at which dynamic thermal tolerance assessments are carried out in the laboratory can have substantial effects on estimation not only of thermal limits, but also on the responses of populations, via plasticity (acclimation), to different conditions (Fig. 4.1a), and on estimates of the heritability of the traits in question (Terblanche *et al.*, 2007; Chown *et al.*, 2009; Mitchell and Hoffmann, 2010). Although some authors suggest that these effects might be as a consequence of the confounding effects of desiccation owing to trial conditions (Rezende *et al.*, 2011), empirical relationships between mass, water loss and metabolic rate suggest that this situation is likely to apply only to the very smallest of species (Fig. 4.1b), and of course not when humidity is maintained at a reasonable level.

While it is useful and often essential to collect information on the microclimates of the species/populations investigated, especially for forecasts of climate change effects, it is not always possible to obtain such information, especially in retrospect. The absence of microclimate information need not be considered a barrier to broader scale, comparative work, which is useful for establishing general principles that can be used to forecast insect responses to environmental change (see Chown and Gaston, 2008). Modelling procedures for downscaling macroclimate to animal body temperatures are now widely available (see Pincebourde *et al.*, 2007; Kearney and Porter, 2009; Helmuth *et al.*, 2010). Moreover, for macrophysiological studies, or investigations of the ecological implications of physiological variation over broad spatial and temporal scales (Chown *et al.*, 2004), much of the important signal remains, despite noise added by variation associated with microscale responses (see discussion in Chown *et al.*, 2003; Hodkinson, 2003; Huey *et al.*, 2009). Therefore understanding patterns in macroclimate variation can assist substantially in elucidating the ways in which insects have evolved to maintain populations in different thermal environments, and how future changes are likely to affect them. Indeed, at least so far as forecast future climates are concerned, downscaling typically refers to making forecasts relevant to cells smaller than the very large grid cells typically used in general circulation models, rather than to assessments of microclimate relevant to insects. For the latter, the tools of biophysics and environmental physiology remain essential (Campbell and Norman, 1998; Porter *et al.*, 2002; Helmuth *et al.*, 2010).

Given that macroclimate variation is useful for understanding the current distribution and evolution of variation in insect physiological traits, and particularly thermal tolerances, several aspects of temperature variation are most noteworthy.

- Variation in the thermal maxima and thermal minima differ substantially across both latitude and altitude. In general, maximum temperatures show much less spatial variation than do minimum temperatures. Thus tropical and high temperate region species may encounter similar thermal maxima, but very different thermal mimima (Addo-Bediako *et al.*, 2000). Mean thermal conditions (ignoring altitudinal variation) are very similar across the tropics, and

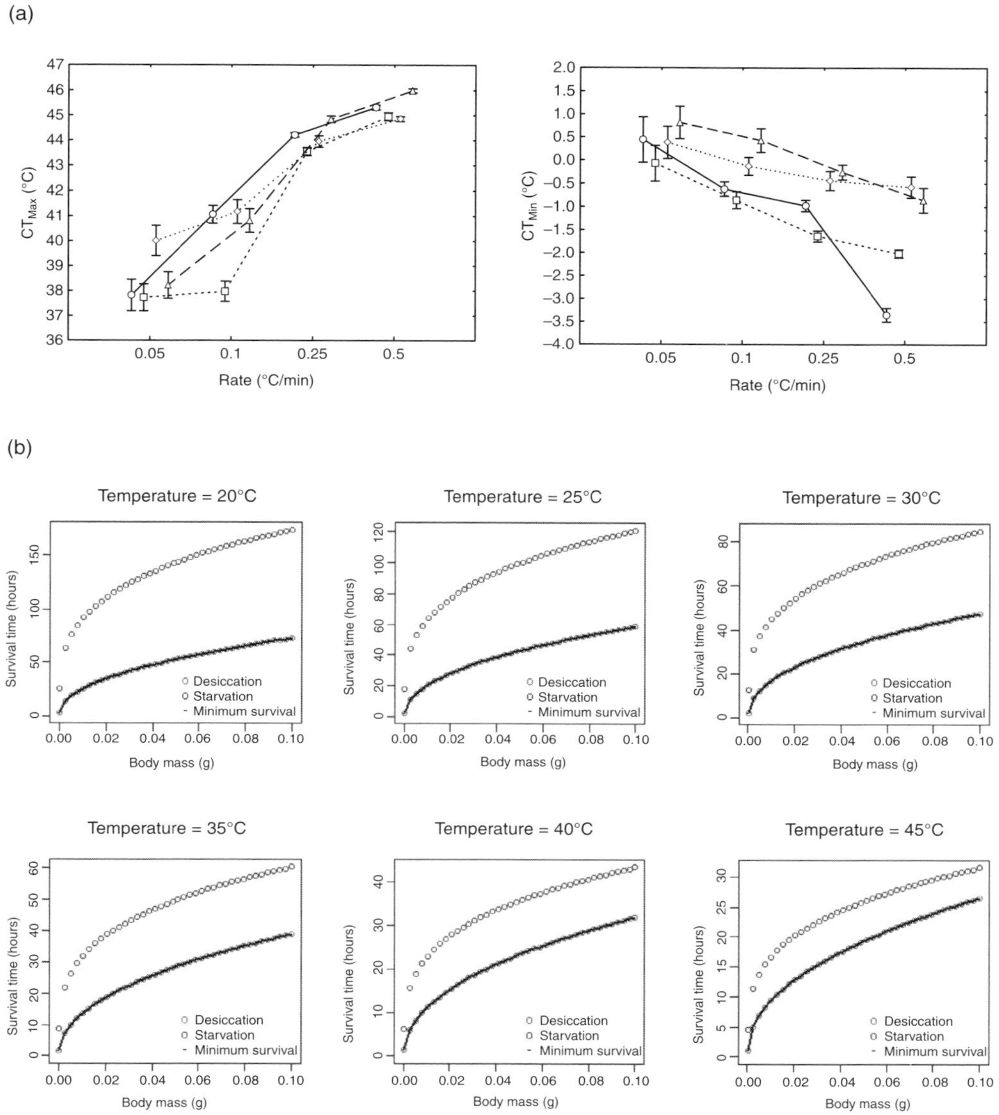

Fig. 4.1. (a) The impacts of different rates of experimental temperature change in critical thermal limit estimations both on the values of these limits and on their response to different acclimation conditions in *Linepithema humile* (Argentine ant) (redrawn from Chown *et al.*, 2009). Acclimation treatments are: circles, 15°C; squares, 20°C; diamonds, 25°C; triangles, 30°C. (b) Modelling based on equations relating metabolic rate to temperature and mass (from Irlich *et al.*, 2009) and desiccation rate to these conditions (unpublished data) indicating that death from starvation or desiccation takes much longer than the duration of a typical critical thermal limits experiment (c. 6 h at the slowest rates) in all but the very smallest of species. Minimum survival indicates whether starvation or desiccation will determine survival, which in all cases is the lower desiccation line.

extreme maximum temperatures are found just outside the tropics, rather than within their bounds (Bonan, 2002). The frequency distribution of thermal conditions also changes markedly with latitude, which also has significant consequences for thermal physiology (Hoffmann, 2010).

- Between 30° and 60° of latitude, minimum temperatures tend to decline to much lower levels in the continent-dominated northern hemisphere than in the ocean-dominated south (Addo-Bediako *et al.*, 2000; Bonan, 2002). Intra- and inter-annual variation in temperature is also considerably larger in the high northern than in the high southern hemisphere, acknowledging that little land is found between 50° and 60° S. Across the year at the higher southern latitudes, the likelihood of encountering sub-zero temperatures is quite high, and this is true also of higher elevation, southern hemisphere areas further to the north (Sinclair and Chown, 2005b). The predictability of thermal conditions may explain both patterns in cold hardiness and in the extent of acclimation, though additional tests of these ideas are needed to establish their broad generality. None the less, they draw attention to the asymmetry in the relationship between variability and predictability. While unvarying conditions are always predictable, variable conditions may range between being highly predictable and completely unpredictable. The latter are important because highly variable environments are also those most prone to extreme events (Katz and Brown, 1992), and these are forecast to become more common as climates continue to change. Methods for analysing changes in the nature of variation, such as wavelet analysis, are well developed (see Chown and Terblanche, 2007).
- Tropical and extra-tropical mountains show very different patterns of annual temperature variation and its change with elevation (Janzen, 1967; McCain, 2009). In tropical regions, temperatures decline with elevation in a constant way across the year owing to minimal seasonal differences. By contrast, in higher latitude areas, temperature variation is much larger seasonally and the overlap in temperatures between lower and higher elevations over the course of a year is also much greater. Similarly, diurnal variation in temperature is much greater at higher than at lower elevations, but this elevational variation may be much less pronounced in higher latitude areas. The interactions between thermal variation across latitude and elevation have important consequences for the physiology and distribution of animals (Ghalambor *et al.*, 2006).
- Care should be taken when calculating means and variances for use in environmental physiology owing to the very different behaviour of different measures (Gaston and McArdle, 1994), Jensen's inequality (Ruel and Ayres, 1999) and the fallacy of the averages (Savage, 2004).

4.2.2 Physiological adaptations, plasticity and behaviour

Physiological responses to a varying thermal environment have often been characterized as resistance and capacity adaptations (e.g. Cossins and Bowler, 1987), though both short-term (plasticity) and evolutionary responses may be involved. The clear distinction between these categories is also blurred by the way in which ultimate limits might be set by changing performances with temperature, such as in the case of oxygen limitation of thermal tolerance (Pörtner, 2001). Similarly, behavioural and physiological responses have strong reciprocal influences (Clusella-Trullas *et al.*, 2007, 2008; Marais and Chown, 2008; Angilletta, 2009).

These complexities and the more straightforward responses of individuals and populations to the thermal environment have been widely investigated in insects. The literature on the mechanistic basis of spatial, temporal and phylogenetic patterns in, and the evolution of, insect thermal tolerances is large. In many respects it was already substantial when Keister and Buck (1964) remarked that the effect of temperature on metabolic rate is the most over-confirmed fact in insect physiology, and continues to grow. This growth is not only furthering knowledge of the mechanisms underlying insect responses to the thermal environment (e.g. Morin *et al.*, 2005; Michaud *et al.*, 2008; McMullen and Storey, 2010; Ragland *et al.*, 2010), but also providing new insights into what the consequences of evolutionary constraints and large-scale variation in physiological tolerances mean for the responses of insects to changing environments (Deutsch *et al.*, 2008; Kellermann *et al.*, 2009; Chown *et al.*, 2009, 2010). Providing a full synopsis of these adaptations, their evolution, and their variation over time and space is well beyond the scope of the current overview. Moreover, owing to several recent comprehensive reviews it is not necessary to do so. In the case of lower thermal limits, foundational aspects of the field are dealt with by Lee and Denlinger (1991), and recent advances by Denlinger and Lee (2010). For upper thermal limits a range of reviews are available, including those by Hoffmann *et al.* (2003a) which is focused on *Drosophila*, and by Denlinger and Yocum (1998), Robertson (2004), Chown and Nicolson (2004), and Chown and Terblanche (2007) which are more general in scope. In the case of temperature effects on physiological rates, a general overview is available in Chown and Nicolson (2004) and most of the theory and many empirical examples are available in Angilletta (2009). The significance and evolution of various forms of plasticity (such as acclimation) have typically been dealt with in a comprehensive fashion by these works, although the importance of plasticity for insects in general is also reviewed by Whitman and Ananthakrishnan (2008). In consequence, this chapter will highlight several areas that are relevant particularly to insect responses to changing climates, or areas that require further work in this context.

First, insects pass through several different stages, which often have very different ecological roles, and also differ among taxa as to which stages endure the most stressful conditions (e.g. adults versus eggs). These differences are important for understanding responses to environmental change (Bale, 1987; Hoffmann, 2010; Woods, 2010), and in many cases investigations of physiological responses have concerned the majority of the stages (see discussion in Bowler and Terblanche, 2008). However, it is much more difficult to find macrophysiological investigations, at least of thermal limits, which deal with non-adult stages. Because immature stages are often found only during the most favourable periods (see Hoffmann, 2010), the absence of such studies might not be problematic. Moreover, thermal responses of immatures are captured in some investigations, such as of large-scale variation in development rate (Irlich *et al.*, 2009) and population growth rate (Deutsch *et al.*, 2008). None the less, any comprehensive understanding of the likely impacts of changing climates must take into account variation in responses among all stages, if only to confirm that the immatures might be discounted (which in some cases they clearly cannot be – see, for example, Marais and Chown, 2008; Fig. 4.2), or that general relationships (such as those applied by Dillon *et al.* (2010) in the case of metabolic rate responses to warming) hold across all stages. None the less, such general relationships are unlikely to be useful for forecasts of the responses of particular populations (Crozier, 2004; see also Crozier and Dwyer, 2006) and here information across all stages will be required. While it might seem unlikely that particular species, rather than broad groups, will be the focus of research, the opposite is true in many instances. Specific species or populations are typically the focus of pest management interventions (Régnière and Bentz, 2007; Bale and Hayward, 2010), including for biocontrol agents and their hosts (Samways, 1989), of concerns regarding the spread and likely dynamics of invasive species and disease vectors (Rogers and Randolph, 2006; Kearney *et al.*, 2009a; Hartley *et al.*, 2010), and of conservation interventions (e.g. Chown *et al.*, 1995; Carroll *et al.*, 2009).

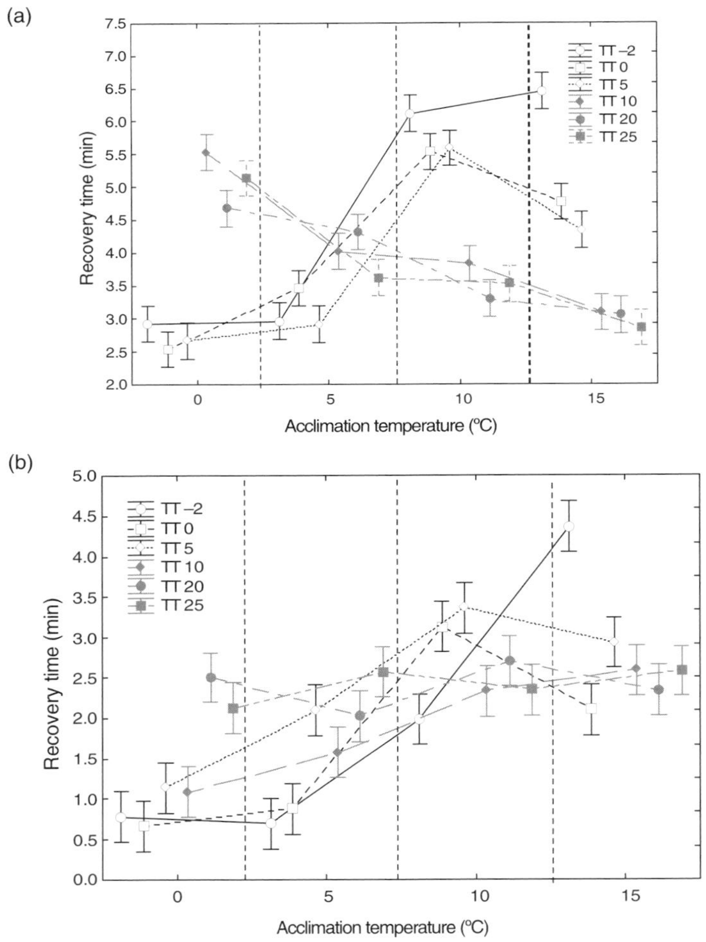

Fig. 4.2. Mean (± SE) chill coma recovery times (minutes) in *Paractora dreuxi* larvae (a) and adults (b) after a 7-day acclimation at 0°C, 5°C, 10°C and 15°C with 2 h treatments (−2°C, 0°C, 5°C, 10°C, 20°C and 25°C). The acclimation temperatures are provided along the x-axis, and recovery times at each of the test temperatures (TT) indicated using a different symbol. Lower test temperature responses are indicated with open symbols and higher test temperature responses with closed symbols. Panel (a) shows that if larvae are held at a low temperature and then pre-treated at a low temperature they recover quickly, but if held at a high temperature and pre-treated at a low one they take much longer to recover. The converse also applied. This is beneficial acclimation. Panel (b) shows that adults recover quickly only if held at low acclimation temperature, irrespective of the test temperature. Thus performance is best at low temperatures (colder is better). Redrawn from Marais and Chown (2008).

Second, just as it is clear that maximum temperatures vary less across large spatial gradients than do minimum temperatures, so too is it becoming clear that thermal maxima in insects vary much less than thermal minima. The typically more limited variation in the critical thermal minima is found not only over small and large spatial scales (Addo-Bediako *et al.*, 2000; Chown, 2001; Hoffmann *et al.*, 2003a; Hazell *et al.*, 2010a,b, though see also Calosi *et al.*, 2010), but is also characteristic of short-term acclimation responses which tend to be more restricted at the upper end of the thermal tolerance range. These differences may well reflect very different mechanisms underlying upper and lethal limits and different genetic underpinnings, though common mechanisms and some cross-tolerance have also been identified (reviewed in Chown and Nicolson, 2004; Chown and Terblanche, 2007). An important consequence of what appears to be less variation in upper lethal limits across space, and a lower capacity for plasticity, is that the warming tolerance (i.e. the difference between critical thermal maxima and habitat temperature) for tropical species may well be lower than those for species from more temperate areas (Deutsch *et al.*, 2008). However, because thermal maxima are often most pronounced just outside the tropics, it is in these areas where insects may be most at risk (Chown *et al.*, 2010). This risk also carries over to changes in the costs of living. Owing to the form of the response in most biological rates to temperature, increases at higher temperatures are likely to have more pronounced effects than those at lower temperatures (e.g. Ruel and Ayres, 1999). In consequence, small warming increments have larger effects on rates in areas of high average temperature than large increments in areas of low average temperature. At least for metabolic rate this appears to be the case (Dillon *et al.*, 2010) (Fig. 4.3). Warming has generally had a greater influence on insects in tropical systems than on those in more temperate systems. This effect may be particularly pronounced because of what appears to be a strong relationship between optimum temperature and maximum population growth rate (Frazier *et al.*, 2006). It should be recognized that any rate-sensitive process will experience similar effects; thus the eventual ecosystem-wide outcome will depend on the differential sensitivity of these effects among plants, their herbivores, and higher trophic levels. If, as Berggren *et al.* (2009) propose, rate–temperature relationships are steeper in predators than herbivores, and optima are broader and somewhat lower for plants than animals, the differential effects may be substantial.

Third, although the mechanisms underlying the major cold-hardiness strategies are well researched (Denlinger and Lee, 2010), and the environmental correlates thereof more certain now than they have been in the past (Chown and Sinclair, 2010), the life history benefits and evolution of each of these strategies have still not been fully explored (Sinclair and Renault, 2010). At least one major life history model for the evolution of these strategies has been proposed (Voituron *et al.*, 2002), but it has not been widely examined. In addition, Makarieva *et al.* (2006) suggested that

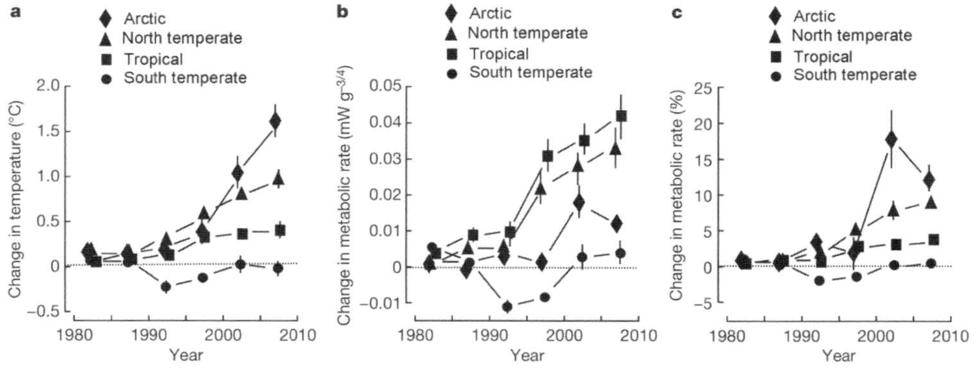

Fig. 4.3. Global temperature changes in ectotherm metabolic rates between 1980 and 2010. The panels indicate changes in (a) temperature, (b) mass-normalized metabolic rate and (c) relative changes in mass-normalized metabolic rate, respectively. Redrawn, with permission, from Dillon *et al.* (2010).

freezing tolerance represents 'abandoned metabolic control' under sub-zero conditions, while freeze intolerance represents 'minimum metabolic control'. Although some evidence exists in favour of these ideas (summarized in Chown and Terblanche, 2007) they have not been widely assessed. None the less, both of these concepts, which are related in several ways, deserve further scrutiny in the context of changing climates. In particular the hypotheses of Makarieva *et al.* (2006) imply that freezing-tolerant species (i.e. with abandoned metabolic control) will be highly susceptible to periods of elevated temperature owing to a reasonably high thermal sensitivity of damage accumulation, whereas in freeze-intolerant species energy is utilized from specific storage organs and periodic repair is undertaken. Certainly the susceptibility of the freezing-tolerant *Eurosta solidaginis* to warm periods has been documented (Irwin and Lee, 2003). Thus, if periodic warm periods become more common in areas previously characterized by low temperatures throughout winter, such as in far north temperate and Arctic areas, then freezing-tolerant species may be more affected than freeze-intolerant ones. The relative demographic merits of the two major cold-hardiness strategies (setting cryoprotective dehydration aside for the moment), and indications that repeated low temperature events may have significant impacts on survival and growth (Brown *et al.*, 2004; Sinclair and Chown, 2005a) also raise the question of the significance of sublethal versus lethal events (see also Dillon *et al.*, 2007). To some extent this depends on the colour of environmental noise (Vasseur and Yodzis, 2004) and the nature of the response of the population in question (Schwager *et al.*, 2006). As yet these questions have not been widely explored for insect responses to sublethal and lethal thermal extremes and the ways in which these might also influence cold-hardiness strategies. Doing so could improve forecasts of the effects of extreme events and of increasing thermal variability, and provide insight into the extent to which plasticity will assist or limit responses to changing environments (see Ghalambor *et al.*, 2007).

Fourth, as was noted in the Introduction, temperature changes will not only be accompanied by other changes in the landscape (for example, in tropical regions forest removal may reduce water availability and increase microclimate temperatures; see Nair *et al.*, 2003; Webb *et al.*, 2006), but will also differentially affect several aspects of organismal physiology and hence demographic rates (e.g. Kingsolver and Woods, 1997). Perhaps most significantly, recent theoretical and empirical work has shown that water loss is a universal cost of gas exchange in animals and plants (Woods and Smith, 2010). In consequence, any increase in metabolic rates, such as those documented by Dillon *et al.* (2010), and forecast to continue with rising temperatures, will have profound impacts on water loss in all ectotherms (including plants). If water stress is not especially significant in a given local environment then the impacts might not be pronounced. However, if the organisms occur in areas that experience seasonal or longer-term drought, as is typical of many tropical and extratropical areas (Sanderson *et al.*, 2011), survival may decline as a consequence of increased water loss rates associated with elevated metabolism. What is missing from the arguments of Dillon *et al.* (2010), and which may therefore be grounds for being less pessimistic about actual changes in metabolic and water loss rates, is the fact that rate–temperature relationships can show substantial plastic responses – in the laboratory and across seasons (Cossins and Bowler, 1987; Davis *et al.*, 2000; Terblanche *et al.*, 2010) – and also vary substantially among taxa (Irlich *et al.*, 2009). Another important interaction is found between food quality and the thermal environment available to and chosen by individuals to maximize growth (Coggan *et al.*, 2011). These interactions are only now beginning to be explored (though see Woods *et al.*, 2003), but joint considerations of biophysical ecology, dynamic energy budget theory, and the geometric framework for nutritional ecology hold much promise (Kearney *et al.*, 2010).

4.3 Responses to Climate Change

In previous sections several of the responses to climate change by insects and the outcomes thereof have been discussed. Here, further consideration is given to these responses at two different levels: that of assemblage richness per se, and whether this might be useful for forecasting; and at the population level, dealing specifically with likely limits to responses and how these might further be comprehended. In the final section, the largely neglected area of the potential impacts of bioengineering and biogeoengineering to mitigate climate change (see Royal Society, 2009; Woodward *et al.*, 2009) are also briefly examined.

4.3.1 Diversity

While much variation in the specifics of the relationships exist, the abundance and/or distribution of many insect species can readily be modelled using just a few variables related to temperature and water availability (e.g. Andrewartha and Birch, 1954; Jeffree and Jeffree, 1996; Todd *et al.*, 2002; Seely *et al.*, 2005; Duncan *et al.*, 2009). It is these relationships that are often used for forecasting responses to change (reviewed in Kearney and Porter, 2009; Buckley *et al.*, 2010). Perhaps unsurprisingly, it is also commonly found that a few thermal and water availability variables are strong predictors of insect diversity across reasonably large spatial scales (e.g. Turner *et al.*, 1987; Kerr and Packer, 1999; Kaspari *et al.*, 2000a; Lobo *et al.*, 2002; Rodriguero and Gorla, 2004; Botes *et al.*, 2006; Dunn *et al.*, 2009). These variables are often encapsulated in composite measures of environmental energy availability (either productive or ambient energy, see Clarke and Gaston (2006) for discussion), most commonly some measure of evapotranspiration or net primary productivity (see, for example, Kaspari *et al.*, 2000b; Hawkins and Porter, 2003). While the influence of energy may be indirect in some instances, such as via changes to habitat heterogeneity (Kerr *et al.*, 2001), energy availability remains a strong correlate of richness.

Energy availability is also among the variables most commonly captured in databases through, for example, measurements of temperature and precipitation and their change though time at ground-based climate stations, or through remote sensing (RS). Productivity measures are also readily available through RS (see, for example, Kerr and Ostrovsky, 2003) as are measures of habitat variability (Kerr *et al.*, 2001). Therefore, the ready documentation of these variables and their change might not only be a way of determining likely broad-scale changes to insect diversity, but also of forecasting ongoing change. This has been done for butterflies in Canada between 1900 and 1999, showing that increases in richness are associated with increases in temperature, while declines are associated with increases in human population density (White and Kerr, 2006). Thus, general forecasts should be possible for other areas. However, these kinds of approaches will not be useful for conditions that lie outside empirically determined relationships, for obvious statistical reasons. At some point, thermal conditions will also exceed the tolerances of most species in an assemblage, irrespective of how much energy is available (Deutsch *et al.*, 2008). None the less, species–energy relationships could be useful for initial forecasting of assemblage level changes where data are sparse, and where substantial alterations to energy availability are forecast. This is true, for example, for much of sub-Saharan Africa (Thornton *et al.*, 2011).

4.3.2 Populations

Because of their significance as agricultural pests, pollinators, biological control agents, and disease vectors, population-level models of insect responses to the environment, and to ongoing climate change, will continue to form a major component of entomological research (see, for example, Gray, 2010). While the various strengths and weaknesses of the correlative and mechanistic approaches available have been widely discussed, and much progress will clearly be made in improving the methods, three areas require further attention. Perhaps most significant among these are investigations of the role of evolutionary change in affecting species responses. By and large, both mechanistic and correlative approaches ignore evolutionary change. However, not only can small amounts of evolutionary change lead to dramatically different model outcomes, such as the evolution of additional desiccation resistance in *Aedes aegypti* eggs (Kearney *et al.*, 2009a), but in some cases adaptive constraints (for whatever reason – see Hoffmann and Willi, 2008), might limit the extent to which species can respond to environmental change. For example, in *Drosophila birchii* little opportunity exists for evolving additional desiccation resistance (Hoffmann *et al.*, 2003b). Moreover, the extent of evolutionary ability may vary in predictable ways. Among *Drosophila* species in Australia, for example, limited genetic variability and minimal resistance to low temperature and desiccation stress in tropical specialists compared with extra-tropical, more broadly distributed species, means that evolutionary responses are constrained in the former (Kellerman *et al.*, 2009).

The extent to which populations are able to evolve also depends on their size and on extraneous factors. Again, using *Drosophila* as a model organism, Willi and Hoffmann (2009) showed that under directional selection for thermal stress resistance, small populations succumbed because of low growth rates and susceptibility to demographic stochasticity. Larger populations performed better and had higher additive genetic variance, but only the largest populations responded well to directional selection. Empirical work on *Tribolium castaneum* populations has also shown that the extinction risks associated with demographic stochasticity have probably been underestimated (Melbourne and Hastings, 2008), while modelling work has also demonstrated that adaptive responses can be substantially constrained by the presence of other species (de Mazancourt *et al.*, 2008; Angilletta, 2009). The role of evolution in affecting species responses to climate change, mediated through their physiological traits, requires further exploration (see also Chevin *et al.*, 2010).

Finally, any investigations of population responses simply cannot afford to ignore the human context of the environment. In essence, the world is now a human-dominated one (Ellis and Ramankutty, 2008). Thus, human demographics and human interventions now play much larger roles in influencing diversity than may previously have been the case. In consequence, human activities will have to be included in studies seeking to understand current and future responses to change (White and Kerr, 2006; Lafferty, 2009; Hill *et al.*, 2011).

4.3.3 Geoengineering

Owing to the failure of governments to agree on the kinds of emissions reductions that will limit climate change to below 2°C (New *et al.*, 2011) (and some authors such as Hansen *et al.* (2008) rightly consider even this level of change of critical concern), the possibilities of geoengineering the climate are being seriously considered (Schneider (2008) provides an historical overview). There is much discussion now on the ethics of this, and on the difficulty of maintaining the sorts of international arrangements that might make the most intrusive forms problematic, such as the use of stratospheric sulphate aerosols to increase planetary albedo (e.g. Hamilton, 2010; Stafford Smith *et al.*, 2011). Recently, the efficacy and likely impacts of the various options for geoengineering have also come under scrutiny (Royal Society, 2009). Two classes of methods exist: carbon dioxide removal from the atmosphere, and solar radiation management by reflecting a small percentage of solar radiation back into space. Both classes can involve either strictly engineered solutions, or what is also known as biogeoengineering (Woodward *et al.*, 2009) which involves some manipulation of the biosphere. Examples of the former include carbon capture and storage, or the injection of sulphate aerosols into the lower stratosphere. The latter includes manipulations such as fertilization of the oceans to enhance nutrient uptake, or planting of crops with enhanced reflectivity.

Few detailed investigations of the likely impacts of geoengineering on insects have been made, or indeed of impacts on biodiversity generally (see Woodward *et al.*, 2009 for a notable exception). None the less, the likely impacts have been generally considered in the comprehensive report prepared by the Royal Society (Royal Society, 2009) (Fig. 4.4). Given the strong relationships between insect biodiversity (here interpreted to mean variation from the mechanistic functioning level to the level of assemblages) and both temperature and water availability, the safety ratings provided by the Royal Society and in the discussions by Woodward *et al.* (2009) seem applicable to insects. For example, carbon capture at source will deal directly with the problem of CO_2 emissions. In so doing it will address increases in temperature and changing C:N ratios in plants, which are likely to affect herbivorous insects and the food webs within which they are embedded (Lindroth, 2010). It will also influence changing growth patterns of C_3 and C_4 plants, which appear to be leading to shrub encroachment and the disappearance of grasslands in many areas, with downstream effects on insect assemblages (see Bond, 2008; Chown, 2010). By contrast, biochar (≈ charcoal, buried in the soil) seems a risky option because of its impacts on soil ecosystems, of which insects are such an important component (Wardle *et al.*, 2004). The slowing of deforestation and an increase in afforestation in previously deforested areas seem low risk options for insects. The former has the benefits of reducing habitat alteration, which is a major cause of change in insect diversity (Samways, 2007), and also ensuring the maintenance not only of local climatic conditions (Nair *et al.*, 2003; Webb *et al.*, 2006), but also of larger-scale gradients in rainfall (see Makarieva and Gorshkov, 2007). Both have substantial implications for ectotherms (see above and also Kearney *et al.*, 2009b). However, widespread afforestation, especially in grasslands or shrub-dominated areas (such as the Fynbos in South Africa) will have

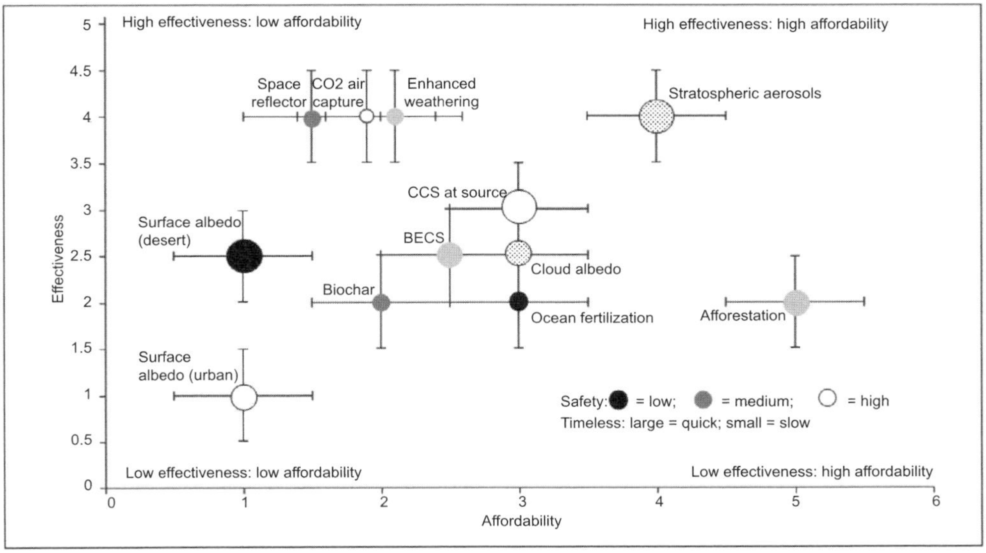

Fig. 4.4. Graphical summary of the Royal Society's (2009) evaluation of the affordability, effectiveness and risk of geoengineering techniques considered in its report. The colours indicate high risk (black) to lower risk (white). Redrawn, with permission, from Royal Society (2009).

substantially negative effects on insects simply by altering habitats and microclimates and so leading to the extirpation of many species adapted to these conditions. For example, work on grasslands and open savannahs invaded by alien thickets has shown that substantial changes in the average body size of insect assemblages take place in the latter (Chown and Steenkamp, 1996; Coetzee *et al.*, 2007). Many physiological and ecological traits vary strongly with size (Chown and Gaston, 2008), and at least part of the reason for the impacts of alien thickets is impairment of flight ability of beetles in dense stands.

How other proposed geoengineering solutions are likely to affect insects is not clear. Many of the proposals are likely to affect local and regional climates (Royal Society, 2009), which may have unpredictable consequences for insect population dynamics via alterations in temperature and water availability. Moreover, large-scale planting of particular trees or crops, especially if in monoculture, may bring all the associated pest problems, which would not only have food web effects on insects (see Samways, 1994), but also affect the utility of these interventions. That said, these kinds of problems are well within the scope of those currently encountered in agriculture and forestry. Perhaps least clear are the effects that changes in polarization of the sky, associated with stratospheric aerosols, might have on insect navigation. Insects certainly make use of polarization for navigation, and more local impacts on insects of changes in polarization have been documented (Hegedüs *et al.*, 2007), as well as large-scale changes in polarization associated with volcanic eruptions, which are accompanied by the ejection of aerosols into the stratosphere.

4.4 Conclusions

Research about the biological impacts of climate change is fast becoming one of the largest areas of environmental science. However, with notable exceptions and most frequently in less formal considerations of climate change and its impacts, a sense of urgency is rarely felt in formal papers. Rather, what is known and how further responses and outcomes can be forecast are usually discussed dispassionately. This measured approach is misleading as to the gravity of the current climate crisis. Although politicians discuss limiting climate change to 2°C, it is a missed target – the least optimistic scenario now indicates such warming will be reached by 2045 (i.e. in 34 years' time) (New *et al.*, 2011). What seems probable is a +4°C scenario by the end of the century. A wide range of research shows that our world will be profoundly different by then, and will be more difficult to live in. Much of what is currently taken for granted, including a spectacular diversity of insects that helps make the world habitable, will have disappeared. Therefore, while our science should continue to be soberly judged and measured to ensure the validity of its conclusions, we should not only be more responsible about the climate change impacts of conducting it, but should also be more prepared to take appropriate advocacy positions to ensure that its messages resonate widely. Changes in the earth system that have already taken place and that are forecast are truly alarming. Only with profound action will further transformation be slowed.

4.5 Acknowledgements

Ken Storey and Karen Tanino are thanked for their invitation to contribute this chapter and their patience while waiting for it. Melodie McGeoch provided useful comments on a previous version, and Charlene Janion assisted with the figures. This work was partially supported by an NRF SANAP Research Grant; Working for Water; and by the HOPE project of Stellenbosch University.

References

Addo-Bediako, A., Chown, S.L. and Gaston, K.J. (2000) Thermal tolerance, climatic variability and latitude. *Proceedings of the Royal Society B* 267, 739–745.

Andrewartha, H.G. and Birch, L.C. (1954) *The Distribution and Abundance of Animals.* University of Chicago Press, Chicago.

Angilletta, M.J.J. (2009) *Thermal Adaptation. A Theoretical and Empirical Synthesis.* Oxford University Press, Oxford.

Archer, D. and Rahmstorf, S. (2010) *The Climate Crisis. An Introductory Guide to Climate Change.* Cambridge University Press, Cambridge.

Ashworth, A.C. and Kuschel, G. (2003) Fossil weevils (Coleoptera: Curculionidae) from latitude 85°S Antarctica. *Palaeogeography, Palaeoclimatology, Palaeoecology* 191, 191–202.

Bale, J.S. (1987) Insect cold hardiness: freezing and supercooling - an ecophysiological perspective. *Journal of Insect Physiology* 33, 899–908.

Bale, J.S. and Hayward, S.A.L. (2010) Insect overwintering in a changing climate. *Journal of Experimental Biology* 213, 980–994.

Bale, J.S., Masters, G.J., Hodkinson, I.D., Awmack, C., Bezemer, T.M., Brown, V.K., Butterfield, J., *et al.* (2002) Herbivory in global climate change research: direct effects of rising temperature on insect herbivores. *Global Change Biology* 8, 1–16.

Battisti, A., Stastny, M., Buffo, E. and Larsson, S. (2006) A rapid altitudinal range expansion in the pine processionary moth produced by the 2003 climatic anomaly. *Global Change Biology* 12, 662–671.

Berggren, A., Björkman, C., Bylund, H. and Ayres, M.P. (2009) The distribution and abundance of animal populations in a climate of uncertainty. *Oikos* 118, 1121–1126.

Bonan, G. (2002) *Ecological Climatology. Concepts and Applications.* Cambridge University Press, Cambridge.

Bond, W.J. (2008) What limits trees in C-4 grasslands and savannas? *Annual Review of Ecology, Evolution, and Systematics* 39, 641–659.

Botes, A., Mcgeoch, M.A., Robertson, H.G., Van Niekerk, A., Davids, H.P. and Chown, S.L. (2006) Ants, altitude and change in the northern Cape Floristic Region. *Journal of Biogeography* 33, 71–90.

Both, C., van Asch, M., Bijlsma, R.G., van den Burg, A.B. and Visser, M.E. (2009) Climate change and unequal Phenological constraints across four trophic levels: constraints or adaptations? *Journal of Animal Ecology* 78, 73–83.

Bowler, K. and Terblanche, J.S. (2008) Insect thermal tolerance: what is the role of ontogeny, ageing and senescence? *Biological Reviews* 83, 339–355.

Brown, C.L., Bale, J.S. and Walters, K.F.A. (2004) Freezing induces a loss of freeze tolerance in an overwintering insect. *Proceedings of the Royal Society B* 271, 1507–1511.

Buckley, L., Urban, M.C., Angilletta, M.J., Crozier, L.G., Rissler, L.J. and Sears, M.W. (2010) Can mechanism inform species distribution models? *Ecology Letters* 13, 1041–1054.

Calosi, P., Bilton, D.T., Spicer, J.I., Votier, S.C. and Atfield, A. (2010) What determines a species' geographical range? Thermal biology and latitudinal range size relationships in European diving beetles (Coleoptera, Dytiscidae). *Journal of Animal Ecology* 79, 194–204.

Campbell, G.S. and Norman, J.M. (1998) *An Introduction to Environmental Biophysics.* Springer, Berlin.

Carroll, M.J., Anderson, B.J., Brereton, T.M., Knight, S.J., Kudrna, O. and Thomas, C.D. (2009) Climate change and translocations: The potential to re-establish two regionally-extinct butterfly species in Britain. *Biological Conservation* 142, 2114–2121.

Casey, T.M. (1988) Thermoregulation and heat exchange. *Advances in Insect Physiology* 20, 119–146.

Cheng, L. (1976) *Marine Insects.* North Holland Publishing, Amsterdam.

Chevin, L.M., Lande, R.L. and Mace, G.M. (2010) Adaptation, plasticity, and extinction in a changing environment: towards a predictive theory. *PLoS Biology* 8, e1000357, doi:10.1371/journal.pbio.1000357.

Chown, S.L. (2001) Physiological variation in insects: hierarchical levels and implications. *Journal of Insect Physiology* 47, 649–660.

Chown, S.L. (2002) Respiratory water loss in insects. *Comparative Biochemistry and Physiology A* 133, 791–804.

Chown, S.L. (2010) Temporal biodiversity change in transformed landscapes: a southern African perspective. *Philosophical Transactions of the Royal Society B* 365, 3729–3742.

Chown, S.L. and Gaston, K.J. (2008) Macrophysiology for a changing world. *Proceedings of the Royal Society B* 275, 1469–1478.

Chown, S.L. and Nicolson, S.W. (2004) *Insect physiological ecology. Mechanisms and patterns.* Oxford University Press, Oxford.

Chown, S.L. and Sinclair, B.J. (2010) The macrophysiology of insect cold hardiness. In: Denlinger, D.L. and Lee Jr, R.E. (eds) *Low Temperature Biology of Insects*. Cambridge University Press, Cambridge, pp. 191–222.

Chown, S.L. and Steenkamp, H.E. (1996) Body size and abundance in a dung beetle assemblage: optimal mass and the role of transients. *African Entomology* 4, 203–212.

Chown, S.L. and Terblanche, J.S. (2007) Physiological diversity in insects: ecological and evolutionary contexts. *Advances in Insect Physiology* 33, 50–152.

Chown, S.L., Scholtz, C.H., Klok, C.J., Joubert, F.J. and Coles, K.S. (1995) Ecophysiology, range contraction and survival of a geographically restricted African dung beetle (Coleoptera: Scarabaeidae). *Functional Ecology* 9, 30–39.

Chown, S.L., Addo-Bediako, A. and Gaston, K.J. (2003) Physiological diversity: listening to the large-scale signal. *Functional Ecology* 17, 562–572.

Chown, S.L., Gaston, K.J. and Robinson, D. (2004) Macrophysiology: large-scale patterns in physiological traits and their ecological implications. *Functional Ecology* 18, 159–167.

Chown, S.L., Jumbam, K.R., Sørenson, J.G. and Terblanche, J.S. (2009) Phenotypic variance, plasticity and heritability estimates of critical thermal limits depend on methodological context. *Functional Ecology* 23, 133–140.

Chown, S.L., Hoffmann, A.A., Kristensen, T.N., Angilletta, M.J., Stenseth, N.C. and Pertoldi, C. (2010) Adapting to climate change: a perspective from evolutionary physiology. *Climate Research* 43, 3–15.

Clarke, A. and Fraser, K.P.P. (2004) Why does metabolism scale with temperature? *Functional Ecology* 18, 243–251.

Clarke, A. and Gaston, K.J. (2006) Climate, energy and diversity. *Proceedings of the Royal Society B* 273, 2257–2266.

Clusella-Trullas, S., Van Wyk, J.H. and Spotila, J.R. (2007) Thermal melanism in ectotherms. *Journal of Thermal Biology* 32, 235–245.

Clusella-Trullas, S., Terblanche, J.S., Blackburn, T.M. and Chown, S.L. (2008) Testing the thermal melanism hypothesis: a macrophysiological approach. *Functional Ecology* 22, 232–238.

Clusella-Trullas, S., Blackburn, T.M. and Chown, S.L. (2011) Climatic predictors of temperature performance curve parameters in ecotherms imply complex responses to climate change. *American Naturalist*, 177, 738–751.

Coetzee, B.W.T., Van Rensburg, B.J. and Robertson, M.P. (2007) Invasion of grasslands by silver wattle, *Acacia dealbata* (Mimosaceae), alters beetle (Coleoptera) assemblage structure. *African Entomology* 15, 328–339.

Coggan, N., Clissold, F. and Simpson, S.J. Locusts use dynamic thermoregulatory behaviour to optimise nutritional outcomes. *Proceedings of the Royal Society B*, 278, 2745–2752.

Cossins, A.R. and Bowler, K. (1987) *Temperature Biology of Animals*. Chapman and Hall, London.

Crozier, L. (2004) Warmer winters drive butterfly range expansion by increasing survivorship. *Ecology* 85, 231–241.

Crozier, L. and Dwyer, G. (2006) Combining population-dynamic and ecophysiological models to predict climate-induced insect range shifts. *American Naturalist* 167, 853–866.

Davis, A.L.V., Chown, S.L., Mcgeoch, M.A. and Scholtz, C.H. (2000) A comparative analysis of metabolic rate in six *Scarabaeus* species (Coleoptera: Scarabaeidae) from southern Africa: further caveats when inferring adaptation. *Journal of Insect Physiology* 46, 553–562.

de Jong, G. and van der Have, T.M. (2008) Temperature dependence of development rate, growth rate and size: from biophysics to adaptation. In: Whitman, D.W. and Ananthakrishnan, T.N. (eds) *Phenotypic Plasticity of Insects. Mechanisms and Consequence*. Science Publishers, Enfield, New Hampshire, pp. 461–526.

De Mazancourt, C., Johnson, E. and Barraclough, T.G. (2008) Biodiversity inhibits species' evolutionary responses to changing environments. *Ecology Letters* 11, 380–388.

Denlinger, D.L. and Lee, R.E. (2010) *Low temperature biology of insects*. Cambridge University Press, Cambridge.

Denlinger, D.L. and Yocum, G.D. (1998) Physiology of heat sensitivity. In: Hallam, G.J. and Denlinger, D.L. (eds) *Temperature Sensitivity in Insects and Application in Integrated Pest Management*. Westview Press, Boulder, Colorado, pp. 7–53.

Deutsch, C.A., Tewksbury, J.J., Huey, R.B., Sheldon, K.S., Ghalambor, C.K., Haak, D.C. and Martin, P.R. (2008) Impacts of climate warming on terrestrial ectotherms across latitude. *Proceedings of the National Academy of Sciences of the USA* 105, 6668–6672.

Dillon, M.E., Cahn, L.R.Y. and Huey, R.B. (2007) Life history consequences of temperature transients in *Drosophila melanogaster*. *Journal of Experimental Biology* 210, 2897–2904.

Dillon, M.E., Wang, G. and Huey, R.B. (2010) Global metabolic impacts of recent climate warming. *Nature* 467, 704–707.

Duncan, R.P., Cassey, P. and Blackburn, T.M. (2009) Do climate envelope models transfer? A manipulative test using dung beetle

introductions. *Proceedings of the Royal Society B* 276, 1449–1457.
Dunn, R.R. (2005) Modern insect extinctions, the neglected majority. *Conservation Biology* 19, 1030–1036.
Dunn, R.R., Agosti, D., Andersen, A.N., Arnan, X., Bruhl, C.A., Cerdá, X., Ellison, A.M., *et al.* (2009) Climatic drivers of hemispheric asymmetry in global patterns of ant species richness. *Ecology Letters* 12, 324–333.
Ellis, E.C. and Ramankutty, N. (2008) Putting people in the map: anthropogenic biomes of the world. *Frontiers in Ecology and the Environment* 6, 439–447.
Feder, M.E., Blair, N. and Figueras, H. (1997) Oviposition site selection: unresponsiveness of *Drosophila* to cues of potential thermal stress. *Animal Behaviour* 53, 585–588.
Fonseca, C.R. (2009) The silent mass extinction of insect herbivores in biodiversity hotspots. *Conservation Biology* 23, 1507–1515.
Frazier, M.R., Huey, R.B. and Berrigan, D. (2006) Thermodynamics constrains the evolution of insect population growth rates: "Warmer Is Better". *American Naturalist* 168, 512–520.
Frenot, Y., Chown, S.L., Whinam, J., Selkirk, P.M., Convey, P., Skotnicki, M. and Bergstrom, D.M. (2005) Biological invasions in the Antarctic: extent, impacts and implications. *Biological Reviews* 80, 45–72.
Gaines, S.D. and Denny, M.W. (1993) The largest, smallest, highest, lowest, longest, and shortest: extremes in ecology. *Ecology* 74, 1677–1692.
Gaston, K.J. (1996) Biodiversity - latitudinal gradients. *Progress in Physical Geography* 20, 466–476.
Gaston, K.J. and Chown, S.L. (1999) Elevation and climatic tolerance: a test using dung beetles. *Oikos* 86, 584–590.
Gaston, K.J. and McArdle, B.H. (1994) The temporal variability of animal abundances: measures, methods and patterns. *Philosophical Transactions of the Royal Society B* 345, 335–358.
Gaston, K.J., Chown, S.L. and Evans, K.L. (2008) Ecogeographical rules: elements of a synthesis. *Journal of Biogeography* 35, 483–500.
Ghalambor, C.K., Huey, R.B., Martin, P.R., Tewksbury, J.J. and Wang, G. (2006) Are mountain passes higher in the tropics? Janzen's hypothesis revisited. *Integrative and Comparative Biology* 46, 5–17.
Ghalambor, C.K., McKay, J.K., Carroll, S.P. and Reznick, D.N. (2007) Adaptive versus non-adaptive phenotypic plasticity and the potential for contemporary adaptation in new environments. *Functional Ecology* 21, 394–407.
Gratz, N.G. (1999) Emerging and resurging vector-borne diseases. *Annual Review of Entomology* 44, 51–75.
Gray, D.R. (2010) Hitchhikers on trade routes: a phenology model estimates the probabilities of gypsy moth introduction and establishment. *Ecological Applications* 20, 2300–2309.
Grimaldi, D. and Engel, M.S. (2005) *Evolution of the Insects*. Cambridge University Press, Cambridge.
Hamilton, C. (2010) *Requiem for a Species. Why We Resist the Truth About Climate Change.* Earthscan, London.
Hansen, J., Sato, M., Kharecha, P., Beerling, D., Berner, R., Masson-Delmotte, V., Pagani, M., *et al.* (2008) Target atmospheric CO2: where should humanity aim? *The Open Atmospheric Science Journal* 2, 217–231.
Harrison, J.F., Kaiser, A. and Vandenbrooks, J.M. (2010) Atmospheric oxygen level and the evolution of insect body size. *Proceedings of the Royal Society B* 277, 1937–1946.
Hartley, S., Krushelnycky, P.D. and Lester, P.J. (2010) Integrating physiology, population dynamics and climate to make multi-scale predictions for the spread of an invasive insect: the Argentine ant at Haleakala National Park, Hawaii. *Ecography* 33, 83–94.
Hawkins, B.A. and Holyoak, M. (1998) Transcontinental crashes of insect populations? *American Naturalist* 152, 480–484.
Hawkins, B.A. and Porter, E.E. (2003) Water-energy balance and the geographic pattern of species richness of western Palearctic butterflies. *Ecological Entomology* 28, 678–686.
Hazell, S.P., Groutides, C., Neve, B.P., Blackburn, T.M. and Bale, J.S. (2010a) A comparison of low temperature tolerance traits between closely related aphids from the tropics, temperate zone, and Arctic. *Journal of Insect Physiology* 56, 115–122.
Hazell, S.P., Neve, B.P., Groutides, C., Douglas, A.E., Blackburn, T.M. and Bale, J.S. (2010b) Hyperthermic aphids: Insights into behaviour and mortality. *Journal of Insect Physiology* 56, 123–131.
Hegedüs, R., Åkesson, S. and Horváth, G. (2007) Anomalous celestial polarization caused by forest fire smoke: why do some insects become visually disoriented under smoky skies? *Applied Optics* 46, 2717–2726.
Heinrich, B. (1993) *The Hot-blooded Insects: Strategies and Mechanisms of Thermo-regulation.* Harvard University Press, Cambridge, Massachusetts.
Helmuth, B., Broitman, B.R., Yamane, L., Gilman, S.E., Mach, K., Mislan, K.A.S. and Denny, M.W.

(2010) Organismal climatology: analyzing environmental variability at scales relevant to physiological stress. *Journal of Experimental Biology* 213, 995–1003.

Hill, J.K., Griffiths, H.M. and Thomas, C.D. (2011) Climate change and evolutionary adaptations at species' range margins. *Annual Review of Entomology* 56, 143–159.

Hillebrand, H. (2004) On the generality of the latitudinal diversity gradient. *American Naturalist* 163, 192–211.

Hodkinson, I.D. (2003) Metabolic cold adaptation in arthropods: a smaller-scale perspective. *Functional Ecology* 17, 562–567.

Hoffmann, A.A. (2010) Physiological climatic limits in *Drosophila*: patterns and implications. *Journal of Experimental Biology* 213, 870–880.

Hoffmann, A.A., Sørensen, J.G. and Loeschcke, V. (2003a) Adaptation of *Drosophila* to temperature extremes: bringing together quantitative and molecular approaches. *Journal of Thermal Biology* 28, 175–216.

Hoffmann, A.A., Hallas, R.J., Dean, J.A. and Schiffer, M. (2003b) Low potential for climatic stress adaptation in a rainforest *Drosophila* species. *Science* 301, 100–102.

Hoffmann, A.A. and Willi, Y. (2008) Detecting genetic responses to environmental change. *Nature Reviews Genetics* 9, 421–432.

Huey, R.B., Deutsch, C.A., Tewksbury, J.J., Vitt, L.J., Hertz, P.E., Ãlvarez Pérez, H., J. and Garland, T. (2009) Why tropical forest lizards are vulnerable to climate warming. *Proceedings of the Royal Society B* 276, 1939–1948.

Irlich, U.M., Terblanche, J.S., Blackburn, T.M. and Chown, S.L. (2009) Insect rate-temperature relationships: environmental variation and the metabolic theory of ecology. *American Naturalist* 174, 819–835.

Irwin, J.T. and Lee, R.E. (2003) Cold winter microenvironments conserve energy and improve overwintering survival and potential fecundity of the goldenrod gall fly, *Eurosta solidaginis*. *Oikos* 100, 71–78.

Janzen, D.H. (1967) Why mountain passes are higher in the tropics. *American Naturalist* 101, 233–247.

Jeffree, C.E. and Jeffree, E.P. (1996) Redistribution of the potential geographical ranges of Mistletoe and Colorado Beetle in Europe in response to the temperature component of climate change. *Functional Ecology* 10, 562–577.

Jepsen, J.U., Hagen, S.B., Ims, R.A. and Yoccoz, N.G. (2008) Climate change and outbreaks of the geometrids *Operophtera brumata* and *Epirrita autumnata* in subarctic birch forest: evidence of a recent outbreak range expansion. *Journal of Animal Ecology* 77, 257–264.

Jetz, W. and Rubenstein, D.R. (2011) Environmental uncertainty and the global biogeography of cooperative breeding in birds. *Current Biology* 21, 1–7.

Jetz, W., Sekercioglu, C.H. and Böhning-Gaese, K. (2008) The worldwide variation in avian clutch size across species and space. *PloS Biology* 6, 2650–2657.

Kaspari, M., Alonso, L. and O'Donnell, S. (2000a) Three energy variables predict ant abundance at a geographical scale. *Proceedings of the Royal Society B* 267, 485–489.

Kaspari, M., O'Donnell, S. and Kercher, J.R. (2000b) Energy, density, and constraints to species richness: ant assemblages along a productivity gradient. *American Naturalist* 155, 280–293.

Katz, R.W. and Brown, B.G. (1992) Extreme events in a changing climate: variability is more important than averages. *Climatic Change* 21, 289–302.

Kearney, M. and Porter, M. (2009) Mechanistic niche modelling: combining physiological and spatial data to predict species' ranges. *Ecology Letters* 12, 334–350.

Kearney, M., Porter, W.P., Williams, C., Ritchie, S. and Hoffmann, A.A. (2009a) Integrating biophysical models and evolutionary theory to predict climatic impacts on species' ranges: the dengue mosquito *Aedes aegypti* in Australia. *Functional Ecology* 23, 528–538.

Kearney, M., Shine, R. and Porter, W.P. (2009b) The potential for behavioral thermoregulation to buffer "cold-blooded" animals against climate warming. *Proceedings of the National Academy of Sciences of the USA* 106, 3835–3840.

Kearney, M., Simpson, S.J., Raubenheimer, D. and Helmuth, B. (2010) Modelling the ecological niche from functional limits. *Philosophical Transactions of the Royal Society B* 365, 3469–3483.

Keister, M. and Buck, J. (1964) Some endogenous and exogenous effects on rate of respiration. In: Rockstein, M. (ed.) *Physiology of Insecta Volume 3*. Academic Press, New York, pp. 617–658.

Kellermann, V., Van Heerwaarden, B., Sgro, C.M. and Hoffmann, A.A. (2009) Fundamental evolutionary limits in ecological traits drive *Drosophila* species distributions. *Science* 325, 1244–1246.

Kerr, J.T. and Ostrovsky, M. (2003) From space to species: ecological applications for remote sensing. *Trends in Ecology and Evolution* 18, 299–305.

Kerr, J.T. and Packer, L. (1999) The environmental basis of North American species richness patterns among *Epicauta* (Coleoptera:

Meloidae). *Biodiversity and Conservation* 8, 617–628.
Kerr, J.T., Southwood, T.R.E. and Cihlar, J. (2001) Remotely sensed habitat diversity predicts butterfly species richness and community similarity in Canada. *Proceedings of the National Academy of Sciences of the USA* 98, 11365–11370.
Kingsolver, J.G. (1989) Weather and the population dynamics of insects: integrating physiological and population ecology. *Physiological Zoology* 62, 314–334.
Kingsolver, J.G. and Woods, H.A. (1997) Thermal sensitivity of growth and feeding in *Manduca sexta* caterpillars. *Physiological Zoology* 70, 631–638.
Klok, C.J., Sinclair, B.J. and Chown, S.L. (2004) Upper thermal tolerance and oxygen limitation in terrestrial arthropods. *Journal of Experimental Biology* 207, 2361–2370.
Kukalová-Peck, J. (1985) Ephemeroid wing venation based upon new gigantic Carboniferous mayflies and basic morphology, phylogeny, and metamorphosis of pterygote insects (Insecta, Ephemerida). *Canadian Journal of Zoology* 63, 933–955.
Labandeira, C.C. and Sepkoski, J.J. (1993) Insect diversity in the fossil record. *Science* 261, 310–315.
Lafferty, K.D. (2009) The ecology of climate change and infectious diseases. *Ecology* 90, 888–900.
Larsen, T.H., Williams, N.M. and Kremen, C. (2005) Extinction order and altered community structure rapidly disrupt ecosystem functioning. *Ecology Letters* 8, 538–547.
Lawton, J.H. (1992) There are not 10 million kinds of population dynamics. *Oikos* 63, 337–338.
Leather, S.R., Walters, K.F.A. and Bale, J.S. (1993) *The Ecology of Insect Overwintering*. Cambridge University Press, Cambridge.
Lee, R.E., Jr and Denlinger, D.L. (eds) (1991) *Insects at Low Temperature*. Chapman and Hall, New York.
Lighton, J.R.B. (2007) Hot hypoxic flies: whole-organism interactions between hypoxic and thermal stressors in *Drosophila melanogaster*. *Journal of Thermal Biology* 32, 134–143.
Lindroth, R.L. (2010) Impacts of elevated atmospheric CO_2 and O_3 on forests: phytochemistry, trophic interactions, and ecosystem dynamics. *Journal of Chemical Ecology* 36, 2–21.
Lobo, J.M., Lumaret, J.-P. and Jay-Robert, P. (2002) Modelling the species richness distribution of French dung beetles (Coleoptera, Scarabaeidae) and delimiting the predictive capacity of different groups of explanatory variables. *Global Ecology and Biogeography* 11, 265–277.
Lovegrove, B.G. (2000) The zoogeography of mammalian basal metabolic rate. *American Naturalist* 156, 201–219.
Mahlman, J.D. (1998) Science and nonscience concerning human-caused global warming. *Annual Review of Energy and the Environment* 23, 83–105.
Makarieva, A.M. and Gorshkov, V.G. (2007) Biotic pump of atmospheric moisture as driver of the hydrological cycle on land. *Hydrology and Earth System Sciences* 11, 1013–1033.
Makarieva, A.M., Gorshkov, V.G., Li, B.-L. and Chown, S.L. (2006) Size- and temperature-independence of minimum life-supporting metabolic rates. *Functional Ecology* 20, 83–96.
Marais, E. and Chown, S.L. (2008) Beneficial acclimation and the Bogert effect. *Ecology Letters* 11, 1027–1036.
Marshall, K.E. and Sinclair, B.J. (2010) Repeated stress exposure results in a survival-reproduction trade-off in *Drosophila melanogaster*. *Proceedings of the Royal Society B* 277, 963–969.
McCain, C.M. (2009) Vertebrate range sizes indicate that mountains may be 'higher' in the tropics. *Ecology Letters* 12, 550–560.
McMullen, D.C. and Storey, K.B. (2010) In cold-hardy insects, seasonal, temperature, and reversible phosphorylation controls regulate sarco/endoplasmic reticulum Ca^{2+}-ATPase (SERCA). *Physiological and Biochemical Zoology* 8, 677–686.
Melbourne, B.A. and Hastings, A. (2008) Extinction risk depends strongly on factors contributing to stochasticity. *Nature* 454, 100–103.
Michaud, M.R., Benoit, J.B., Lopez-Martinez, G., Elnitsky, M.A., Lee, R.E. and Denlinger, D.L. (2008) Metabolomics reveals unique and shared metabolic changes in response to heat shock, freezing and desiccation in the Antarctic midge, *Belgica antarctica*. *Journal of Insect Physiology* 54, 645–655.
Mitchell, K. and Hoffmann, A.A. (2010) Thermal ramping rate influences evolutionary potential and species differences for upper thermal limits in *Drosophila*. *Functional Ecology* 24, 694–700.
Morin, P.J., McMullen, D.C. and Storey, K.B. (2005) HIF-1α involvement in low temperature and anoxia survival by a freeze tolerant insect. *Molecular and Cellular Biochemistry* 280, 99–106.
Nair, U.S., Lawton, R.O., Welch, R.M. and Pielke Sr, R.A. (2003) Impact of land use on Costa Rican tropical montane cloud forests: sensitivity of cumulus cloud field characteristics to lowland deforestation. *Journal of Geophysical Research* 108, D74206, doi: 10.1029/2001JD001135.
New, M., Liverman, D., Schroder, H. and Anderson,

K. (2011) Four degrees and beyond: the potential for a global temperature increase of four degrees and its implications. *Philosophical Transactions of the Royal Society A* 369, 6–19.

Nichols, E., Spector, S., Louzada, J., Larsen, T., Amezquita, S. and Favila, M.E. (2008) Ecological functions and ecosystem services provided by Scarabaeinae dung beetles. *Biological Conservation* 141, 1461–1474.

Parmesan, C., Root, T.L. and Willig, M.R. (2000) Impacts of extreme weather and climate on terrestrial biota. *Bulletin of the American Meteorological Society* 81, 443–450.

Pelini, S.L., Keppel, J.A., Kelley, A.E. and Hellmann, J.J. (2010) Adaptation to host plants may prevent rapid insect responses to climate change. *Global Change Biology* 16, 2923–2929.

Pincebourde, S. and Casas, J. (2006a) Leaf miner-induced changes in leaf transmittance cause variations in insect respiration rates. *Journal of Insect Physiology* 52, 194–201.

Pincebourde, S. and Casas, J. (2006b) Multitrophic biophysical budgets: thermal ecology of an intimate herbivore insect-plant interaction. *Ecological Monographs* 76, 175–194.

Pincebourde, S., Sinoquet, H., Combes, D. and Casas, J. (2007) Regional climate modulates the canopy mosaic of favourable and risky microclimates for insects. *Journal of Animal Ecology* 76, 424–438.

Porter, W.P., Sabo, J.L., Tracy, C.R., Reichman, O.J. and Ramankutty, N. (2002) Physiology on a landscape scale: Plant-animal interactions. *Integrative and Comparative Biology* 42, 431–453.

Pörtner, H.O. (2001) Climate change and temperature-dependent biogeography: oxygen limitation of thermal tolerance in animals. *Naturwissenschaften* 88, 137–146.

Potter, K., Davidowitz, G. and Woods, H.A. (2009) Insect eggs protected from high temperatures by limited homeothermy of plant leaves. *Journal of Experimental Biology* 212, 3448–3454.

Ragland, G.J., Denlinger, D.L. and Hahn, D.A. (2010) Mechanisms of suspended animation are revealed by transcript profiling of diapause in the flesh fly. *Proceedings of the National Academy of Sciences of the USA* 107, 14909–14914.

Régnière, J. and Bentz, B. (2007) Modeling cold tolerance in the mountain pine beetle, *Dendroctonus ponderosae*. *Journal of Insect Physiology* 53, 559–572.

Rezende, E.L., Tejedo, M. and Santos, M. (2011) Estimating the adaptive potential of critical thermal limits: methodological problems and evolutionary implications. *Functional Ecology* 25, 111–121; doi:10.1111/j.1365-2435.2010.01778.x.

Ricklefs, R.E. (2008) Disintegration of the ecological community. *American Naturalist* 172, 741–750.

Robertson, R.M. (2004) Thermal stress and neural function: adaptive mechanisms in insect model systems. *Journal of Thermal Biology* 29, 351–358.

Rodriguero, M.S. and Gorla, D.E. (2004) Latitudinal gradient in species richness of the New World Triatominae (Reduviidae). *Global Ecology and Biogeography* 13, 75–84.

Rogers, D.J. and Randolph, S.E. (2006) Climate change and vector-borne diseases. *Advances in Parasitology* 62, 345–381.

Royal Society (2009) *Geoengineering the Climate. Science, Governance and Uncertainty*. Royal Society Publishing, London.

Ruel, J.M. and Ayres, M.P. (1999) Jensen's inequality predicts effects of environmental variation. *Trends in Ecology and Evolution* 14, 361–366.

Samways, M.J. (1989) Climate diagrams and biological control: an example from the areography of the ladybird *Chilocorus nigritus* (Fabricius,1798) (Insecta, Coleoptera, Coccinellidae). *Journal of Biogeography* 16, 345–351.

Samways, M.J. (1994) *Insect Conservation Biology*. Chapman and Hall, London.

Samways, M.J. (2007) Insect conservation: a synthetic management approach. *Annual Review of Entomology* 52, 465–487.

Sanderson, M.G., Hemming, D.L. and Betts, R.A. (2011) Regional temperature and precipitation changes under high-end (≥4°C) global warming. *Philosophical Transactions of the Royal Society A* 369, 85–98.

Savage, V.M. (2004) Improved approximations to scaling relationships for species, populations, and ecosystems across latitudinal and elevational gradients. *Journal of Theoretical Biology* 227, 525–534.

Schneider, S.H. (2008) Geoengineering: could we or should we make it work? *Philosophical Transactions of the Royal Society A* 366, 3843–3862.

Schwager, M., Johst, K. and Jeltsch, F. (2006) Does red noise increase or decrease extinction risk? Single extreme events versus series of unfavorable conditions. *American Naturalist* 167, 879–888.

Scriber, J.M. (2002) Latitudinal and local geographic mosaics in host plant preferences as shaped by thermal units and voltinism in *Papilio* spp. (Lepidoptera). *European Journal of Entomology* 99, 225–239.

Seely, M., Henschel, J.R. and Hamilton, W.J. III (2005) Long-term data show behavioural fog collection adaptations determine Namib Desert beetle abundance. *South African Journal of Science* 101, 570–572.

Sinclair, B.J. and Chown, S.L. (2005a) Deleterious effects of repeated cold exposure in a freeze-tolerant sub-Antarctic caterpillar. *Journal of Experimental Biology* 208, 969–879.

Sinclair, B.J. and Chown, S.L. (2005b) Climatic variability and hemispheric differences in insect cold tolerance: support from southern Africa. *Functional Ecology* 19, 214–221.

Sinclair, B.J. and Renault, D. (2010) Intracellular ice formation in insects: Unresolved after 50 years? *Comparative Biochemistry and Physiology A* 155, 14–18.

Sinclair, B.J., Addo-Bediako, A. and Chown, S.L. (2003) Climatic variability and the evolution of insect freeze tolerance. *Biological Reviews* 78, 181–195.

Stafford Smith, M. Horrocks, L., Harvey, A. and Hamilton, C. (2011) Rethinking adaptation for a 4°C world. *Philosophical Transactions of the Royal Society A* 369, 196–216.

Stevens, M.M., Jackson, S., Bester, S.A., Terblanche, J.S. and Chown, S.L. (2010) Oxygen limitation and thermal tolerance in two terrestrial arthropod species. *Journal of Experimental Biology* 213, 2209–2218.

Stillwell, R.C., Morse, G.E. and Fox, C.W. (2007) Geographic variation in body size and sexual size dimorphism of a seed-feeding beetle. *American Naturalist* 170, 358–369.

Tenow, O. and Nilssen, A. (1990) Egg cold hardiness and topoclimatic limitations to outbreaks of *Epirrita autumnata* in Northern Fennoscandia. *Journal of Applied Ecology* 27, 723–734.

Terblanche, J.S., Deere, J.A., Clusella-Trullas, S., Janion, C. and Chown, S.L. (2007) Critical thermal limits depend on methodological context. *Proceedings of the Royal Society B* 274, 2935–2942.

Terblanche, J.S., Clusella-Trullas, S. and Chown, S.L. (2010) Phenotypic plasticity of gas exchange pattern and water loss in *Scarabaeus spretus* (Coleoptera: Scarabaeidae): deconstructing the basis for metabolic rate variation. *Journal of Experimental Biology* 213, 2940–2949.

Thornton, P.K., Jones, P.G., Ericksen, P.J. and Challinor, A.J. (2011) Agriculture and food systems in sub-Saharan Africa in a 4°C+ world. *Philosophical Transactions of the Royal Society A* 369, 117–136.

Todd, M.C., Washington, R., Cheke, R.A. and Kniveton, R. (2002) Brown locust outbreaks and climate variability in southern Africa. *Journal of Applied Ecology* 39, 31–42.

Turner, J.R.G., Gatehouse, C.M. and Corey, C.A. (1987) Does solar energy control organic diversity? Butterflies, moths and the British climate. *Oikos* 48, 195–205.

Turner, J.S. (2000) *The Extended Organism. The Physiology of Animal-Built Structures.* Harvard University Press, Cambridge, Massachusetts.

Unwin, D.M. (1991) *Insects, Plants and Microclimate.* The Richmond Publishing Co. Ltd, Slough.

Vannier, G. (1994) The thermobiological limits of some freezing tolerant insects: the supercooling and thermostupor points. *Acta Oecologica* 15, 31–42.

Vasseur, D.A. and Yodzis, P. (2004) The color of environmental noise. *Ecology* 85, 1146–1152.

Voituron, Y., Mouquet, N., De Mazancourt, C. and Clobert, J. (2002) To freeze or not to freeze? An evolutionary perspective on the cold-hardiness strategies of overwintering ectotherms. *American Naturalist* 160, 255–270.

Wallace, J.B. and Webster, J.R. (1996) The role of macroinvertebrates in stream ecosystem function. *Annual Review of Entomology* 41, 115–139.

Wardle, D.A., Bardgett, R.D., Klironomos, J.N., Setälä, H., Van Der Putten, W.H. and Wall, D.H. (2004) Ecological linkages between aboveground and belowground biota. *Science* 304, 1629–1633.

Warren, M.S., Hill, J.K., Thomas, J.A., Asher, J., Fox, R., Huntley, B., Roy, D.B., *et al.* (2001) Rapid responses of British butterflies to opposing forces of climate and habitat change. *Nature* 414, 65–69.

Warren, M., McGeoch, M.A., Nicolson, S.W. and Chown, S.L. (2006) Body size patterns in *Drosophila* inhabiting a mesocosm: interactive effects of spatial variation in temperature and abundance. *Oecologia* 149, 245–255.

Webb, T.J., Gaston, K.J., Hannah, L. and Woodward, F.I. (2006) Coincident scales of forest feedback on climate and conservation in a diversity hotspot. *Proceedings of the Royal Society B* 273, 757–765.

White, P. and Kerr, J.T. (2006) Contrasting spatial and temporal global change impacts on butterfly species richness during the 20th century. *Ecography* 29, 908–918.

Whitman, D.W. and Ananthakrishnan, T.N. (2008) *Phenotypic Plasticity of Insects. Mechanisms and Consequence.* Science Publishers, Enfield, New Hampshire.

Wilf, P., Labandeira, C.C., Johnson, K.R. and Ellis,

B. (2006) Decoupled plant and insect diversity after the end-cretaceous extinction. *Science* 313, 1112–1115.

Williams, J.B., Shorthouse, J.B. and Lee, R.E. (2003) Deleterious effects of mild simulated overwintering temperatures on survival and potential fecundity of rose-galling Diplolepis wasps (Hymenoptera: Cynipidae). *Journal of Experimental Zoology* 298, 23–31.

Willi, Y. and Hoffmann, A.A. (2009) Demographic factors and genetic variation influence population persistence under environmental change. *Journal of Evolutionary Biology* 22, 124–133.

Willmer, P.G. (1982) Microclimate and the environmental physiology of insects. *Advances in Insect Physiology* 16, 1–57.

Woods, H.A. (2010) Water loss and gas exchange by eggs of *Manduca sexta*: Trading off costs and benefits. *Journal of Insect Physiology* 56, 480–487.

Woods, H.A. and Smith, J.N. (2010) Universal model for water costs of gas exchange by animals and plants. *Proceedings of the National Academy of Sciences of the USA* 107, 8469–8474.

Woods, H.A., Makino, W., Cotner, J.B., Hobbie, S.E., Harrison, J.F., Acharya, K. and Elser, J.J. (2003) Temperature and the chemical composition of poikilothermic organisms. *Functional Ecology* 17, 237–245.

Woodward, F.I., Bardgett, R.D., Raven, J.A. and Hetherington, A.M. (2009) Biological approaches to global environment change mitigation and remediation. *Current Biology* 19, R615–R623.

5 Temperature Adaptation in Changing Climate: Marine Fish and Invertebrates

Doris Abele

5.1 Changing Climate Affects Marine Ectothermic Species

Marine invertebrates and fish are important biomass components of oceanic and coastal food webs which link carbon flux between primary producers and top order predators. Global climate change already has profound effects on these food webs. At present, warming in some regions of the earth is causing mismatches between species that affect predator–prey relationships or species competition, and lead to shifts at the population and community level (Pörtner *et al.*, 2001; Somero, 2005; Perry *et al.*, 2005; Parmesan, 2006; Kochmann *et al.*, 2008). Although the principles of adaptation to rising temperatures are universal across marine ectotherms, the response of animals and populations to regional and global warming differs considerably with climatic background (adaptation) and additionally varies with seasonal or regional acclimatization (that is, physiological adjustment to meet thermal/environmental change in nature).

The width of annual temperature windows, as well as the thermal extremes (both high and low) that animals experience in natural habitats, are critical determinants of the thermal tolerance of adult fish and invertebrates (Dahlhoff and Somero, 1993). Thermal adaptation of adult specimens is reflected in the maximum survival temperature of gametes, eggs, and spawn which, once released to the outside world, rely on an inherited level of stress defence (Andronikov, 1975). Thermal tolerance windows of larvae are often significantly narrower than those of the adult forms, and the larval tolerances can often more clearly explain the climatic borders for sustainable colonization in a given species.

Thermal adaptation and acclimatization of individuals or within local populations of marine ectotherms are reflected in the Arrhenius break temperatures (ABT) at which extracted enzymes, isolated organelles (especially mitochondria) or organs, such as the heart, become functionally impaired during *in vitro* testing studies (O'Brien *et al.*, 1991; Somero, 2005). Many studies aimed at defining species-specific thermal tolerance windows have therefore investigated ABT of rate-determining key metabolic enzymes, expression levels, and induction temperatures of heat shock proteins, and other systems that protect tissues against heat stress. Importantly, however, the upper critical temperatures (T_{crit}) that delimit functioning of isolated cells or subcellular systems are usually higher – often considerably (10°C) higher – than the lethal temperature of a complex organism (Bowler, 1963; Somero, 2005). So, while ABT are instrumental in comparing the actual physiological acclimatization of individuals in and among populations (Stillman and Tagmount (2009), for example, showed that upper limits of cardiac thermal performance vary with acclimatization temperature in porcelain crabs), they do not allow predictions with respect to the adaptive capacities of animals for coping with environmental change.

Furthermore, there is a huge discrepancy

© CAB International 2012. *Temperature Adaptation in a Changing Climate: Nature at Risk* (eds K.B. Storey and K.K. Tanino)

between the response of test animals to rapid experimental warming and the actual effect of natural dynamic warming of the same species, or the width of the biogeographic thermal range that a species can colonize (Barnes and Peck, 2008). Rapid experimental warming is often deleterious because it does not give the animals time to adjust their cellular 'hardware' (membranes, organelles, functional proteins), and a critical temperature is reached when cellular structures and molecules become thermally damaged. Accelerated repair mechanisms are energy consuming, and respiration can be limited as ventilation is impaired in rapidly heated animals. Under such conditions, some sensitive, complex animals cross the energetic limits and end up in a thermally reinforced state of oxygen shortage (Pörtner, 2002). If animals cannot rebound upon return to control temperature, the energetic drainage can be fatal. By contrast, climate change-induced global warming implies prolonged seasonal warming periods, including short episodes of extreme heat and subsequent cooling. Furthermore, regional climate scenarios are often down-scaled into diverse small-scale temperature patterns with cooler protective microhabitats (e.g. rock pools, crevices) within spatially complex natural habitats (Helmuth and Hofmann, 2001; Helmuth *et al.*, 2006). Under natural conditions animals are often warmed slowly and dynamically (that is, discontinuously), allowing them – within certain boundaries – to adjust their cellular hardware and acclimatize to the new thermal condition. Discontinuous, dynamic heating can also induce heat hardening of cells and tissues, for example in intertidal species where time-limited non-lethal heat exposure induces cellular defence mechanisms (Hofmann, 1999; Tomanek, 2005). On one hand, these defences can help the organism to survive future periods of stressful warming. On the other hand, stress-hardening responses are costly and can explain different growth rates and population age-structures of the same species in ecologically differing habitats (Helmuth *et al.*, 2006; Stillman and Tagmount, 2009; Basova *et al.*, in press). Thus, while climate change will indeed eradicate various sensitive species in some oceanic and coastal environments, many ectotherms will be able to adapt by changing and adjusting life history parameters such as growth rates and maximal attainable size and age, and by seasonally shifting life cycle events such as onset of reproduction (see also Daufresne *et al.*, 2009). Based on long-term data sets of fish and planktonic communities in mesocosm studies, the authors predicted a decrease in body size at the individual (size at age) and population level under conditions of global warming, and further argued that the highest trophic levels (with large body masses) could be most sensitive to climate warming (Daufresne *et al.*, 2009). In temperate (cod) and tropical (zebrafish) fish species, a direct negative effect of critically elevated temperatures on daily growth rate, muscle protein stores, and condition index has been reported (Pörtner *et al.*, 2001; Vergauwen *et al.*, 2010). By contrast, in most cases where elevated temperatures accelerate growth in ectotherms (such as polar molluscs or fish), this happens at the cost of stress hardiness and damage repair capacity (see Abele *et al.*, 2009 for review).

The most pronounced biochemical and physiological response to temperature change and thermal stress is observed in temperate species, adapted to cooler climate in most marine areas, but prepared to face seasonal fluctuations to sometimes much higher habitat temperatures. Therefore, I will concentrate first on the adaptive physiological and ecological responses to warming in cold–temperate marine ectotherms. To understand centrally important traits of thermal tolerance, it is also important to see how evolutionary adaptation is coined in an *extreme* climatic background. In other words, which adaptive traits restrict the capacities of cold stenotherms from Antarctic environments to acclimatize to the present warming trend in Western Antarctica? Finally I will summarize the present knowledge of regulatory pathways that support metabolic adaptations to warmer temperatures.

5.2 The Cellular Stress Response to Critical Warming in Marine Invertebrates

Exposure of marine ectothermic animals to natural or experimental warming accelerates metabolic energy (ATP) requirements and elevates the oxygen consumption by isolated mitochondria, tissues and whole animals. Experimental studies showed that this causes a corresponding increase in reactive oxygen species (ROS) production. If not tightly controlled by antioxidant systems, ROS damage complex molecules and cellular structural components, and this damage is enhanced as animals are warmed beyond critical temperature (T_{crit}) limits (Abele and Puntarulo, 2004). Below the species or population-specific T_{crit}, oxidative stress is controlled at low levels (sometimes prevented) by the animals' antioxidant defence systems (enzymatic and biochemical ROS scavengers, such as glutathione, and vitamins C, E and A). Above T_{crit}, mitochondrial ROS release increases exponentially (Abele *et al.*, 2002; Heise *et al.*, 2003) as oxygen radical scavengers become exhausted and antioxidant enzymes are thermally impaired. Damage that accumulates impairs cellular functioning (Terman and Brunk, 2006; Philipp and Abele, 2010) and, under stress, repair and removal of oxidative damage are slowed, as cell division and apoptotic cell renewal are protracted (Place *et al.*, 2008). All of the cellular processes that accompany a thermal stress condition apply equally to temperate/eurythermal and polar (often highly stenothermal) species. However, during 20 million years (Ma) of isolation in a permanent cold environment, many polar marine ectotherms in the Antarctic have lost the adaptive plasticity that protects the cells of temperate animals during warming. Thus, in temperate invertebrates and fish, thermally accelerated oxygen consumption and ROS formation is efficiently curtailed by a decrease in mitochondrial densities, whereas polar species lack this adaptive capacity. The mitochondrial proton leak can have an important antioxidant function, as it uncouples the inner mitochondrial membrane potential which alleviates the reduction state of respiratory electron carriers and mitigates ROS formation and oxidative stress (summarized in Buttemer *et al.*, 2010). This 'mild uncoupling mechanism' (Brand, 2000) is adjusted low when higher ATP yields are required in actively moving or swimming animals, and high when activity levels and energy requirements are low, and oxygen accumulates and threatens to pollute cells with ROS. Under conditions of moderately stressful warming at the border of the thermal tolerance range of a species, locomotory performance is curtailed as the animals start to become energetically restricted, and cellular oxygen, not consumed any more by oxidative phosphorylation in muscle tissues, can be removed via futile H^+ cycling before it generates toxic ROS (Keller *et al.*, 2004; Abele *et al.*, 2007). By contrast, critical heating outside the thermal tolerance window increased oxygen consumption, ROS formation and proton leak in thermally uncoupled mitochondria isolated from temperate and Antarctic mud clams (Abele *et al.*, 2002; Heise *et al.*, 2003). Concentrations of the so-called thiobarbituric acid-reactive substances (TBARS), a very crude measure of ongoing lipid peroxidation, increased in bivalves and other ectotherms upon short-term warming beyond T_{crit}; and enzymatic antioxidants, which are sometimes activated during heat shock, usually fail to balance stress when warming is slow, critical and persistent (Abele *et al.*, 2001, 2002; also see below).

Warming above the critical thermal limit, where *in vitro* mitochondrial oxygen demand rises progressively, causes thermal hypoxaemia (summarized in Pörtner and Lannig, 2009) and dramatically enhances energetic costs for cellular oxidative damage repair, protein turnover and new expression/synthesis of stress proteins. Oxidative damage repair and stress protein synthesis slow under energetically limited and oxygen-deficient conditions during heat stress. *In vivo* ROS generation may indeed subside in the hypoxic state (mitochondria in less well-oxygenated cells generate less ROS than organelles having full oxygen supply, in fully oxygenated tissues). Exposing dogwhelks (*Nucella lapillus*) to short-term warming from 16°C to 26.5°C and 30°C, Gardeström *et al.* (2007) showed how high temperatures alter the gene expression patterns, with the number of heat stress-induced genes increasing as long as the animals could maintain their energetic homeostasis. T_{crit} between 26.5°C and 30°C was confirmed by a reduction of the number of expressed temperature-specific (heat stress and antioxidant) genes at 30°C. This limitation could be completely reverted when dogwhelks were exposed to 30°C in a hyperoxic atmosphere, and nicely supports the principle of oxygen-limited survival during acute heat stress. It underlines that fatal heat stress occurs primarily during acute warming. Dogwhelks were collected in summer (June) and adaptive plasticity for response to temperature oscillations was high.

Genes that are up-regulated under heat stress include protein homeostasis genes such as heat shock proteins (Hsp) and ß-tubulin, whereas cell proliferation genes are down-regulated (Place *et al.*, 2008; Stillman and Tagmount, 2009). Stillman and Tagmount further reported that genes of energy metabolism, including cytochrome-*c* oxidase, NADH dehydrogenase, arginine kinase and ATP synthase were down-regulated in cold- (winter) acclimatized porcelain crabs, and induced in warm-acclimatized crabs, or crabs that had been experiencing strong thermal fluctuations. Clearly, higher mitochondrial densities acquired by the crabs during the winter create a problem upon sudden exposure to an experimental heat wave, whereas crabs prepared to tolerate high and fluctuating temperatures synthesized proteins in support of elevated mitochondrial respiration rates. In keeping with possible immune deficiencies under warmer conditions, immune response and bacterial recognition genes (anti-LPS factor) were more highly expressed in cold-acclimatized porcelain crabs and repressed by heat stress (Stillman and Tagmount, 2009).

As climate change proceeds, thermal stress conditions (heat waves) will be more frequent, and these will consume cellular energy reserves in marine ectotherms to cover the costs for damage repair. It should not be forgotten that climate change leads to important sea surface temperature increases in many regions especially during winter, mostly due to climate shift-induced effects on oceanic circulation patterns (Deser and Blackmon, 1993). However, low temperatures in winter are important for regenerative processes and cellular defence mechanisms, which build up as ectotherms enter into a metabolic slowdown, similar to a diapause state. If higher maximum temperatures increase the demand for repair and, at the same time, less winter cold restricts natural regenerative periods, this is bound to render many marine species, especially the long-lived cold water forms, more susceptible to stress and presumably shortens life expectancy. In this context, it is interesting that temperate bivalves that spawn once a year after severe winters show a second spawning when preceding winters were mild (Günther *et al.*, 1998). It is not known how the second spawning affects larval fitness and recruitment success, but it demonstrates that changes in the lower annual temperature exposure limit of a species can have diverse effects on its life history parameters and performance.

5.3 Cold Adaptation is a One-way Road and Biases the Physiological Response to Warming in Antarctic Marine Ectotherms

Permanent cold adaptation and the principle of higher mitochondrial densities in cold-adapted species are best illustrated by comparing populations from the same or closely related species (ecological and morphological comparability and at least from the same family) on biogeographical climate or depth gradients. In addition to an increase in mitochondrial volume density, extension of the inner mitochondrial membrane surface characterizes the long-term adaptation of Antarctic stenotherms to permanent cold. Thus, Lurman *et al.* (2010) found two- to fourfold enhanced volume densities and also higher mitochondrial cristae surface area in the Antarctic limpet *Nacella concinna* compared to temperate and tropical molluscs that did not differ among each other. Laboratory acclimation of *N. concinna* to higher temperature (0°C –> 3°C) for several months reduced cristae surface area, but did not reduce the number of mitochondria per cell in *N. concinna* (Morley *et al.*, 2009). So, contrary to many examples from temperate species that exhibit rapid adjustments of mitochondrial densities to seasonal change or warm acclimation in the laboratory (Sommer and Pörtner, 2002; Keller *et al.*, 2004), Antarctic stenotherms lack the short-term adaptive capacities of their cellular energetic hardware. With the decrease in cristae surface area upon warm acclimation (and a corresponding decrease of citrate synthase activity, representing the aerobic scope of tissues), *N. concinna* may even be the 'one-eyed among the blind' of the Antarctic species with respect to thermal adaptive plasticity. These active crawlers (max speed 10 mm/sec at 0°C, Davenport, 1997) have conserved high activity against the depressing effect of low Antarctic water temperatures, which is intuitively associated with higher metabolic capacities and enhanced metabolic flexibility in *N. concinna* compared to less mobile Antarctic ectotherms (Pörtner, 2002; Morley *et al.*, 2009). *N. concinna* colonizes intertidal and sublittoral shore levels (see below). Intertidal specimens survive in cooled niches such as rock crevices, but are still exposed to higher air and elevated water temperatures in rock pools that can occasionally be warmed to above 4°C during summer. When these intertidal limpets are experimentally heat shocked by exposing them for 24 h or 48 h to 4°C and 9°C, some antioxidant enzyme activities including superoxide dismutase (SOD) and catalase are slightly up-regulated at both temperatures, although this does not prevent a massive increase of oxidative damage markers (lysosomal membrane destabilization and lipofuscin accumulation; Abele *et al.*, 1998). By contrast, in the strictly sublittoral and highly stenothermal mud clam *Yoldia eightsi* from the same Antarctic environment (Potter Cove at King George Island (KGI), South Shetland Archipelago), the activity of the major radical scavenger SOD was thermally inhibited when animals were maintained for 5 days above 2°C (Abele *et al.*, 2001). Of course lipid peroxidation markers increased correspondingly, reflecting oxidative damage in the tissue lipid fraction. The mud clam *Laternula elliptica* (from near Rothera Station, Adelaide Island, West Antarctic) also shows no real acclimatory response to long-term mild experimental warming (0°C –>3°C during several months), and Morley *et al.* (2009) concluded that sluggish Antarctic stenotherms, such as infaunal clams, will respond less flexibly to thermal challenge. And while this is certainly true, the story might be even more complex.

More specific information comes from well-studied areas around Antarctic stations, such as Potter Cove at KGI, where yearly mean temperatures within the whole water column have been rising by on average 0.6°C within the last 20 years (0.3°C per decade, Schloss *et al.*, in preparation). Here, the overall abundance of the mud clam *L. elliptica* is still increasing, and the clams colonize areas that were previously not colonized by *Laternula* (Sahade et al., 2008). In certain areas of Potter Cove, however, changes in population structure and thinning of densities are observable. Populations under severe ice impact (coastal ice blocks are a major product of local glacier disintegration) are less dense than at neighboring (1–2 km distance) or deeper sites without or with less scouring impact. Furthermore, smaller individuals were observed to be more resistant to injury and rebury more rapidly after being unearthed (e.g. by icebergs) than larger clams (Philipp *et al.*, 2011). Peck *et al.* (2004) reported that *L. elliptica* suffer 50% failure in reburying as temperatures rise to 2°C–3°C and thereby confirmed the interacting effect of two factors directly related to climate change. The forcing effect by icebergs on population size structure has also been observed for the clam *Y. eightsi* at Signy Island (Peck and Bullough, 1993) and for Antarctic bryozoans at Rothera (Brown *et al.*, 2004), areas under long-term investigation by scientists from the British Antarctic Survey (BAS). Furthermore, in Potter Cove, Sahade *et al.* (2008) observed a diminishment of stalked ascidians in areas with increased sediment import from coastal sedimentary runoff. As climate warming proceeds, the indirect effects of aerial warming on marine coastal populations, not only in the Antarctic, are becoming clearly visible and it is hoped they will be better studied in the future.

Nacella concinna is one of the best-studied macroinvertebrates and a model for successful colonization of the Antarctic intertidal zone and, thus, for ongoing adaptation of a stenothermal organism to a broader thermal window and a generally more variable abiotic environment. The limpets are reported to migrate to the intertidal zone only during the summer season, although recent evidence suggests that specimens also overwinter in rock crevices under the ice (Waller *et al.*, 2006). Whether this habitat extension into the intertidal zone is new and relates to the ongoing rapid regional air warming and the reduction of winter coastal ice duration at the Antarctic Peninsula is currently disputed. Limpets that colonize the intertidal zone have taller shells and less wet mass at the same shell size than their sublittoral congeners. The taller shells allow the animals to store more shell water during tidal emersion, which buffers them thermally (keeps them colder) and prevents desiccation (Weihe and Abele, 2008). Intertidal Antarctic limpets systematically display higher antioxidant activities in gill tissues and maintain higher adenylate levels in foot muscle. Moreover, they have developed a behavioural strategy that involves interspersed air gaping and slowdown of metabolism to delay onset of anaerobic metabolism during prolonged air exposure. This behaviour helps them to survive during tidal emersion when the limpets clamp their shells down to the rock surfaces to prevent water loss and desiccation and the shell water of the intertidal limpets becomes hypoxic. By contrast, sublittoral limpets rapidly lose shell water and switch to anaerobic metabolism when experimentally exposed to air and desiccation (Weihe and Abele, 2008; Weihe *et al.*, 2009). So far, no evidence exists for a genetic separation of intertidal and sublittoral Antarctic limpets and, indeed, phenotypic plasticity seems to allow the species to colonize high intertidal shores on the Antarctic Peninsula. Further to the north and into the warmer climate of the South American continent, the closely related limpets *Nacella deaurata* and *N. magellanica* colonize the Patagonian rocky shores. All three limpet species are genetically very closely related and *N. concinna* is thought to have crossed the Drake Passage less than 20 Ma ago to colonize the Antarctic Peninsula. Recent investigations using rapidly evolving microsatellite markers suggest that *N. concinna* has evolved into an independent species separated from gene flow from South America for approximately 7 Ma (Pöhlmann and Held, in review). Thus, the genus *Nacella* with its Patagonian and West Antarctic subspecies may be a perfect indicator system for studying the effects of relatively recent adaptation to Antarctic cold conditions. In other words, the stenothermal Antarctic limpet, *N. concinna*, may, to some little extent, have preserved genetic features of eurythermality that supports its successful colonization of Antarctic intertidal rock pools along the WAP and the adjacent islands. Some features between the three limpet species show a tiered response pattern apparently linked to climatic conditions on the geographic gradient between Puerto Montt in the north and KGI, south of the Antarctic circumpolar current (ACC): Patagonian limpets from Puerto Montt have significantly higher shell height/length ratios, adaptive to higher environmental stress conditions (Weihe and Abele, 2008). Furthermore, constitutive levels and the intensity to which a heat shock response can be induced increases and becomes much more variable towards the northern warmer regions, with KGI (Antarctica) < Punta Arenas < Puerto Montt. At Puerto Montt (41°30´S) the limpets have reached their northernmost habitat extension and presumably their thermal limit and are unable to colonize shallow intertidal and splash zones as they do at the colder Magellan (Punta Arenas) and KGI sites (Pöhlmann *et al.*, 2011). Metabolic rate measurements performed with *Nacella concinna* from the shallow subtidal area (5–7 m) at KGI, Antarctica (Abele *et al.* 1998, measuring temperature = 1°C) and with intertidal *N. magellanica* from the Magellan Strait at Ushuaia (Malanga *et al.*, 2007, measured at seasonal water temperature between 3°C in winter and 10°C in summer) indicate higher tissue-specific respiration (1 µmol O_2/g fresh mass (fm)/h at mean shell length 29.3 ± 3 mm) in Antarctic compared to Patagonian specimens (0.1–0.3 µmol O_2/g fm/h). Although part of the difference can be attributed to scaling (Patagonian specimens had >50 mm shell length and were thus nearly twice as large as the Antarctic specimens), this suggests complete metabolic cold compensation in the Antarctic limpets. This is not a typical response to low temperatures in Antarctic stenotherms (see below) and, if it should hold true, corroborates relatively recent adaptation to permanent cold. Further comparative measurements of metabolic rates and cellular aerobic capacities of similarly sized limpets on a gradient between Puerto Montt and Adelaide Island (Rothera) are needed to obtain a consistent picture of limpet evolutionary history and climatic adaptation.

Most Antarctic ectotherms with longer evolutionary history in the permanently cold climate feature reduced locomotory scope in spite of higher mitochondrial densities. Their muscle cells have less space for contractile elements, which reduces contractile power (Weibel, 1985; Seibel *et al.*, 2007; O'Brien and Sidell, 2000) and cuts down on the capacity for burst swimming. Many Antarctic fishes and even squids are indeed so slow moving that divers can often catch them by hand. This principle applies to such diverse models as red and white muscle of temperate and polar fish (Guderley and St-Pierre, 2002; Verde *et al.*, 2006) and high locomotory species such as pteropods ('sea angels') from Antarctic and temperate seas (Seibel *et al.*, 2007). The Antarctic pteropod, *Clione antarctica*, has lost neuromuscular components (fast twitch anaerobic muscle and large motoneurons) to negative selection in the cold and, in spite of having twice as many mitochondria per cell, is unable to perform fast swim cycles, comparable to its temperate congener *C. limacina* (Rosenthal *et al.*, 2009). Again, adaptive loss of phenotypic plasticity in permanent cold environments limits the possibility for an adaptive response to a thermal stress stimulus, a phenomenon observed more often since genetic sequencing of polar organisms has been intensified. It resembles the loss of erythropoiesis in Antarctic ice fish, as well as their reduced capacity for producing an inducible heat shock response (Hofmann *et al.*, 2000). It may not necessarily be a question of the time duration of the adaptation history of an Antarctic species or of its genetic isolation (see the example of incomplete stenothermality in *N. concinna*). Mutation is a contingent process and, once accepted and selected for under prevailing environmental extreme conditions, it may represent a one-way road that may lead to extinction in regions where Antarctic waters are warming, especially in the West Antarctic Peninsula region (Meredith and King, 2005).

5.4 Cellular Stress Signals in Response to Warming

Recent advances in marine genomics and molecular studies of non-biomedical model organisms allow us to distinguish two phases in the response to thermal/environmental stress in marine ectotherms. In a first rapid response (2–4 days) to critical (and finally lethal) heating, animals increase the levels of chaperones, mainly Hsp. No sophisticated signalling pathway is necessary to induce transcription of *hsp* genes. Instead, increasing amounts of denatured proteins cause a release of heat shock factor (HSF1) previously bound to Hsp, which in turn interact with the denatured proteins. When released, HSF1 binds to heat shock elements in the promoter regions of *hsp* genes and induces the transcriptional response.

Besides HSF1, other regulatory factors presumably fine-tune the heat shock response (Hofmann, 1999; Buckley *et al.*, 2001). The early stress response to warming is often characterized by a depletion of energy stores in order to rapidly limit and repair damage (Vergauwen *et al.*, 2010). Damaged proteins can affect other macromolecules, such as DNA, and also inhibit cellular pathways.

Upon prolonged thermal acclimation, a regulatory response for compensation of metabolism starts (Somero, 2005). The first transcriptome approaches applied to analyse the responses by freshwater and marine fish and ectotherms to warming and cooling have highlighted possible signalling pathways involved in these metabolic adjustments. For zebrafish, Vergauwen *et al.* (2010) reported that the acclimatory stress response is indeed similar for warming and cooling, and thus independent of the direction of temperature change. It involves induction of haemopoiesis genes and angiogenesis factors, as well as proteins related to oxidative stress or antitumour activity. Taken together, the acclimation response is aimed at maintaining cellular integrity and regulation of cell turnover, whereas genes supporting oogenesis are down-regulated during high temperature stress, for energy-saving purposes. Haemopoiesis induction points at an involvement of hypoxia-inducible factor- (HIF) regulated pathways during stressful warming in zebrafish and to transient heat-induced hypoxemia in zebrafish warmed from 18°C to 34°C. Thus, thermal hypoxemia is a critical condition but can also have a protective signalling function, since it induces metabolic down-regulation in hypoxia-tolerant animals and cell types.

A HIF signal was observed in North Sea eelpout liver within 24 h of acute heating from 12°C (control temperature) to 18°C, a benevolent thermal challenge. The HIF signal consisted of enhanced binding of eelpout HIF protein to the (human) erythropoietin- (EPO) enhancer sequence in an electromobility shift assay (EMSA, Heise *et al.*, 2006a). The response in eelpout aligns with the short-term heat shock response in the zebrafish study of Vergauwen *et al.* (2010), with the difference that 18°C for eelpout is not a lethal temperature. Heating to 22°C (which is above the critical temperature in eelpout) reduced the positive HIF response in the EMSA assay, and we related it to increased tissue oxygenation and ROS formation entailing more oxidized tissue redox potential (ΔE_{mV}) in eelpout liver as the fish suffered critical heat exposure. The HIF-binding signal (seen by EMSA) was also observed when eelpout were exposed to short-term severe cooling to 1°C and tissue redox potential was correspondingly reduced (Heise *et al.*, 2006b). Contrary to the situation in mammals, the oxygen-sensitive alpha subunit of the HIF transcription factor (Fandrey *et al.*, 2006; Hoogewijs *et al.*, 2007) seems to be constitutively present in hypoxia-tolerant fish such as eelpout and crucian carp (Nikinmaa and Rees, 2005; Rissanen *et al.*, 2006; Heise *et al.*, 2007) and further stabilizes at much higher cellular oxygen concentration in hypoxia-sensitive rainbow trout (Soitamo *et al.*, 2001). The HIF signal (HIF-1α protein levels or DNA binding) further increases during cold exposure and hypoxia or during subcritical warming (Rissanen, 2006; Heise 2006a,b). It seems very likely that in hypoxia-tolerant ectotherms cellular oxygen concentration might not be the only regulator of HIF-1α levels, and that ROS or NO could act as additional modulators that fine-tune HIF stability, possibly in concert with molecular mechanisms related to HIF-1ß/ARNT availability (Randall and Yang, 2004). ROS (especially H_2O_2) and NO can both reduce prolylhydroxylase (PHD) activity (that is, cellular hydroxylases that initiate HIF-1α degradation under normoxic cellular conditions), either through direct interaction/oxidation with the PHD proteins or by oxidizing the PHD cofactor Fe^{2+} to Fe^{3+} and inhibiting PHDs (for review see Murphy, 2009). In North Sea eelpout caught during summer, iron reduction rates were twice as high as when the same fish were caught in winter, and lower HIF-binding activity in warm-adapted summer fish might relate to a higher Fe^{2+}/Fe^{3+} ratio during summer (Heise *et al.*, 2007). Long-term acclimation of polar eelpout, *Pachycara brachicephalum*, to sub-critically elevated temperatures (0°C –> 5°C over 3 months) caused increased HIF signals, although several oxidative stress parameters were simultaneously increased and the same effect was found when common eelpout were cooled to normal winter temperature in the North Sea for several months (Heise *et al.*, 2007). Thus, HIF-signalling seems to form part of the adaptive response of ectotherms under environmental/thermal change conditions within the tolerance boundaries of a given species, and possibly this includes a tiered effect of ROS that modulates the signal. Low ROS levels seem to be supportive (Benarroch, 2009), whereas high ROS levels (oxidative burst conditions) may abrogate the HIF signal and thereby terminate basic adaptive and protective mechanisms. In keeping with the idea that a limited amount of cellular ROS may have stabilizing effects on HIF levels during thermal challenge, over-expression of the uncoupling protein UCP2 was found to accompany HIF stabilization in both eelpout species under the applied mild thermal exposure conditions (Mark *et al.*, 2006). At least in fish, the major HIF targets seem to be the same genes as in mammals and include EPO, glucose transporters, vascular endothelial growth factor, and possibly nitric oxide synthase isoforms. However, HIF targets in hypoxia-tolerant, non-haemoglobin-carrying marine invertebrates still need to be verified. HIF and its regulatory network may indeed turn out to be centrally important for the adaptive response of marine animal ectotherms to changing thermal windows due to ongoing climate change.

5.5 Conclusions

Climate warming will not only cause a rise of annual and seasonal temperature means, but, in polar and temperate regions, a widening of the window of temperature fluctuations, and generally higher variability of environmental conditions. By contrast, in the tropics, climate warming may actually narrow the thermal envelopes between the minimum temperature and the species-specific lethal extreme. Broader thermal windows will raise costs of stress hardening in cold temperate and polar species, including an increased expression of stress proteins such as antioxidants, Hsp, and cellular autophagic and apoptotic mechanisms in marine invertebrates and fish. Costs of cellular and mitochondrial turnover increase as ROS formation exacerbates during warming, reducing the energy availability for growth. For some slow-growing species, a reduction of maximum body size and also of maximum attainable lifespan can already be observed in stressful environments (*Arctica islandica* in the Baltic Sea, *L. elliptica* in coastal waters of the Antarctic Peninsula), and a decrease in maximum individual sizes in ectotherms is more generally expected as global warming proceeds. Thus, the acclimatory response involves transcription of genes supporting cellular maintenance and turnover, whereas growth and reproduction are down regulated. The interplay of temperature and oxygen, with oxygen solubility decreasing at higher temperatures, is of central importance in this context. Hyperoxic conditions were shown to alleviate thermal stress under experimental protocols, but in nature the chances that oxygen availability will increase under warming conditions is virtually nil. Thus, a major adaptive pathway in temperature acclimation involves stabilization of HIF and expression of its target genes that stabilize tissue oxygenation under thermal and other forms of stress. Polar animals, especially in the Antarctic, long-term adapted to constant cold climate, have evolved a stenothermal phenotype. This phenotype includes a tendency to attain larger body sizes, high mitochondrial densities and loss of genes centrally important for warm acclimation. Further research in Antarctic animal physiology, especially with species from the rapidly warming Western Antarctic, needs to take into account these adaptive regulatory genes and pathways, as well as to study the evolutionary history and genetic separation time (genetic distance) between related animal species, to predict their adaptive genetic and phenotypic capabilities to survive climate change.

References

Abele, D. and Puntarulo, S. (2004) Formation of reactive species and induction of antioxidant defense systems in polar and temperate marine invertebrates and fish. *Comparative Biochemistry and Physiology A* 138, 405–415.

Abele, D., Großpietsch, H. and Pörtner, H.O. (1998) Temporal fluctuations and spatial gradients of environmental PO_2, temperature, H_2O_2 and H_2S in its intertidal habitat trigger enzymatic antioxidant protection in the capitellid worm *Heteromastus filiformis*. *Marine Ecology Progress Series* 163, 179–191.

Abele, D., Tesch, C., Wencke, P. and Pörtner, H.O. (2001) How does oxidative stress relate to thermal tolerance in the Antarctic bivalve *Yoldia eightsi*? *Antarctic Science* 13, 111–118.

Abele, D., Heise, K., Pörtner, H.O. and Puntarulo, S. (2002) Temperature-dependence of mitochondrial function and production of reactive oxygen species in the intertidal mud clam *Mya arenaria*. *Journal of Experimental Biology* 205, 1831–1841.

Abele, D., Philipp, E., Gonzalez, P. and Puntarulo, S. (2007) Marine invertebrate mitochondria and oxidative stress. *Frontiers in Biochemistry* 12, 933–946.

Abele, D., Brey, T. and Philipp, E.E.R. (2009) Bivalve models of aging and the determination of molluscan lifespans. *Experimental Gerontology* 44, 307–315.

Andronikov, V.B. (1975) Heat resistance of gametes of marine invertebrates in relation to temperature conditions under which the species exist. *Marine Biology* 30, 1–11.

Barnes, D.K.A. and Peck, L.S. (2008) Vulnerability of Antarctic shelf biodiversity to predicted regional warming. *Climate Research* 37, 149–163.

Basova, L., Begum, S., Strahl, J., Sukhotin, A., Brey, T., Philipp, E. and Abele, D. Age dependent patterns of antioxidants in *Arctica islandica* from six regionally separate populations with different life spans. *Aquatic Biology*, in press.

Benarroch, E. (2009) Hypoxia-induced mediators and neurologic disease. *Neurology* 73, 560–565.

Bowler, K. (1963) A study of the factors involved in acclimatization to temperature and death at high temperatures in *Astacus pallipes*. I. Experiments on intact animals. *Journal of Cellular and Comparative Physiology* 62, 119–132.

Brand, M.D. (2000) Uncoupling to survive? The role of mitochondrial inefficiency in ageing. *Experimental Gerontology* 35, 811–820.

Brown, K.M., Fraser, K.P.P., Barnes, D.K.A. and Peck, L.S. (2004) Links between the structure of an Antarctic shallow-water community and ice-scour frequency. *Oecologia* 141, 121–129.

Buckley, B.A., Owen, M.E. and Hofmann, G.E. (2001) Adjusting the thermostat: the threshold induction temperature for the heat-shock response in intertidal mussels (genus *Mytilus*) changes as a function of thermal history. *Journal of Experimental Biology* 204, 3571–3579.

Buttemer, W.A., Abele, D. and Costantini, D. (2010) From bivalves to birds: Oxidative stress and longevity. *Functional Ecology*, 24, 971–983.

Dahlhoff, E. and Somero, G.N. (1993) Effects of temperature on mitochondria from abalone (genus *Haliotis*): adaptive plasticity and its limits. *Journal of Experimental Biology* 185, 151–168.

Daufresne, M., Lengfellner, K. and Sommer, U. (2009) Global warming benefits the small in aquatic ecosystems. *Proceedings of the National Academy of Sciences of the USA* 106, 12788–12793.

Davenport, J. (1997) Comparison of the biology of the intertidal subantarctic limpets *Nacella concinna* and *Kerguelenella lateralis*. *Journal of Molluscan Studies* 63, 39–48.

Deser, C. and Blackmon, M. (1993) Surface climate variations over the North Atlantic Ocean during winter 1900–1989. *Journal of Climate* 6, 1743–1753.

Fandrey, J., Gorr, T.A. and Gassmann, M. (2006) Regulating cellular oxygen sensing by hydroxylation. *Cardiovascular Research* 71, 642.

Gardeström, J., Elfwing, T., Löf, M., Tedengren, M., Davenport, J.L. and Davenport, J. (2007) The effect of thermal stress on protein composition in dogwhelks (*Nucella lapillus*) under normoxic and hyperoxic conditions. *Comparative Biochemistry and Physiology A* 148, 869.

Guderley, H. and St-Pierre, J. (2002) Going with the flow or life in the fast lane: contrasting mitochondrial responses to thermal change. *Journal of Experimental Biology* 205, 2237–2249.

Günther, C.-P., Boysen-Ennen, E., Niesel, V., Hasemann, C., Heuers, J., Bittkau, A., Fetzer, I., *et al.* (1998) Observations of a mass occurrence of *Macoma balthica* larvae in midsummer. *Journal of Sea Research* 40, 347–351.

Heise, K., Puntarulo, S., Pörtner, H.O. and Abele, D. (2003) Production of reactive oxygen species by isolated mitochondria of the Antarctic bivalve *Laternula elliptica* (King and Broderip) under heat stress. *Comparative Biochemistry and Physiology C* 134, 79–90.

Heise, K., Puntarulo, S., Nikinmaa, M., Abele, D. and Pörtner, H.O. (2006a) Oxidative stress during stressful heat exposure and recovery in the North Sea eelpout (*Zoarces viviparus*). *Journal of Experimental Biology* 209, 353–363.

Heise, K., Puntarulo, S., Nikinmaa, M., Pörtner, H.O. and Abele, D. (2006b) Functional hypoxia and reoxygenation upon cold exposure effects oxidative stress and hypoxic signalling in the North Sea eelpout (*Zoarces viviparus*). *Comparative Biochemistry and Physiology* A 143, 494–503.

Heise, K., Estevez, M., Puntarulo, S., Galleano, M., Nikinmaa, M., Pörtner, H. and Abele, D. (2007) Effects of seasonal and latitudinal cold on oxidative stress parameters and activation of hypoxia inducible factor (HIF-1) in zoarcid fish. *Journal of Comparative Physiology B* 177, 765–777.

Helmuth, B., Broitman, B.R., Blanchette, C.A., Gilman, S., Halpin, P., Harley, C.D.G., O'Donnell, M.J., *et al.* (2006) Mosaic patterns of thermal stress in the rocky intertidal zone: implications for climate change. *Ecological Monographs* 76, 461–479.

Helmuth, B.S.T. and Hofmann, G.E. (2001) Microhabitats, thermal heterogeneity, and patterns of physiological stress in the rocky intertidal zone. *Biological Bulletin* 201, 374–384.

Hofmann, G., Buckley, B., Airaksinen, S., Keen, J. and Somero, G.N. (2000) Heat-shock protein expression is absent in the Antarctic fish *Trematomus bernacchii* (family Nototheniidae). *Journal of Experimental Biology* 203, 2331–2339.

Hofmann, G.E. (1999) Ecologically relevant variation in induction and function of heat shock proteins in marine organisms. *American Zoology* 39, 889–900.

Hoogewijs, D., Terwilliger, N.B., Webster, K.A., Powell-Coffman, J.A., Tokishita, S., Yamagata, H., Hankeln, T., *et al.* (2007) From critters to cancers: bridging comparative and clinical research on oxygen sensing, HIF signaling, and adaptations towards hypoxia. *Integrative and Comparative Biology*, 1–26.

Keller, M., Sommer, A.M., Pörtner, H.O. and Abele, D. (2004) Seasonality of energetic functioning and production of reactive oxygen species by lugworm (*Arenicola marina*) mitochondria exposed to acute temperature changes. *Journal of Experimental Biology* 207, 2529–2538.

Kochmann, J., Buschbaum, C., Volkenborn, N. and Reise, K. (2008) Shift from native mussels to alien oysters: Differential effects of ecosystem engineers. *Journal of Experimental Marine Biology and Ecology* 364, 1–10.

Lurman, G., Blaser, T., Lamare, M., Tan, K.-S., Poertner, H., Peck, L. and Morley, S. (2010) Ultrastructure of pedal muscle as a function of temperature in nacellid limpets. *Marine Biology* 157, 1705–1712.

Malanga, G., Estevez, M.S., Calvo, J., Abele, D. and Puntarulo, S. (2007) The effect of seasonality on oxidative metabolism in *Nacella* (*Patinigera*) *magellanica*. *Comparative Biochemistry and Physiology A* 146, 551–558.

Mark, F., Lucassen, M. and Pörtner, H.O. (2006) Thermal sensitivity of uncoupling protein expression in polar and temperate fish. *Comparative Biochemistry and Physiology-Part D* 1, 365–374.

Meredith, M. and King, J. (2005) Climate change in the ocean to the west of the Antarctic peninsula during the second half of the 20th century. *Geophysical Research Letters* 32, L19604, doi:10.1029/2005GL024042.

Morley, S.A., Lurman, G.J., Skepper, J.N., Pörtner, H.-O. and Peck, L.S. (2009) Thermal plasticity of mitochondria: A latitudinal comparison between Southern Ocean molluscs. *Comparative Biochemistry and Physiology A* 152, 423–430.

Murphy, M.P. (2009) How mitochondria produce reactive oxygen species. *Biochemical Journal* 417, 1–13.

Nikinmaa, M. and Rees, B. (2005) Oxygen-dependent gene expression in fishes. *American Journal of Physiology* 288, R1079–1090, doi:10.1152/ajpregu.00626.2004.

O'Brien, J., Dahlhoff, E. and Somero, G.N. (1991) Thermal resistance of mitochondrial respiration: hydrophobic interactions of membrane proteins may limit thermal resistance. *Physiological Zoology* 64, 1509–1526.

O'Brien, K.M. and Sidell, B. (2000) The interplay among cardiac ultrastructure, metabolism and the expression of oxygen-binding proteins in Antarctic fishes. *Journal of Experimental Biology* 203, 1287–1297.

Parmesan, C. (2006) Ecological and evolutionary responses to recent climate change. *Ecology, Evolution and Systematics* 37, 637–669.

Peck, L. and Bullough, L.W. (1993) Growth and population structure in the infaunal bivalve *Yoldia eightsi* in relation to iceberg activity at Signy Island, Antarctica. *Marine Biology* 117, 235–241.

Peck, L.S., Webb, K.E. and Bailey, D.M. (2004) Extreme sensitivity of biological function to temperature in Antarctic marine species. *Functional Ecology* 18, 625–630.

Perry, A.L., Low, P.L., Ellis, J.R. and Reynolds, J.D. (2005) Climate change and distribution shifts in marine fishes. *Science* 308, 1912–1915.

Philipp, E.E.R. and Abele, D. (2010) Masters of longevity: Lessons from long-lived bivalves. *Gerontology* 56, 55–65.

Philipp, E.E.R., Husmann, G. and Abele, D. (2011) The impact of sediment deposition and iceberg scour on the Antarctic soft shell clam *Laternula elliptica* at King George Island, Antarctica. *Antarctic Science* doi:10.1017/S0954102010000970.

Place, S.P., O'Donnell, M.J. and Hofmann, G.E. (2008) Gene expression in the intertidal mussel *Mytilus californianus*: physiological response to environmental factors on a biogeographic scale. *Marine Ecology Progress Series* 356, 1–14.

Pöhlmann, K., Koenigstein, S., Alter, K., Abele, D. and Held, C. (2011) Heat-shock response and antioxidant defense during air exposure in Patagonian shallow-water limpets from different climatic habitats. *Cell Stress & Chaperones*, DOI 10.1007/s12192-011-0272-8.

Pörtner, H.O. (2002) Climate variations and the physiological basis of temperature dependent biogeography: systematic to molecular hierarchy of thermal tolerance in animals. *Comparative Biochemistry and Physiology A* 132, 739–761.

Pörtner, H.O. and Lannig, G. (2009) Oxygen and capacity limited thermal tolerance. In: Richards, J.G., Farrell, A.P. and Brauner, C.J. (eds) *Hypoxia*. Fish Physiology series 27, Academic Press, London, pp. 143–191.

Pörtner, H.O., Berdal, B., Blust, R., Brix, O., Colosimo, A., De Wachter, B., Giuliani, A., *et al.* (2001) Climate induced temperature effects on growth performance, fecundity and recruitment in marine fish: developing a hypothesis for cause and effect relationships in Atlantic cod (*Gadus morhua*) and common eelpout (*Zoarces viviparus*). *Continental Shelf Research* 21, 1975–1997.

Randall, D. and Yang, H. (2004) The role of hypoxia, starvation, beta–naphthoflavone and the aryl hydrocarbon receptor nuclear translocator in the inhibition of reproduction in fish. In: Rupp, G.G. and White, M.D. (eds) *Fish Physiology, Toxicology and Water Quality*. Proceedings of the seventh international symposium, Tallinn, Estonia, May 12–25, 2003, pp. 253–261.

Rissanen, E., Tranberg, H.K., Sollid, J., Nilsson, G.E. and Nikinmaa, M. (2006) Temperature regulates hypoxia-inducible factor-1 (HIF-1) in a poikilothermic vertebrate, crucian carp (*Carassius carassius*). *Journal of Experimental Biology* 209, 994–1003.

Rosenthal, J.J.C., Seibel, B.A., Dymowska, A. and Bezanilla, F. (2009) Trade-off between aerobic capacity and locomotor capability in an Antarctic pteropod. *Proceedings of the National Academy of Sciences of the USA* 106, 6192–6196.

Sahade, R., Tarantelli, S., Tatian, M. and Mercuri, G. (2008) Benthic community shifts: a possible linkage of climate change? In: Wiencke, C., Ferreyra, G.A., Abele, D. and Marenssi, S. (eds) *The Antarctic Ecosystem of Potter Cove, King-George Island (Isla 25 de Mayo). Berichte zur Polar- und Meeresforschung* 571, pp. 331–337.

Seibel, B.A., Dymowska, A. and Rosenthal, J. (2007) Metabolic temperature compensation and coevolution of locomotory performance in pteropod molluscs. *Integrative and Comparative Biology* 47, 880–891.

Soitamo, A.J., Rabergh, C.M.I., Gassmann, M., Sistonen, L. and Nikinmaa, M. (2001) Characterization of a hypoxia-inducible factor (HIF-1α) from rainbow trout. *Journal of Biological Chemistry* 276, 19677–19705.

Somero, G.N. (2005) Linking biogeography to physiology: Evolutionary and acclimatory adjustments of thermal limits. *Frontiers in Zoology* 2, 1–9.

Sommer, A.M. and Pörtner, H.O. (2002) Metabolic cold adaptation in the lugworm *Arenicola marina*: comparison of a North Sea and a White Sea population. *Marine Ecology Progress Series* 240, 171–182.

Stillman, J.H. and Tagmount, A. (2009) Seasonal and latitudinal acclimatization of cardiac transcriptome responses to thermal stress in porcelain crabs, *Petrolisthes cinctipes*. *Molecular Ecology* 18, 4206–4226.

Terman, A. and Brunk, U.T. (2006) Oxidative stress, accumulation of biological 'garbage', and aging. *Antioxidants & Redox Signaling* 8, 197–204.

Tomanek, L. (2005) Two-dimensional gel analysis of heat-shock response in marine snails (genus *Tegula*): interspecific variation in protein expression and acclimation ability. *Journal of Experimental Biology* 208, 3133–3143.

Verde, C., Parisi, E. and di Prisco, G. (2006) The evolution of thermal adaptation in polar fish. *Gene* 385, 137–145.

Vergauwen, L., Benoot, D., Blust, R. and Knapen, D. (2010) Long-term warm or cold acclimation elicits a specific transcriptional response and affects energy metabolism in zebrafish. *Comparative Biochemistry and Physiology - Part A: Molecular & Integrative Physiology* 157, 149–157.

Waller, C., Worland, M., Convey, P. and Barnes, D. (2006) Ecophysiological strategies of Antarctic intertidal invertebrates faced with freezing stress. *Polar Biology* 29, 1077–1083.

Weibel, E. (1985) Design and performance of

muscular systems: an overview. *Journal of Experimental Biology* 115, 405–412.

Weihe, E. and Abele, D. (2008) Differences in the physiological response of inter- and subtidal Antarctic limpets (*Nacella concinna*) to aerial exposure. *Aquatic Biology* 4, 155–166.

Weihe, E., Kriews, M. and Abele, D. (2009) Differences in heavy metal concentrations and in the response of the antioxidant system to hypoxia and air exposure in the Antarctic limpet *Nacella concinna*. *Marine Environmental Research* 69, 127–135.

6 Fish: Freshwater Ecosystems

Tommi Linnansaari and Richard A. Cunjak

6.1 Introduction

All freshwater fish are ectothermic. That is, their body temperature and metabolic processes are directly regulated by the ambient water temperature. This feature makes them particularly susceptible to one of the main effects of climate change: the increased variability in temperature (Meehl *et al.*, 2007). Changes in temperature are predicted to be larger over land than in the ocean (Meehl *et al.*, 2007), implying larger effects in freshwater than marine habitats if temperature effects are considered independently. For example, the increase in air temperature will inevitably lead to warming of water temperatures and therefore will have a direct effect on freshwater fish both at the individual/physiological and species/population levels.

Climate change will also have a strong effect on precipitation patterns (Meehl *et al.*, 2007) with a direct link to freshwater fish populations by affecting quantity and suitability of their habitat. The predicted changes in these two primary environmental variables will cause a number of other physico-chemical changes that will affect freshwater fish directly or indirectly, as conceptualized in Fig. 6.1. Among the most important cascading effects are the changes in flow regime (timing and intensity) and reductions in the duration and amount of ice cover in lakes and rivers (Lemke *et al.*, 2007; Meehl *et al.*, 2007). Flow and ice regime are primary determinants of the structure of riverine communities (Poff *et al.*, 1997; Prowse, 2001) and significant effects on fish are to be expected when these parameters are altered due to climate change.

To understand the magnitude of climate change and the various future predictions, we will provide an overview of the different climate change scenarios. The common (national) climate models are summarized and their main ensemble predictions described with regard to changes in air temperature and precipitation as a basis for forecasting freshwater fish responses. Our approach is to predict the outcome of climate change and fish interactions by reviewing examples from the established literature; where such sources were unavailable, we used reasonable, albeit subjective, inference to predict the likely outcome. Climate change and the effects of summer warming and extreme high temperatures, even in high latitudes, have been the topic of numerous recent reviews (Reist *et al.*, 2006; Ficke *et al.*, 2007; McCullough *et al.*, 2009; Prowse *et al.*, 2009). Our focus here is on *cold adaptation in freshwater fish*, and especially how predicted climate change may alter the winter environment and conditions in streams and lakes, and how selected freshwater fish are likely to respond. Consequently, we have restricted our geographical research to high latitude regions in northern North America, northern Eurasia, and the southernmost areas of South America (Patagonia).

Taxonomically, we have focused on select groups of fish to make our points for the various geographically constrained climate scenarios. Fish have traditionally been grouped into different functional categories (i.e. guilds). In the temperate–polar regions,

 © CAB International 2012. *Temperature Adaptation in a Changing Climate: Nature at Risk* (eds K.B. Storey and K.K. Tanino)

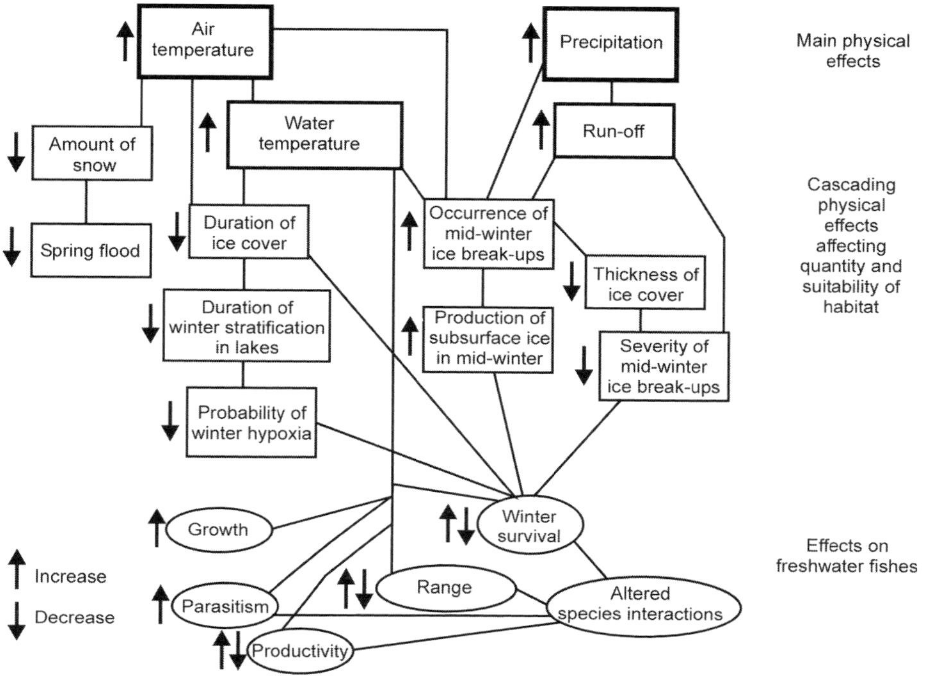

Fig. 6.1. Conceptual framework of the predicted effects of climate change on winter environmental conditions and the likely response of freshwater fishes. Physical effects are shown as boxes and the biological effects as ovals. Note that, to maintain the clarity of presentation, not all possible connections between the different variables are drawn.

fish guilds have been categorized according to optimal physiological performance and temperature selection (Magnuson *et al.*, 1979). These thermal guilds are specifically described as coldwater, coolwater and warmwater fish with corresponding optimum water temperatures of 11°C–15°C, 21°C–25°C, and 27°C–31°C, respectively (Magnuson *et al.*, 1979). Parallel with this classification, fish are also distinguished by the breadth of their thermal niche as stenothermal (narrow temperature tolerance; 'specialist') and eurythermal (wide temperature tolerance; 'generalist'). We have adopted these classifications in the chapter in order to generalize predictions from a single species to groups of similar fish.

It must be remembered that the effects of climate change are acting *in addition* to anthropogenic stressors (e.g. hydropower development, land use changes, species introductions) that are simultaneously affecting freshwater ecosystems and their fish fauna irrespective of climate change. In reality, it will be difficult to tease out causal connections to specific perturbations. Accurate interpretation of the synergistic effects of multiple stressors on freshwater fish is impossible, especially in areas where climate change is predicted to be relatively small (e.g. Patagonia) in comparison with high latitudes in the north. Still, cumulative effects, even involving small incremental changes, may prove to be biologically significant.

6.2 What do the Models Predict?

6.2.1 Climate change scenarios and models

In order to predict how climate change will affect the cold adaptation of freshwater fish, it is necessary to first understand what that future climate will probably be like. When referring to global climate change, climatic responses vary widely between geographic regions such that predicting change is not a trivial task. For our review, we have adopted the template set forward by the Fourth Assessment Report (AR4) (Solomon *et al.*, 2007) of the Intergovernmental Panel on Climate Change (IPCC) as a basis of climatic change predictions.

There are two fundamental sources of uncertainty in predictions of the future climate. First, it is obvious that the amount of greenhouse gas emissions in the future is unknown, especially as the climate predictions are often made for the decadal mean change during the 21st century. The IPCC has taken this uncertainty into account by introducing different *scenarios* about future globalization, economy and environmental development (with a direct effect on greenhouse gas emissions). Overall, the IPCC has drafted 40 equally probable scenarios outlining how the future might unfold, all of which are within four alternative 'storylines' (Nakicenovic and Swart, 2000). The four storylines, A1, B1, A2 and B2, describe a world with different future weighting on two dimensions: sustainability and globalization (Table 6.1). In addition to these two primary dimensions, the IPCC has described three subgroups within the A1 family (A1F1, A1B and A1T) in an attempt to better discern the effects of alternative energy technologies on future greenhouse gas (GHG) emissions (Nakicenovic and Swart, 2000). The A1F1 scenario describes a future where energy dependency remains with fossil fuels. The A1B scenario assumes equal development across all energy supply technologies and results in a balanced energy solution where the world is not relying heavily on any one particular energy source. In the A1T scenario, the energy supply and end use technologies are predominantly relying on non-fossil fuel energy sources. In order to adequately capture the uncertainty associated with future GHG emissions, the IPCC has recommended that consideration of future climate predictions be based on at least these six 'marker' scenarios (Nakicenovic and Swart, 2000). It should be noted that no IPCC scenario assumes implementation of any explicit climate change mitigation policies, such as ratification of the Kyoto Protocol (Nakicenovic and Swart, 2000). All scenarios, and their fundamentals, are described in a special report on emissions scenarios (SRES) (Nakicenovic and Swart, 2000).

The second fundamental source of variability in the future climate predictions relates to the multitude of national or

Table 6.1. Description of the main characteristics of the four IPCC storylines. These describe alternative futures in the 21st century with a direct effect on the amount of greenhouse gas emissions and thus on climate change (modified from Nakicenovic and Swart, 2000).

Storyline	Sustainability	Globalization	Global population
A1[a]	Focus on economy Very rapid economic growth	Convergent world Global solutions	Peak mid-century, decline thereafter
A2	Focus on regional economy Intermediate and uneven economic growth	Local and regional focus	Continuously increasing
B1	Focus on environmental sustainability Rapid economic growth	Convergent world Global solutions	Peak mid-century, decline thereafter
B2	Focus on environmental sustainability Intermediate economic growth	Local and regional focus	Continuously increasing, slower than A2

[a] The A1 storyline is further divided into three alternative scenarios based on energy dependencies, in an attempt to form six 'marker' scenarios that capture a wide range of uncertainty of future greenhouse gas emissions.

institutional climate *models* used to simulate the future climate. Large-scale climatic predictions are carried out using coupled atmosphere–ocean general circulation models (AOGCMs) that provide predictions of climate variables typically in the range of 200 km × 200 km horizontal grid units (Randall *et al.*, 2007). Because each AOGC model typically results in somewhat different representations of the future climate within each scenario, the approach adopted by the IPCC is to utilize an ensemble of plausible models (23 AOGC models are described in detail in Randall *et al.* (2007)) ensuring that the between-model uncertainty is addressed. As the ensemble AOGCM predictions for air temperature and precipitation variables are readily available for each continent courtesy of IPCC (Christensen *et al.*, 2007; Meehl *et al.*, 2007), we used these predictions as the basis of our assessment of cold adaptation for freshwater fish. Of the two variables, precipitation has the higher uncertainty (i.e. larger disagreement between different AOGCMs; Christensen *et al.*, 2007). For further details about different AOGC models, their assumptions and predictions, readers are referred to Randall *et al.* (2007).

6.2.2 Model predictions for temperature and precipitation in selected regions

Predictions for changes in temperature and precipitation for four different high latitude (seasonally cold) land masses resulting from an ensemble of AOGCMs are shown in Table 6.2, with the corresponding regions highlighted in Fig. 6.2. The AOGCMs consistently predict that the warming throughout the northern North America and northern Europe will be highest during the winter months (December–February), whereas southern South America is likely to experience similar average warming throughout the year (Table 6.2). The increase in average air temperatures is also predicted to be much higher in the high latitudes of the northern hemisphere than the increase in southern South America (Table 6.2). Similarly, the largest increase in precipitation

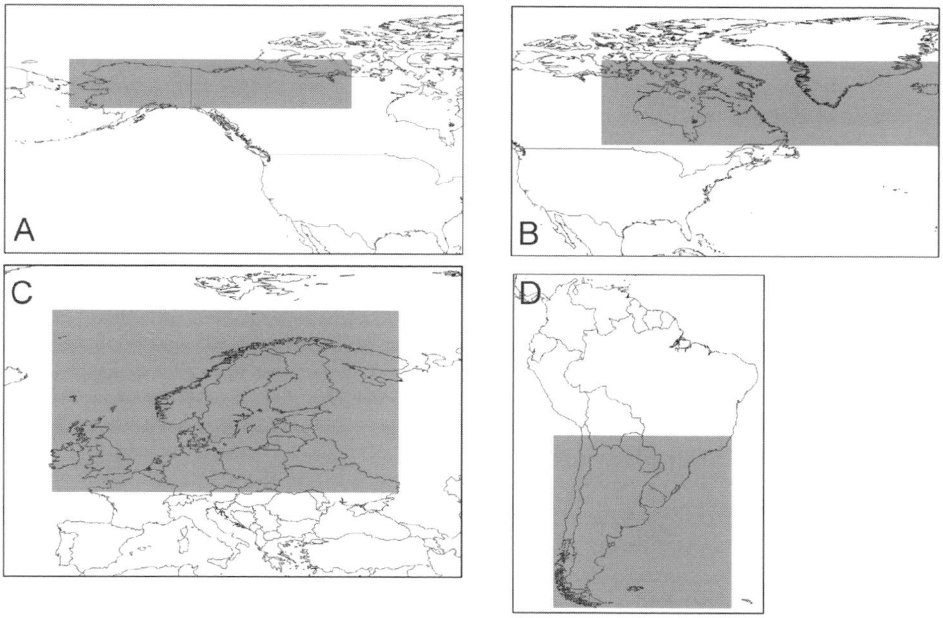

Fig. 6.2. The four high latitude (seasonally cold) land masses modelled in Table 6.2. A, North-western North America; B, North-eastern North America; C, Northern Europe; D, Patagonia, Tierra del Fuego.

Table 6.2. Regional averages of temperature and precipitation projections from a set of 21 atmosphere–ocean global climate models (AOGCM) in the IPCC multi-model data set for the A1B scenario. Values are for the predicted change by the 2080–2099 period relative to values in 1980–1999.

Geographic region	Temperature (°C) A1B	Precipitation (%) A1B
North-western North America (60N, 170W–72N, 103W, Fig. 6.2A)		
DJF	**6.3** (4.4–11)	**28** (6–56)
MAM	**3.5** (2.3–7.7)	**17** (2–38)
JJA	**2.4** (1.3–5.7)	**14** (1–30)
SON	**4.5** (2.3–7.4)	**19** (6–36)
North-eastern North America (50N, 103W–72N, 10W, Fig. 6.2B)		
DJF	**5.9** (3.3–8.5)	**26** (6–42)
MAM	**3.8** (2.4–7.2)	**17** (4–34)
JJA	**2.8** (1.5–5.6)	**11** (0–19)
SON	**4.0** (2.7–7.3)	**16** (7–37)
Northern Europe (48N, 10W–75N, 40E, Fig. 6.2C)		
DJF	**4.3** (2.6–8.2)	**15** (9–25)
MAM	**3.1** (2.1–5.3)	**12** (0–21)
JJA	**2.7** (1.4–5.0)	**2** (–21–16)
SON	**2.9** (1.9–5.4)	**8** (–5–13)
Patagonia, Tierra del Fuego (56 S, 76W–20S, 40W, Fig. 6.2D)		
DJF	**2.7** (1.5–4.3)	**1** (–16–10)
MAM	**2.6** (1.8–4.2)	**1** (–11–7)
JJA	**2.4** (1.7–3.6)	**0** (–20–17)
SON	**2.7** (1.7–3.9)	**1** (–20–11)

Values in bold font are the median values for the average change within the 21 AOGCM ensemble, with the minimum and maximum average values in brackets. DJF, December, January, February; MAM, March, April, May; JJA, June, July, August; SON, September, October, November. The corresponding geographic regions are shown in Fig. 6.2. (Data based on Solomon *et al.* (2007), Table 11.1.)

is expected during the winter months in the northern parts of the northern hemisphere, whereas the average change from the median AOGC model in the multi-model data set indicates a very subtle change in precipitation during any part of the year for southern South America (Table 6.2). The different AOGCMs all agree about the direction of the change in precipitation patterns during the winter months in the northern hemisphere (i.e. increase), whereas there is between-model disagreement on the direction of the precipitation change for southern South America under the A1B scenario. The increased precipitation in the high latitudes of the northern hemisphere will lead to increased annual runoff in the order of 10%–30% by 2050 under the A1B emission scenario, whereas reductions in annual streamflow of 5%–20% are predicted for Patagonia (Milly *et al.*, 2005). Although Milly *et al.* (2005) considered changes in annual flows only, it is predicted that the increase in annual flows for northern areas is largely attributable to flow increases in winter because:

1. The amount of precipitation will increase in winter;
2. Precipitation will fall as rain, not snow; and
3. Evapotranspiration is lower in winter than in summer.

6.3 Redistribution of Freshwater Fish – Range Extensions, Contractions and Shifts

For freshwater fish, one of the most profound effects of the warming climate will be the change in the thermal isolines that largely determine the distribution of the suitable thermal niches for different fish species (Wrona *et al.*, 2005; Reist *et al.*, 2006; Ficke *et al.*, 2007). The ranges of freshwater fish are predicted to both extend and contract depending on the thermal guild of the species. Generally, the thermal habitat for stenothermic coldwater species is predicted to contract, and the habitat for cool and warmwater species to extend towards the poles (Lehtonen, 1996; Gómez *et al.*, 2004; Reist *et al.*, 2006; Sharma *et al.*, 2007; Jonsson and Jonsson, 2009; Cussac *et al.*, 2009).

It appears that the range contractions may be exclusively dictated by summer warming, such that the upper tolerance temperatures for coldwater species are exceeded resulting in the eventual local extinctions of metapopulations (Ficke *et al.*, 2007; McCullough *et al.*, 2009). Winter warming may have negative effects for some coldwater fish in which low temperatures are essential for spawning success (e.g. salmonids and burbot, *Lota lota*) (McPhail and Paragamian, 2000; Ficke *et al.*, 2007), but the predicted poleward range contractions even for these species are more likely to be affected by the summer conditions becoming too warm or hypoxic.

In contrast to range contractions, range extensions of freshwater fish are predicted to be directly affected by the warming of *winter* temperatures. Winter warming will directly benefit coldwater species in very high latitude (e.g. arctic) waters that are currently marginally suitable or devoid of fish due to extreme ice conditions (i.e. frozen solid in winter). Some warmwater species will benefit directly from winter warming if their poleward distribution is limited by their lower incipient lethal temperatures (Ortubay *et al.*, 1997; Cussac *et al.*, 2009). Several Patagonian native fish (ranging from less resistant *Callichthys callichthys* to more resistant *Jenynsia lineata*), for example, show clear correlations between the southern limit of their geographic distribution and lower lethal temperature (Gómez, 1996). In lakes, the shorter ice-covered period is predicted to reduce the period of winter stratification (Fang and Stefan, 2009). This, in turn, may lower the probability of hypoxia (or anoxia) that can lead to winter kill of fish (Fang and Stefan, 2009; Magnuson, 2010). Therefore, typical winter-kill lakes may become available for fish that cannot tolerate low oxygen conditions (Rahel and Olden, 2008).

For many freshwater fish of temperate and high latitudes, survival is not a function of the acute effects of cold water temperature per se. Rather, the current distribution is limited by the length of winter during which energy deficits may develop (Cunjak *et al.*, 1987; Cunjak, 1988) and lead to significant mortality (Shuter and Post, 1990; Miranda and Hubbard, 1994). Winter mortality is strongly size selective, mostly affecting the smallest individuals with concomitantly low levels of stored energy reserves (i.e. typically the young-of-the-year fish) (Johnson and Evans, 1990; Shuter and Post, 1990; Miranda and Hubbard, 1994; Lappalainen *et al.*, 2000). In North America and Europe, the species for which the distribution is most strongly limited by winter mortality typically belong to the cool- and warmwater guilds (e.g. Centrarchidae, Percidae, Cyprinidae). These fish overwinter in a non-feeding passive (low activity) state compared with coldwater species (e.g. Salmonidae) that remain active, feed and can more effectively meet the energy demands of overwintering at low temperatures (Cunjak, 1988,1996; Finstad *et al.*, 2004a; Graham and Harrod, 2009). The situation for the cool- and warmwater fish is predicted to improve due to climate warming because:

1. The growing season will be longer leading to larger size, and thus more energy reserves heading into the first winter;
2. The winter fasting (starvation) period will be shorter; and
3. Winters will be milder (Lappalainen and Lehtonen, 1997; Wrona *et al.*, 2005; Reist *et al.*, 2006; but see Lehtonen and Lappalainen, 1995).

Predicted distributional shifts in North America and northern Europe involve biologically similar groups of fish as the two continents share many fish species (or biological equivalents), and the predicted effects of climate change are similar (Table 6.2). Lehtonen (1996) compiled predicted distributional outcomes (i.e. positive or negative) for a number of common fish species in northern Europe, and these results are applicable for North America at the genus and family level. As expected, the most vulnerable species with respect to climate change in northern America and Europe are the stenothermic coldwater fish (Salmonidae: *Salmo*, *Salvelinus*, *Hucho*, *Coregonus*, *Stedonus*, and *Thymallus*; Osmeridae: *Osmerus* and *Hypomesus*; as well as individual species such as the Alaska blackfish *Dallia pectoralis* and burbot) (Lehtonen, 1996; Reist *et al.*, 2006). These species will benefit to some extent from the habitat that will become available in areas in the very highest latitudes; however, some of them already occupy the northernmost areas of land masses and simply have no more room to redistribute further north. One such example is the arctic char (*Salvelinus alpinus*) that has the most northerly distribution of any freshwater fish (including the arctic archipelago surrounding the north pole) (Scott and Crossman, 1973). In the case of arctic char, a range extension further north is not possible. Another socially important salmonid species, the lake trout (*Salvelinus namaycush*) may expand its range in the high latitudes, at least initially, as lakes that freeze solid under the current climate may become available. Recent evidence of this species' ability to forage in coastal marine areas may facilitate its range extension throughout the north (Swanson *et al.*, 2010). In the long run, however, it is predicted that thermally suitable habitat for lake trout will be reduced because the hypolimnion, especially in temperate latitudes, will experience a 'temperature-oxygen squeeze' (Ficke *et al.*, 2007). The squeeze is a result of warming epilimnion and increasing volume of hypoxic conditions at the bottom of lakes, resulting in reduced amounts of suitable hypolimnetic water. Such habitat reduction in lakes could also impact other stenothermic fish, for example arctic char, whitefish (*Coregonus* spp.), striped bass (*Morone saxatilis*) and northern pike (*Esox lucius*).

The current distributions for many North American and European fish are well approximated by the mean July isotherms, and examination of these values reveals the potential for range extensions under predicted climate change scenarios (Lehtonen, 1996). For example, the northern distribution of Eurasian perch (*Perca fluviatilis*) is governed by the July 10°C isoline, making further northward expansion on the European continent physically impossible (Shuter and Post, 1990; Lehtonen, 1996). However, the abundance of Eurasian perch in northern waters is certainly predicted to increase based on the positive effects of temperature on growth and population dynamics (Lehtonen, 1996). By contrast, the distribution of zander (*Sander lucioperca*) in northern Europe is governed by the July 15°C isoline, allowing significant increase in the potentially available thermal habitat in the future (Lehtonen, 1996) and a similar expansion is probable for sauger (*S. canadensis*) and walleye (*S. vitreus*) in North America. Smallmouth bass (*Micropterus dolomieu*), with a northern distribution limit associated with July 16–18°C isotherms is also predicted to expand its range northward in North America (Sharma *et al.*, 2007) with possible deleterious effects on many native cyprinid species (Jackson and Mandrak, 2002) and lake trout (Sharma *et al.*, 2009). Northern distribution limit for cyprinids should also be controlled, theoretically, by the need to grow above a threshold size to survive the first winter (Shuter and Post,

1990). However, there is evidence to suggest that the pattern for cyprinids is disconnected from such simple mechanistic models, and that predator–prey dynamics play a greater role in determining year-class size for this group (Lehtonen and Lappalainen, 1995; Jackson and Mandrak, 2002).

Predicted changes in Patagonia are distinct from the northern hemisphere due to differences in the freshwater fish fauna and the amplitude of climate change (Table 6.2). In Patagonia there are 26 native freshwater species and a number of introduced species, of which salmonids are dominant (for a list of species, see Pascual *et al.*, 2007; Aigo *et al.*, 2008). The exotic salmonids have established self-sustaining populations, both resident and anadromous, although the populations are supplemented with vigorous stocking (Pascual *et al.*, 2007). Economic importance of inland fisheries for native species is negligible in Patagonia (Junk, 2007) whereas the recreational fishery for exotic salmonids is one of the best in the world, with significant local and regional economic impact (Pascual *et al.*, 2007).

There is a clear decreasing trend in fish diversity along the latitudinal gradient poleward in South America such that only four native species (*Galaxias* spp.) and the introduced salmonids occur on the island of Tierra del Fuego, the southernmost area of the world with a freshwater fish fauna (Cussac *et al.*, 2009). In Patagonia, southward range expansions are expected for fish that are limited by their tolerance to low temperatures such as the Argentinian silverside (*Odontesthes bonariensis*). Poleward expansion for the introduced salmonid fish is limited, as even the Tierra del Fuego region is already occupied by these coldwater exotics (Pascual *et al.*, 2002).

Some changes in freshwater fish fauna, attributable to climate change, have already been observed in the pampas region of southern South America (latitude 34° 58' S). More rain has been recorded in this area in recent years (Gómez *et al.*, 2004; Christensen *et al.*, 2007), in contrast to the predicted reduction in precipitation in the higher latitudes to the south (Table 6.2). In these areas, fish communities have been established in areas that have previously been dry (Gómez *et al.*, 2004). Further south, it is predicted that the future distribution of native and exotic fish species will be largely negatively affected by the reductions in river flow due to lower precipitation, and the trend will be exaggerated by the construction of the planned large-scale hydroelectric projects (Pascual *et al.*, 2007). Further predictions of possible effects of climate change on range shifts in Patagonia are complicated by two factors:

1. The current distribution of fish fauna, especially in riverine habitats, is not well known (Pascual *et al.*, 2007).
2. The fish fauna may not be in equilibrium due to the recent (and ongoing) introductions of exotic salmonids and the changes in population dynamics that may result irrespective of climate change (Habit *et al.*, 2010).

Exotic salmonids have caused declines in native galaxiids which are now restricted to marginal habitats (Habit *et al.*, 2010). In particular, the introduced salmonids outcompete the native galaxiids in cool lotic habitats but the native fish seem to find refugia in the northern (warmer) parts of their range, and in the higher altitudes of the southern part of their range (Habit *et al.*, 2010). With such a distribution, the effects of climate change will probably benefit the native fauna in northern Patagonia where warming may make habitats less suitable for salmonids. Future warming may, however, make the hostile, high altitude areas suitable for salmonids at the expense of native galaxiids. In lakes, most interaction between salmonids and native species occurs in the littoral zone where warming has already been shown to negatively affect the salmonid fish and has led to an increase in the native Creole perch (*Percichthys trucha*; Aigo *et al.*, 2008).

When considering freshwater fish, the question of future shifts in a species' range is not a simple extrapolation of present-day physico-chemical requirements of the species in relation to future conditions under a set of projected climate change scenarios (the 'bioclimatic envelope' approach; Heino *et al.*, 2009). Unlike, for example, marine fish and most of terrestrial fauna, dispersal opportunities for freshwater fish are limited by their current hydrographic networks (interconnected catchments). This means that climate-induced changes in a species' range are not exclusively governed by the thermal niche requirements; fish species cannot necessarily disperse to new areas even though suitable thermal habitat may become available (Grant *et al.*, 2007; Sharma *et al.*, 2007; Buisson *et al.*, 2008). Rather, the dispersal of freshwater fish is limited by their containment in accessible hydrographic networks. One important determinant of future distributions of freshwater fish is the direction of the drainage systems (Sharma *et al.*, 2007). Generally, the rate of expansion in range can be predicted to be faster in systems with a north–south gradient, where fish can move freely along latitudinal gradients into new thermally suitable habitats (e.g. the large Russian rivers Lena, Ob and Yenisey) in contrast to systems flowing east–west where movement over continental divides would be required for range expansion (Wrona *et al.*, 2005).

Predicting species range shifts due to climate warming as a product of altered thermal niche and the available distributional route(s) is too simplistic. Multiple biotic interactions will strongly affect whether a particular fish species is to become established (or succumb) in a region after suitable thermal habitat becomes available due to climate warming (Jackson *et al.*, 2001). System-specific predictions of such outcomes are impossible, but some of the more likely interactions can be envisaged conceptually. The various interactions potentially affecting future fish distributions are considered in Section 6.6.

6.4 Effects of Altered Ice-processes and Winter Flow on Freshwater Fish

There is already compelling evidence that climate change will have a major effect on winter conditions in freshwater systems by reducing the duration of ice cover in lakes and rivers (Magnuson *et al.*, 2000; Prowse and Beltaos, 2002). An altered ice cover regimen can significantly affect the freshwater fish community because ice has a strong influence on light, temperature and dissolved oxygen conditions in lakes and rivers (Prowse, 2001). On the positive side, warming due to climate change is predicted to make more habitat available, at least initially, in the very high arctic where the winter conditions are currently too hostile to support a freshwater fish fauna. Some arctic rivers and ponds freeze to the bottom (Power *et al.*, 1993), and less severe ice conditions in these streams may make them available for colonization by cold stenothermic species such as Arctic char, lake trout and whitefish (Ficke *et al.*, 2007). In lakes, the shorter ice-covered period will lead to a shorter period of winter stratification (Wrona *et al.*, 2005; Magnuson, 2010). This can be predicted to lead to improved dissolved oxygen conditions in the hypolimnion, reduced probability of hypoxia/anoxia, and thus less frequent winter kills in shallow ice-covered lakes (Fang and Stefan, 2009).

The shortening of the ice-covered period is also predicted to enhance survival for many cool- and warmwater guild fish. The prediction is based on the assumption that these fish are unable to maintain a positive (or neutral) energy balance as they either do not feed, or their assimilation capacity is reduced at near freezing temperatures causing size-dependent winter mortality (Conover, 1992). For coldwater species able to supplement their energy reserves even under ice (Finstad *et al.*, 2004b), significant winter mortality has been shown to be more episodic (e.g. Cunjak *et al.*, 1998) rather than a continuous process (Finstad *et al.*, 2004a). For example, Linnansaari and Cunjak (2010) showed that apparent survival of juvenile Atlantic salmon (*Salmo salar*) during the stable, ice-covered period can be relatively high in comparison with the period prior to ice formation when both water temperature and flow are changing markedly. This implies that no net positive effects are likely in response to a shorter ice-covered period for juvenile salmonids in streams. Indeed, Linnansaari (2009) hypothesized that apparent survival could be reduced in the future if winters with a stable ice-cover are replaced by winters with no ice but greater variability in environmental conditions. In support, Finstad (2005) showed that the metabolic costs for arctic char and Atlantic salmon are lower under ice cover when compared with no ice. Winter warming with an associated reduction in ice cover due to climate change is therefore predicted to cause significant negative effects for the energy budget of juvenile stream salmonids in northern populations, and may cause lower winter survival (Finstad *et al.*, 2004c).

Reduction in surface ice cover may also alter the vulnerability of fish to predation, especially by homeothermic mammals such as mink (*Mustela vison*) and river otters (*Lutra canadensis*). Reduced ice cover may increase accessibility to water for these semi-aquatic predators that are otherwise restricted to infrequent open water patches in the ice along streams and rivers (Power *et al.*, 1993). In addition to the altered *amount* of predation, fish populations may also be affected by the change in the *potential risk* of predation if the period with surface ice is reduced. Surface ice provides in-stream cover for fish (Cunjak and Power, 1986; Linnansaari *et al.*, 2009). Even under a warming scenario, winter water temperatures in temperate and high latitude streams are still low enough that the ability of a poikilotherm to avoid predation by a homeotherm is much reduced, especially if suitable instream cover is lacking. Linnansaari *et al.* (2008) observed that winter daytime activity by juvenile Atlantic salmon increased in the presence of surface ice, presumably because the perceived risk of predation was lower under ice where low light reduced the efficiency of hunting by predators. However, with no direct quantitative estimates of predation-induced mortality in winter, it is impossible accurately to assess winter predation risk for fish populations should ice cover be reduced in the future.

Mid-winter ice break-ups are expected to occur more frequently in the future (Prowse and Beltaos, 2002). This is due to the synergistic effects of reduced ice strength (from warmer temperatures) and the increased winter flows as more precipitation will fall as rain instead of snow (Prowse and Beltaos, 2002). The effects of mid-winter ice break-ups on stream fish has been shown to be devastating, especially when mechanically strong or thick ice scours the streambed where juvenile fish or eggs overwinter (Doyle *et al.*, 1993; Cunjak *et al.*, 1998). However, it is expected that warmer winters will also produce less ice (Beltaos and Burrell, 2003) which may reduce the severity of ice break-up effects on aquatic biota. Linnansaari and Cunjak (2010) observed that thermal ice break-ups early in the winter did not significantly affect the apparent survival of juvenile Atlantic salmon. Prowse (2001) hypothesized that moderate ice break-up events may actually contribute to increasing overall biological diversity and productivity in rivers.

In rivers, surface ice serves to stabilize winter hydraulic conditions by precluding the formation of more dynamic subsurface ice (frazil and anchor ice). The predicted increased frequency of mid-winter ice break-ups and subsequent refreezing may lead to increased production of subsurface ice (frazil and anchor ice) in the future. Frazil and anchor ice accumulation can drastically change the physical habitat of rivers and streams over short temporal scales (Cunjak and Caissie, 1993; Stickler *et al.*, 2010). The effects of frazil and anchor ice on large-bodied stream fish that do not seek winter cover within the substrate have been shown to be generally negative. Subsurface ice has been shown to displace fish from preferred microhabitats, potentially leading to stranding of fish, or suffocation (Maciolek and Needham, 1952; Brown *et al.*, 1993, 2001; Cunjak and Caissie, 1993). Recent evidence has shown that smaller-bodied fish are able to find refuge from ice by hiding within the stream substratum and may be relatively unaffected by anchor ice dynamics (Linnansaari, 2009), although this may vary depending on the specific nature of the anchor ice (Stickler and Alfredsen, 2009).

Increased winter runoff may also impact

freshwater fish. Winter is typically a stable low-flow period in many temperate and arctic systems where much of the water is tied up in its solid phase (ice and snow). The swimming capacity of fish is closely tied to their metabolism and at cold temperatures swimming performance is markedly reduced (Rimmer *et al.*, 1985; Graham *et al.*, 1996). It is possible that some riverine fish will not be able to tolerate higher flows at cold winter temperatures in lotic habitats such that more areas may become unsuitable for fish as their critical swimming capacity is exceeded. Further, swimming against higher flows may increase winter-time energy consumption and possibly affect the capability of fish to survive the cold season.

6.5 Changes in Winter Growth and Productivity

Climate change is predicted to have a direct effect on the productivity of freshwater systems in the regions considered in this chapter. The impacts on productivity are likely to be positive in the high latitude areas where a shorter winter and consequent longer growing period is expected in the future (Reist *et al.*, 2006). For example, fish production in all Canadian lakes is predicted to increase by 80.7% by 2080 under the A2 scenario, *if* climate warming was to act independently of biotic interactions (Minns, 2009). However, considering the effects of future anthropogenic impacts (e.g. human population growth and demands) in addition to climate change and the potential impacts from biotic interactions such as species displacements and adaptation to new conditions, a reduction in production of 31.4% was estimated (Minns, 2009). This example highlights the difficulty in projecting changes in fish productivity in response to future climate scenarios because multiple interactions, physical and biological, act simultaneously with complex consequences for fish production.

The impact of climate change on growth per se will vary seasonally and will be species specific. Generally, growth will increase due to warming as long as species-specific upper thresholds for growth are not exceeded. Warmer winter temperatures may lead to positive somatic growth, especially in coldwater species that remain relatively active during winter (Finstad *et al.*, 2004b). However, warmer water in winter necessitates an accelerated metabolism so that more energy (food) will be required to sustain basal functions. Availability of food items has long been identified as a potential constraint for growth opportunities in winter (Hynes, 1970) and might be further complicated if the increased winter water temperature leads to shifts in the emergence times of macroinvertebrates, causing a mismatch between food availability and the nutritional requirements of freshwater fish. For example, Connolly and Petersen (2003) showed that winter warming, coupled with limitations in food availability, can cause energy depletion in young-of-the-year rainbow trout. The magnitude of any temperature increase also plays a critical role in determining whether increased food production is able to offset the increased metabolic demand and therefore lead to somatic growth. Ries and Perry (1995), in a case study using brook char (*Salvelinus fontinalis*), showed that whereas a 2°C rise in water temperature would be likely to increase growth rates in spring and autumn, a 4°C temperature rise would require a 30%–40% increase in food consumption, and would be unlikely to be realized, thereby resulting in reduced growth. The outcome – positive or negative fish growth in a future climate – will thus be a fine balance between increased metabolic demands, food availability and thermal performance (physiological capacity, foraging activity, efficacy of assimilation, etc.) of each fish species.

In autumn-spawning species such as many salmonids, warmer water temperatures in winter will implicitly lead to faster development of eggs and an earlier emergence of the alevins as the early development is directly related to accumulation of degree days (Crisp, 1981,1988). The positive effect of such faster 'intra-gravel growth' is that less time is spent at a stage where the fish is immobile and thus vulnerable to hyporheic water quality stressors such as low dissolved oxygen (Youngson *et al.*, 2004). Accelerated development due to warmer winter temperatures can, however, lead to emergence of fish during unsuitable environmental conditions (e.g. high spring streamflow) or during inadequate food availability, as the life cycle of many freshwater invertebrates is partly controlled by day length (Hynes, 1970). De-coupling of fish emergence timing from suitable environmental conditions can significantly reduce young-of-the-year survival, but accurate predictions of such an ecological 'mismatch' are difficult given the complex role of water temperature on biological and environmental functions.

6.6 Intra- and Interspecific Interactions and Population Dynamics

A great number of changes in population dynamics and species interactions are predicted as a result of climate change (Reist *et al.*, 2006; Ficke *et al.*, 2007; Rahel and Olden, 2008). Alterations in competition, food availability, parasitism and life-history traits are expected. We discuss below a few common responses predicted to occur due to climate change.

Competitive interactions, both at the intra- and interspecific level are predicted to change as alterations in environmental conditions occur. Increased winter survival, for example, may lead to crowding and subsequent decreased growth due to increased intraspecific competition for food (that is, a population self-thinning response; Reist *et al.*, 2006). Changes in the ecological balance between species living in sympatry are expected where the altered environmental conditions may subsequently allow one species to gain a competitive edge. For example, arctic char living in sympatry with brown trout (*Salmo trutta*) in northern European lakes are able to maintain positive growth under ice, and so benefit from long northern winters (Helland *et al.*, 2011). Logically, with a shorter winter, the balance between the two sympatric species is likely to shift in favour of brown trout – a species better adapted to warm water. Similarly, in North America where bull trout (*Salvelinus confluentus*) seem to have a competitive advantage over brook char in cold streams, a warming climate may reverse the interspecific balance (Rahel and Olden, 2008). Lake trout populations, especially in small Canadian lakes, are expected to be negatively affected by the northward dispersal of smallmouth bass due to competition for food resources (Sharma *et al.*, 2009).

Predation will play a significant role in shaping the ichthyofauna under future climate scenarios. Range extensions due to better winter survival are predicted for large predatory fish such as smallmouth bass (Sharma *et al.*, 2007). The presence of smallmouth bass has been associated with marked reductions in native cyprinid species and it has been estimated that over 25,000 populations of small cyprinids may disappear in the aftermath of bass invasions when the climate warms and the species' range expands northward (Jackson and Mandrak, 2002). Altered predatory interactions do not necessarily require an introduction of a new predator to a system, and changes in water temperature alone may have major effects on population dynamics. For example, coexistence of northern pike and brown trout is possible in small lakes if they are relatively cold, presumably because the ability of pike to capture brown trout is reduced below a threshold temperature of 8°C (Öhlund *et al.*, 2010). With the predicted warming in northern Europe, it seems likely that pike predation efficiency will improve during the 'shoulder periods' before and after winter, to the detriment of brown trout. Some of the seemingly positive effects of winter warming may also turn out to be negative for some species.

Population dynamics of all freshwater fish will be altered by complex host–parasite dynamics. Winter temperatures are the major limiting factor for parasite prevalence in the regions considered in this chapter (Marcogliese, 2001). It is possible, therefore, that warmer winters will lead to year-round infections due to improved thermal niches for parasites, but milder winters may also lower infection rates if the fish are in better condition and less stressed (Marcogliese, 2001; Ficke *et al.*, 2007).

Climate change is also predicted to have population-level effects irrespective of interactions between, or within, species. Such effects are due to altered physiology or metabolism of the population, or productivity of the system, and affect the life history strategies of fish. For example, population dynamics of New World silversides (family Atherinpsidae) will be directly affected by climate warming due to their temperature-dependent sex determination (Kopprio *et al.*, 2010; Strüssman *et al.*, 2010). In the Patagonian region, the male-to-female (M:F) ratio for Argentinian and Patagonian silverside (*Odontesthes hatcheri*) is affected to a different extent by an increase of 1°C in temperature. Even in allopatry, subtle changes in water temperature may have significant effects on the population whereby the abundance of males can increase sharply due to higher temperatures (Kopprio *et al.*, 2010). Depending on the ambient water temperature regime, an increase of 1.5°C could change the M:F ratio from 1:1 to 3:1 in Argentinian silversides (Ospina-Álvarez and Piferrer, 2008). Where the two silversides live in sympatry (due to stockings of *O. bonariensis* in the Patagonian lakes; Cussac *et al.*, 2009), the interactions between the species can be altered in unpredictable ways when the M:F ratio changes disproportionately between species.

Warming temperature is also expected directly to affect the age at reproduction, with a shift to younger fish maturing earlier due to improved growth opportunities in the cool- and warmwater guilds. For anadromous salmonids, the timing of smolt migration is predicted to occur earlier in the spring. Smolt migration is typically initiated when a certain sum of degree days has been accumulated in association with a triggering increase in river flow (i.e. spring freshet) (Zydlewski *et al.*, 2005). With warmer winters, the suitable degree-day sum and snowmelt flood will occur earlier, advancing the timing when the juvenile salmon head seaward. This may be problematic in terms of survival at sea, as the physiological processes that precondition juvenile salmonids for life at sea are strongly controlled by photoperiod. Earlier initiation of migration may result in smolts that are physiologically maladapted for acclimation to seawater entry, and may cause fish to miss their 'physiological smolt window', a time frame for sea entry that has been shown to be pivotal in determining sea survival (McCormick *et al.*, 1998).

Another potential change in fish population dynamics that can have important biological and societal impacts is the potential shift in the degree of anadromy. The prevalence of anadromy in various fish species generally increases towards higher latitudes in both hemispheres (McDowall, 2008; Cussac *et al.*, 2009). This is presumed to be due either to low productivity of the freshwater habitats towards the poles (Gross *et al.*, 1988) or possibly because of harsh winter (ice) conditions in high latitude freshwaters (McDowall, 2008). As the productivity in northern freshwaters is predicted to increase and ice conditions are likely to be less severe due to climate change, it is possible that a reduction in anadromy will result (Reist *et al.*, 2006). Anadromous individuals typically attain a much larger body size than do their non-anadromous counterparts, and therefore the anadromous fish are more valued in recreational, commercial and subsistence fisheries. Admittedly, this line of logic also depends on the relative productivity of marine areas for winter feeding by anadromous fish. Still, if anadromy was to be reduced because of climate change, significant negative effects in the fisheries would be experienced (Reist *et al.*, 2006).

6.7 Summary and Conclusions

Climate change is likely to lead to significant negative effects for native freshwater fish by altering both the quantity and suitability of their habitat. The predicted changes in air (and thus water) temperature and precipitation patterns lead to a number of cascading physico-chemical changes with both direct and indirect effects on fish species distribution and assemblage composition, as well as metabolism, growth, production and survival. Whereas range contractions for freshwater fish will largely be caused by summer warming, the predicted warming of winter temperatures may permit range extensions for a number of species in northern North America, northern Eurasia and Patagonia in South America. These range extensions are especially expected where new habitat becomes available at very high latitudes, and where the threat of experiencing incipient lethal low temperatures is reduced. Range extensions are likely to be dictated, however, by the synergistic effects of improved summer growth and shortening of the winter period when energetic deficiencies typically develop. In general, stenothermic coldwater species are predicted to be negatively affected whereas eurythermic cool- and warmwater species are likely to extend their distributions towards the poles. Species-level extinctions are unlikely during the next 50–100 years in the high latitudes, but local metapopulations may disappear. As the dispersal of freshwater fish is limited by the interconnected hydrographic networks where they live, redistribution of many species is likely to be enhanced by introductions (either intentional or accidental) when the thermal habitat changes with the new climate. The future fish distribution will also be largely dependent on altered species interactions affecting competition, food availability, predation and parasitism. Altered ice processes, expected under the new climate, will have both positive and negative effects on freshwater fish. A shorter ice-covered period translates into a longer growth period and shorter winter stratification in many lakes. In fluvial habitats, reduction in ice may lead to changes in predation pressure and increased subsurface ice production that may affect some fish species negatively. How and when (if ever) an 'ecological equilibrium' will be realized for freshwater fish communities under the changing climate is unpredictable given the complexity of variables involved. Comprehensive ecosystem monitoring, coupled with intelligent, multidisciplinary management that balances the needs of human society and the natural environment, will increase the probability that the outcome will have positive aspects for both.

6.8 Acknowledgements

The Secretariat of the Intergovernmental Panel on Climate Change granted the kind permission to reproduce the data shown in Table 6.2. We also thank O.K. Berg, S. Gómez, E. Habit, C. Harrod, J. Heino, J. Lappalainen, C. Milly, O. Ugedal and A. Vøllestad for their correspondence.

References

Aigo, J., Cussac, V., Peris, S., Ortubay, S., Gómez, S., López, H., Gross, M., Barriga, J., *et al.* (2008) Distribution of introduced and native fish in Patagonia (Argentina): patterns and changes in fish assemblages. *Reviews in Fish Biology and Fisheries* 18, 387–408.

Beltaos, S. and Burrell, B.C. (2003) Climatic change and river ice breakup. *Canadian Journal of Civil Engineering* 30, 145–155.

Brown, R.S., Stanislawski, S.S. and Mackay, W.C. (1993) Effects of frazil ice on fish. In: Prowse, T.D. (ed.) *Proceedings of the Workshop on Environmental Aspects of River Ice*. NHRI Symposium Series No. 12, Environment Canada National Hydrology Research Institute, Saskatoon, Saskatchewan, Canada, pp. 261–278.

Brown, R.S., Power, G. and Beltaos, S. (2001) Winter movements and habitat use of riverine brown trout, white sucker and common carp in relation to flooding and ice break-up. *Journal of Fish Biology* 59, 1126–1141.

Buisson, L., Thuiller, W., Lek, S., Lim, P. and Grenouillet, G. (2008) Climate change hastens the turnover of stream fish assemblages. *Global Change Biology* 14, 2232–2248.

Christensen, J.H., Hewitson, B., Busuioc, A., Chen, A., Gao, X., Held, I., Jones, R., *et al.* (2007) Regional climate projections. In: Solomon, S., Qin, D., Manning, M., Chen, Z., Marquis, M., Averyt, K.B., Tignor, M., *et al.* (eds) *Climate Change 2007: The Physical Science Basis. Contribution of Working Group I to the Fourth Assessment Report of the Intergovernmental*

Panel on Climate Change. Cambridge University Press, Cambridge, pp. 847–940.

Connolly, P.J. and Petersen, J.H. (2003) Bigger is not always better for overwintering young-of-year steelhead. *Transactions of the American Fisheries Society* 132, 262–274.

Conover, D.O. (1992) Seasonality and the scheduling of life history at different latitudes. *Journal of Fish Biology* 41, 161–178.

Crisp, D.T. (1981) A desk study of the relationship between temperature and hatching time for the eggs of five species of salmonid fishes. *Freshwater Biology* 11, 361–368.

Crisp, D.T. (1988) Prediction, from temperature, of eyeing, hatching and "swing-up" times for salmonid embryos. *Freshwater Biology* 19, 41–48.

Cunjak, R.A. (1988). Physiological consequences of overwintering in streams: the cost of acclimatization? *Canadian Journal of Fisheries and Aquatic Sciences* 45, 443–452.

Cunjak, R.A. (1996). Winter habitat of selected stream fishes and potential impact from land-use activities. *Canadian Journal of Fisheries and Aquatic Sciences* 53, 267–282.

Cunjak, R.A. and Caissie, D. (1993) Frazil ice accumulation in a large salmon pool in the Northwest Miramichi River, New Brunswick: Ecological implications for overwintering fishes. In: Prowse, T.D. (ed.) *Proceedings of the Workshop on Environmental Aspects of River Ice*. NHRI Symposium Series No. 12, Environment Canada National Hydrology Research Institute, Saskatoon, Saskatchewan, Canada, pp. 279–295.

Cunjak, R.A. and Power, G. (1986) Winter habitat utilization by stream resident brook trout (*Salvelinus fontinalis*) and brown trout (*Salmo trutta*). *Canadian Journal of Fisheries and Aquatic Sciences* 43, 1970–1981.

Cunjak, R.A., Curry, R.A. and Power, G. (1987) The seasonal energy budget of brook trout in a small river: evidence of a winter deficit. *Transactions of the American Fisheries Society* 116, 817–828.

Cunjak, R.A., Prowse, T.D. and Parrish, D.L. (1998) Atlantic salmon (*Salmo salar*) in winter: "The season of parr discontent". *Canadian Journal of Fisheries and Aquatic Sciences* 55, 161–180.

Cussac, V.E., Fernández, D.A., Gómez, S.E. and López, H.L. (2009) Fishes of southern South America: a story driven by temperature. *Fish Physiology and Biochemistry* 35, 29–42.

Doyle, P.F., Kosakoski, G.T. and Costerton, R.W. (1993) Negative effects of freeze-up and breakup on fish in the Nicola River. In: Prowse, T.D. (ed.) *Proceedings of the Workshop on Environmental Aspects of River Ice*. NHRI Symposium Series No. 12, Environment Canada National Hydrology Research Institute, Saskatoon, Saskatchewan, Canada, pp. 299–314.

Fang, X. and Stefan H.G. (2009) Simulations of climate effects on water temperature, dissolved oxygen, ice and snow covers in lakes of the contiguous U.S. under past and future climate scenarios. *Limnology and Oceanography* 54, 2359–2370.

Ficke, A.D., Myrick, C.A. and Hansen, L.J. (2007) Potential impacts of global climate change on freshwater fisheries. *Reviews in Fish Biology and Fisheries* 17, 581–613.

Finstad, A.G. (2005). Salmonid fishes in a changing climate: The winter challenge. PhD thesis, Norwegian University of Science and Technology, Trondheim, Norway.

Finstad, A.G., Næsje, T.F. and Forseth, T. (2004a) Seasonal variation in the thermal performance of juvenile Atlantic salmon (*Salmo salar*). *Freshwater Biology* 49, 1459–1467.

Finstad, A.G., Ugedal, O., Forseth, T. and Næsje, T.F. (2004b). Energy-related juvenile winter mortality in a northern population of Atlantic salmon (*Salmo salar*). *Canadian Journal of Fisheries and Aquatic Sciences* 61, 2358–2368.

Finstad, A.G., Forseth, T., Næsje, T.F. and Ugedal, O. (2004c) The importance of ice cover for energy turnover in juvenile Atlantic salmon. *Journal of Animal Ecology* 73, 959–966.

Gómez, S.E. (1996) Resistance to temperature and salinity in fishes of the province of Buenos Aires (Argentina), with zoogeographical implications. In: *4º Convegno Nazionale A.I.I.A.D.* ATTI Congressuali 2–13 December 1991, Associazione Italiana Ittiologi Acque Dolci, Trento, Italy, pp. 171–192.

Gómez, S.E., Trenti, P.S. and Menni, R.C. (2004) New fish populations as evidence of climate change in former dry areas of the Pampa Region (Southern South America). *Physis* 59, 43–44.

Graham, C.T. and Harrod, C. (2009) Implications of climate change for the fishes of the British Isles. *Journal of Fish Biology* 74, 1143–1205.

Graham, W.D., Thorpe, J.E. and Metcalfe, N.B. (1996) Seasonal current holding performance of juvenile Atlantic salmon in relation to temperature and smolting. *Canadian Journal of Fisheries and Aquatic Sciences* 53, 80–86.

Grant, E.H.C., Lowe, W.H. and Fagan, W.F. (2007) Living in the branches: population dynamics and ecological processes in dendritic networks. *Ecology Letters* 10, 165–175.

Gross, M.R., Coleman, R.M. and McDowall, R.M.

(1988) Aquatic productivity and the evolution of diadromous fish migration. *Science* 239, 1291–1293.

Habit, E., Piedra, P., Ruzzante, D.E., Walde, S.J., Belk, M.C., Cussac, V.E., Gonzalez, J., *et al.* (2010) Changes in the distribution of native fishes in response to introduced species and other anthropogenic effects. *Global Ecology and Biogeography* 19, 697–710.

Heino, J., Virkkala, R. and Toivonen, H. (2009) Climate change and freshwater biodiversity: detected patterns, future trends and adaptations in northern regions. *Biological Reviews* 84, 39–54.

Helland, I.P., Finstad, A.G., Forseth, T., Hesthagen, T. and Ugedal, O. (2011) Ice-cover effects on competitive interactions between two fish species. *Journal of Animal Ecology* 80, 539–547.

Hynes, H.B.N. (1970) *The Ecology of Running Waters*. 1st ed., Blackburn Press, Caldwell, New Jersey.

Jackson, D.A. and Mandrak, N.E. (2002) Changing fish biodiversity: predicting the loss of cyprinid biodiversity due to global climate change. In: McGinn, N.A. (ed.) *American Fisheries Society Symposium 32*. American Fisheries Society, Bethesda, Maryland, pp. 89–98.

Jackson, D.A., Peres-Neto, P.R. and Olden, J.D. (2001) What controls who is where in freshwater fish communities – the roles of biotic, abiotic, and spatial factors. *Canadian Journal of Fisheries and Aquatic Sciences* 58, 157–170.

Johnson, T.B. and Evans, D.O. (1990) Size-dependent winter mortality of young-of-the-year white perch: Climate warming and invasion of the Laurentian Great Lakes. *Transactions of the American Fisheries Society* 119, 301–313.

Jonsson, B. and Jonsson, N. (2009) A review of the likely effects of climate change on anadromous Atlantic salmon Salmo salar and brown trout Salmo trutta, with particular reference to water temperature and flow. *Journal of Fish Biology* 75, 2381–2447.

Junk, W.J. (2007) Freshwater fishes of South America: Their biodiversity, fisheries, and habitats—a synthesis. *Aquatic Ecosystem Health and Management* 10, 228–242.

Kopprio, G.A., Freije, R.H., Strüssmann, C.A., Kattner, G., Hoffmeyer, M.S., Popovich, C.A. and Lara, R.J. (2010) Vulnerability of pejerrey Odontesthes bonariensis populations to climate change in pampean lakes of Argentina. *Journal of Fish Biology* 77, 1856–1866.

Lappalainen, J. and Lehtonen, H. (1997) Temperature habitats for freshwater fishes in a warming climate. *Boreal Environment Research* 2, 69–84.

Lappalainen, J., Erm, V., Kjellman, J. and Lehtonen, H. (2000) Size-dependent winter mortality of age-0 pikeperch (*Stizostedion lucioperca*) in Pärnu Bay, the Baltic Sea. *Canadian Journal of Fish and Aquatic Sciences* 57, 451–458.

Lehtonen, H. (1996) Potential effects of global warming on northern European freshwater fish and fisheries. *Fisheries Ecology and Management* 3, 59–71.

Lehtonen, H. and Lappalainen, J. (1995) The effects of climate on the year-class variations of certain freshwater fish species. *Canadian Special Publication of Fisheries and Aquatic Sciences* 121, 37–44.

Lemke, P., Ren, J., Alley, R.B., Allison, I., Carrasco, J., Flato, G., Fujii, Y., *et al.* (2007) Observations: changes in snow, ice and frozen ground. In: Solomon, S., Qin, D., Manning, M., Chen, Z., Marquis, M., Averyt, K.B., Tignor, M., *et al.* (eds) *Climate Change 2007: The Physical Science Basis. Contribution of Working Group I to the Fourth Assessment Report of the Intergovernmental Panel on Climate Change.* Cambridge. University Press, Cambridge, pp. 337–383.

Linnansaari, T. (2009) Effects of ice conditions on behaviour and population dynamics of Atlantic salmon (*Salmo salar* L.) parr. PhD thesis, University of New Brunswick, Fredericton, New Brunswick, Canada.

Linnansaari, T. and Cunjak, R.A. (2010) Patterns in apparent survival of Atlantic salmon (*Salmo salar*) parr in relation to variable ice conditions throughout winter. *Canadian Journal of Fisheries and Aquatic Sciences* 67, 1744–1754.

Linnansaari, T., Cunjak, R.A. and Newbury, R. (2008) Winter behaviour of juvenile Atlantic salmon *Salmo salar* L. in experimental stream channels: effect of substratum size and full ice cover on spatial distribution and activity pattern. *Journal of Fish Biology* 72, 2518–2533.

Linnansaari, T., Alfredsen, K., Stickler, M., Arnekleiv, J.V., Harby, A. and Cunjak, R.A. (2009) Does ice matter? Site fidelity and movements by Atlantic salmon (*Salmo salar* L.) parr during winter in a substrate enhanced river reach. *River Research and Applications* 25, 773–787.

Maciolek, J.A. and Needham, P.R. (1952) Ecological effects of winter conditions on trout and trout foods in Convict Creek, California, 1951. *Transactions of the American Fisheries Society* 81, 202–217.

Magnuson, J.J. (2010) History and heroes: the

thermal niche of fishes and long-term lake ice dynamics. *Journal of Fish Biology* 77, 1731–1744.

Magnuson, J.J., Crowder, L.B. and Medwick, P.A. (1979) Temperature as an ecological resource. *American Zoologist* 19, 331–343.

Magnuson, J.J., Robertson, D.M., Benson, B.J., Wynne, R.H., Livingstone, D.M., Arai, R.A., Barry, R.G., *et al.* (2000) Historical trends in lake and river ice cover in the Northern Hemisphere. *Science* 289, 1743–1746.

Marcogliese, D.J. (2001) Implications of climate change for parasitism of animals in the aquatic environment. *Canadian Journal of Zoology* 79, 1331–1352.

McCormick, S.D., Hansen, L.P., Quinn, T.P. and Saunders, R.L. (1998) Movement, migration, and smolting of Atlantic salmon (*Salmo salar*). *Canadian Journal of Fisheries and Aquatic Sciences* 55, 77–92.

McCullough, D.A., Bartholomew, J.M., Jager, H.I., Beschta, R.L., Cheslak, E.F., Deas, M.L., Ebersole, J.L., *et al.* (2009) Research in thermal biology: Burning questions for coldwater stream fishes. *Reviews in Fisheries Science* 17, 90–115.

McDowall, R.M. (2008) Why are so many boreal freshwater fishes anadromous? Confronting 'conventional wisdom'. *Fish and Fisheries* 9, 208–213.

McPhail, J.D. and Paragamian, V.L. (2000) Burbot biology and life history. In: Paragamian, V.L. and Willis, D.W. (eds) *Burbot: Biology, Ecology, and Management.* American Fisheries Society, Bethesda, Maryland, pp. 11–23.

Meehl, G.A., Stocker, T.F., Collins, W.D., Friedlingstein, P., Gaye, A.T., Gregory, J.M., Kitoh, A., *et al.* (2007) Global climate projections. In: Solomon, S., Qin, D., Manning, M., Chen, Z., Marquis, M., Averyt, K.B., Tignor, M., *et al.* (eds) *Climate Change 2007: The Physical Science Basis. Contribution of Working Group I to the Fourth Assessment Report of the Intergovernmental Panel on Climate Change.* Cambridge University Press, Cambridge, pp. 747–845.

Milly, P.C.D., Dunne, K.A. and Vecchia, A.V. (2005) Global pattern of trends in streamflow and water availability in a changing climate. *Nature* 438, 347–350.

Minns, C.K. (2009) The potential future impact of climate warming and other human activities on the productive capacity of Canada's lake fisheries: a meta-model. *Aquatic Ecosystem Health and Management* 12, 152–167.

Miranda, L.E. and Hubbard, W.D. (1994) Length-dependent winter survival and lipid composition of age-0 largemouth bass in Bay Springs Reservoir, Mississippi. *Transactions of the American Fisheries Society* 123, 80–87.

Nakicenovic, N. and Swart, R. (eds) (2000) *Intergovernmental Panel on Climate Change Special Report on Emissions Scenarios.* Cambridge University Press, Cambridge, pp. 599.

Öhlund, G., Hein, C.L. and Englund, G. (2010) Temperature alters the outcome of pike-salmonid interactions. In: Harrod, C. and Sims, D. (eds) *Abstracts of the Fisheries Society of the British Isles Annual Symposium 2010: Fish and Climate Change.* Fisheries Society of the British Isles, www.fsbi.org.uk/2010/fsbi-2010-programme.htm.

Ortubay, S.G., Gómez, S.E. and Cussac, V.E. (1997) Lethal temperatures of a neotropical fish relic in Patagonia, the scale-less characinid *Gymnocharacinus bergi* Steindachner 1903. *Environmental Biology of Fishes* 49, 341–350.

Ospina-Álvarez, N. and Piferrer, F. (2008) Temperature-dependent sex determination in fish revisited: Prevalence, a single sex ratio response pattern, and possible effects of climate change. *PLoS ONE* 3, e2837, doi:10.1371/journal.pone.0002837.

Pascual, M., Macchi, P., Urbanski, J., Marcos, F., Rossi, C.R., Novara, M. and Dell'Arciprete, P. (2002) Evaluating potential effects of exotic freshwater fish from incomplete species presence–absence data. *Biological Invasions* 4, 101–113.

Pascual, M.A., Cussac, V., Dyer, B., Soto, D., Vigliano, P., Ortubay, S. and Macchi, P. (2007) Freshwater fishes of Patagonia in the 21st Century after a hundred years of human settlement, species introductions, and environmental change. *Aquatic Ecosystem Health and Management* 10, 212–227.

Poff, N.L., Allan, J.D., Bain, M.B., Karr, J.R., Prestegaard, K.L., Richter, B.D., Sparks, R.E., *et al.* (1997) The natural flow regime. *BioScience* 47, 769–784.

Power, G., Cunjak, R.A., Flannagan, J. and Katopodis, C. (1993) Biological effects of river ice. In: Prowse, T.D. and Gridley, N.C. (eds) *Environmental Aspects of River Ice.* National Hydrology Research Institute, Environment Canada, Saskatoon, Saskatchewan, Canada, pp. 97–125.

Prowse, T.D. (2001) River-ice ecology. II: Biological aspects. *Journal of Cold Regions Engineering* 15, 17–33.

Prowse, T.D. and Beltaos, S. (2002) Climatic control

of river-ice hydrology: a review. *Hydrological Processes* 16, 805–822.

Prowse, T.D., Furgal, C., Wrona, F.J. and Reist, J.D. (2009) Implications of climate change for northern Canada: Freshwater, marine, and terrestrial ecosystems. *Ambio* 38, 282–289.

Rahel, F.J. and Olden, J.D. (2008) Assessing the effects of climate change on aquatic invasive species. *Conservation Biology* 22, 521–533.

Randall, D.A., Wood, R.A., Bony, S., Colman, R., Fichefet, T., Fyfe, J., Kattsov, V., *et al.* (2007) Climate models and their evaluation. In: Solomon, S., Qin, D., Manning, M., Chen, Z., Marquis, M., Averyt, K.B., Tignor, M., *et al.* (eds) *Climate Change 2007: The Physical Science Basis. Contribution of Working Group I to the Fourth Assessment Report of the Intergovernmental Panel on Climate Change.* Cambridge University Press, Cambridge, pp. 589–662.

Reist, J.D., Wrona, F.J., Prowse, T.D., Power, M., Dempson, J.B., Beamish, R.J., King, J.R., *et al.* (2006) General effects of climate change on arctic fishes and fish populations. *Ambio* 35, 370–380.

Ries, R.D. and Perry, S.A. (1995) Potential effects of global climate warming on brook trout growth and prey consumption in central Appalachian streams, USA. *Climate Research* 5, 197–206.

Rimmer, D.M., Saunders, R.L. and Paim, U. (1985) Effects of temperature and season on the position holding performance of juvenile Atlantic salmon (*Salmo salar*). *Canadian Journal of Zoology* 63, 92–96.

Scott, W.B. and Crossman, E.J. (1973) *Freshwater Fishes of Canada.* Bulletin 184, Fisheries Research Board of Canada, Ottawa.

Sharma, S., Jackson, D.A., Minns, C.K. and Shuter, B. (2007) Will northern fish populations be in hot water because of climate change? *Global Change Biology* 13, 2052–2064.

Sharma, S., Jackson, D.A. and Minns, C.K. (2009) Quantifying the potential effects of climate change and the invasion of smallmouth bass on native lake trout populations across Canadian lakes. *Ecography* 32, 517–525.

Shuter, B.J. and Post, J.R. (1990) Climate, population viability, and the zoogeography of temperate fishes. *Transactions of the American Fisheries Society* 119, 314–336.

Solomon, S., Qin, D., Manning, M., Chen, Z., Marquis, M., Averyt, K.B., Tignor, M., *et al.* (eds) (2007) *Contribution of Working Group I to the Fourth Assessment Report of the Intergovernmental Panel on Climate Change.* Cambridge University Press, Cambridge.

Stickler, M. and Alfredsen, K. (2009) Anchor ice formation in streams: a field study. *Hydrological Processes* 23, 2307–2315.

Stickler, M., Alfredsen, K., Linnansaari, T. and Fjeldstad, H.-P. (2010) The influence of dynamic ice formation on hydraulic heterogeneity in steep streams. *River Research and Applications* 26, 1187–1197.

Strüssmann, C.A., Conover, D.O., Somoza, G.M. and Miranda, L.A. (2010) Implications of climate change for the reproductive capacity and survival of New World silversides (family Atherinopsidae). *Journal of Fish Biology* 77, 1818–1834.

Swanson, H.K., Kidd, K.A., Babaluk, J.A., Wastle, R.J., Yang, P.P., Halden, N.M. and Reist, J.D. (2010) Anadromy in Arctic populations of lake trout (*Salvelinus namaycush*): otolith microchemistry, stable isotopes, and comparisons with Arctic char (*Salvelinus alpinus*). *Canadian Journal of Fisheries and Aquatic Sciences* 67, 842–853.

Wrona, F.J., Prowse, T.D., Reist, J.D. (2005) Freshwater ecosystems and fisheries. In: Symon, C., Arris, L. and Heal, B. (eds) *Arctic Climate Impact Assessment.* Cambridge University Press, New York, pp. 353–452.

Youngson, A.F., Malcolm, I.A., Thorley, J.L., Bacon, P.J. and Soulsby, C. (2004) Long-residence groundwater effects on incubating salmonid eggs: low hyporheic oxygen impairs embryo development. *Canadian Journal of Fisheries and Aquatic Sciences* 61, 2278–2287.

Zydlewski, G.B., Haro, A. and McCormick, S.D. (2005) Evidence for cumulative temperature as an initiating and terminating factor in downstream migratory behavior of Atlantic salmon (*Salmo salar*) smolts. *Canadian Journal of Fisheries and Aquatic Sciences* 62, 68–78.

7 Strategies of Molecular Adaptation to Climate Change: The Challenges for Amphibians and Reptiles

Kenneth B. Storey and Janet M. Storey

7.1 Introduction

Over the course of Earth's history, climate has varied widely for numerous reasons, including very recently the actions of humans (Pidwirny, 2006). Through all of this, many organisms adapt and change to the vagaries of climate and at the two 'ends of the spectrum' new species arise whereas others go extinct. In this chapter we consider the lives of ectothermic ('cold-blooded') terrestrial vertebrates – the amphibians and reptiles – and how they are affected by and adapt to changing environmental conditions. We will focus particularly on species that have adapted to life in seasonally cold environments and examine some of the specialized biochemical adaptations that support cold and/or freeze tolerance (Storey and Storey, 1992,2004a; Margesin *et al.*, 2007). We will also survey the range of molecular mechanisms that are available to organisms to make adaptive changes to their biochemistry (and thereby to their viability) in response to environmental stress, with examples from our 30 years of study of the biochemistry of winter cold hardiness of amphibians and reptiles.

All amphibians and the vast majority of reptile species live in terrestrial or freshwater environments, most amphibians being tied to a need for bodies of water or at least damp conditions in their microenvironment. Typically, terrestrial ectotherms have little capacity for migration as a solution when environments become unfavourable (McCallum *et al.*, 2009) and therefore changes in their local environments (temperature, precipitation, pollution, etc.) can put them at risk. They can be particularly sensitive to local thermal conditions that change on daily, seasonal or multi-year time frames and are typically 'sandwiched' between upper and lower critical temperatures in determining their own activity/viability (Hillman *et al.*, 2009). Furthermore, amphibians and reptiles are also affected by secondary consequences of temperature change such as altered availability of food, water (changes in precipitation and evaporation rates), shelter/shade and suitable egg-laying sites. As such, shifting temperature profiles as a result of climate change have already, and will continue to have, a profound impact on the lives of ectothermic vertebrates affecting individuals, local populations and whole species, and undoubtedly causing extirpation of some species from various geographic locations and complete extinction of others.

7.2 Climate Change and Ectothermic Vertebrates

Many amphibian species are in trouble around the world, showing both declining populations and/or extinctions, and this has been the subject of much research over the last 20 years (Stuart *et al.*, 2004; Hayes *et al.*, 2010). Multiple influences have been identified including atmospheric change (temperature, precipitation), habitat destruction, pollution of waters from agricultural and industrial run-offs, invasive species, and pathogens (e.g. chytrid fungus,

© CAB International 2012. *Temperature Adaptation in a Changing Climate: Nature at Risk* (eds K.B. Storey and K.K. Tanino)

parasites). In some studies, all other factors can be eliminated leaving atmospheric (climate) change as the apparent cause of continuing declines in amphibians (and some reptiles) (Wake, 2007). Furthermore, warming temperatures interact with other factors (Hayes *et al.*, 2010) such as by enhancing the spread of chytridiomycosis (Bosch *et al.*, 2007). Some of the problems impacting reptiles and amphibians experiencing sudden climate change are outlined below.

7.2.1 Survival

Reptiles and amphibians are 'cold-blooded' animals and, naively, it may be assumed that rising environmental temperatures due to global warming would be a good thing – higher body temperatures (Tb) should facilitate such features as faster growth and higher activity levels. However, it turns out that ectothermic vertebrates are actually highly adept at managing their Tb and frequently maintain near-constant core Tb values via combinations of basking in the sun and cooling in the shade. Physical performance of these animals is generally greatest at core Tb values of 30°C–35°C and most are heat-stressed above about 40°C; hence, in hot environments, the availability of shade to keep animals from overheating is key to survival (Kearney *et al.*, 2009). Indeed, when the effects of a 3°C increase in environmental temperature were modelled, very serious negative consequences were predicted, especially for tropical reptile species. In particular, food demands to fuel a higher metabolic rate would increase sharply at the same time as the animals would need much more nonforaging time in the shade to keep from overheating (Kearney *et al.*, 2009). Periods of seasonal inactivity (aestivation) due to extreme heat would also have to be extended (requiring greater body fuel reserves to be accumulated in the active season) and the timing of reproduction would need to be altered for optimal egg incubation.

In northern regions, warmer temperatures could have the positive effect of allowing northward range extension of various amphibians and reptiles, but there are also negatives to consider. Higher Tb values directly increase metabolic rate, thereby increasing both daily food requirements as well as the body reserves needed to last through the winter. Warmer temperatures can go hand-in-hand with reduced precipitation and elevated rates of evaporation, creating an overall drier climate not amenable to some species, particularly amphibians that are highly susceptible to water loss across their skin. A reduced thickness of the winter snowpack could also be devastating for amphibian and reptile species that hibernate at or near the soil surface, exposing them to temperatures too low to endure. For example, a thick insulating layer of snow can keep soil surface temperatures from falling below -5°C (survivable by freeze-tolerant frogs) even when air temperature above the snowpack hovers around -30°C (not survivable) (Storey and Storey, 2004a).

Survival under changing climatic conditions also depends on whether animals can adjust effectively to new conditions. This may be difficult, especially if the pace of climatic temperature change is high. For example, McCallum *et al.* (2009) applied a climate change model to predict growth responses of three-toed box turtles over the remainder of this century, forecasting that by 2100 less than 20% of hatchlings would grow in their first year and that they would subsequently show reduced growth rates, lower adult size and reduced fecundity as compared with current populations, all conditions that could trigger an extinction vortex. A recent study of tiger snakes, *Notechis scutatus*, showed that the animals had problems adjusting to a new temperature regime. Aubret and Shine (2010) raised young snakes in different thermal gradients representing cold (19°C–22°C), warm (19°C–26°C) or hot (19°C–37°C) options. They found that all groups adjusted their basking behaviour so that each maintained similar Tb values (24°C–25°C). However, when cold and hot groups were switched in their second year of life, neither group seemed capable of adjusting their length of basking. Animals in the cold group that were used to basking the longest continued to do so and this resulted in a higher mean Tb (26.5°C), whereas those in the hot group continued to bask for the shortest time resulting in a lower Tb (23.5°C). This showed a significant effect of early life experience on thermoregulatory tactics and a limited plasticity to adjust to year-to-year variation in ambient temperatures.

7.2.2 Species range

Warming climates can lead to range extension for some species, for example, expansion northward or to higher altitudes. Other species may see their ranges decline due to competition from southern invaders and limitations to their own range expansion, for example an inability to move above the treeline. Alpine populations could easily become fractured, moving to higher and higher elevations until they are isolated and ultimately extinguished. Indeed, a substantial number of the neotropical amphibian species that have disappeared in recent years are those living in montane zones; animals are 'sandwiched' between inhospitable higher elevations and lowland forests where they cannot compete with the resident species (Stuart *et al.*, 2004). However, the opposite can also occur; studies of Columbia spotted frogs (*Rana luteiventris*) in Montana showed increased survival and breeding probability as the severity of winter decreased, suggesting that ectotherms in alpine or boreal habitats that are living near to their thermal ecological limits would benefit from a warming climate as long as suitable habitats remain intact (McCaffery and Maxell, 2010).

7.2.3 Reproduction

Temperature change associated with global warming may have its most profound effects on breeding. Most amphibians require an aquatic environment for larval development. In temperate regions, warmer temperatures can have positive impacts such as an earlier start to the breeding season and a faster rate of larval growth. Equally, however, there are negative impacts, particularly for species breeding in ephemeral waters. Reduced meltwater (the result of less snow) and higher rates of evaporation at warmer temperatures will decrease the time (hydroperiod) available for larval development. This means that fewer larvae may reach the minimum size needed for metamorphosis before ponds dry out and those that do will be smaller sized when they transform, which is associated with reduced terrestrial survival and fecundity (McMenamin and Hadly, 2010).

Most reptiles rely on the sun to incubate their eggs, with a minimum number of heat days needed to achieve hatching. For species with large north–south distributions this can allow two or more clutches of eggs to be raised per year in the south, whereas the northern range limit is often determined by insufficient heat days to rear a single clutch. For example, this is true of painted turtles at their northern limit in Algonquin Park in Canada, even though this species can deal with another very difficult challenge – survival of whole-body freezing by hatchlings during their first winter (Storey *et al.*, 1988). Therefore, at first glance, global warming could allow beneficial northern range extensions for some species but other factors also determine reproductive success. A critical factor for many reptiles is the phenomenon of temperature-dependent sex determination (TSD) during egg development. In turtles, exposure to lower temperatures during the thermosensitive period produces males, whereas higher temperatures produce females; the opposite pattern occurs in some lizards. In a third pattern, males are produced at intermediate temperatures and females at both higher and lower temperatures; this occurs in crocodilians as well as in some lizards and turtles (Hulin *et al.*, 2009; Mitchell and Janzen, 2010). As a result, rising global temperatures could easily skew the sex ratio of various species to the point of demographic collapse (Hulin *et al.*, 2009). Indeed, as little as a 1.5°C change in temperature can be the difference between

100% female and 100% male progeny in alligators (Lance, 2009). With a 3°C–4°C increase in air temperature, Mitchell *et al.* (2008) predicted that only males would remain in an island population of tuatara (*Sphenodon guntheri*) by about 2050, effectively extirpating the species.

7.3 Amphibian and Reptile Cold Hardiness

In terms of its effects on species biodiversity, various studies predict that climate change will have its greatest effect on tropical species for reasons that include the huge number of species in tropical habitats, the often highly specialized lifestyles, and the high environmental temperatures that are already close to the critical thermal maxima. Dynesius and Jansson (2000) state that climatic oscillations select for vagility (dispersal ability and propensity) and generalism, and this makes tropical species particularly vulnerable to climate change due to high specialization, low vagility and small geographic ranges despite the high species diversity of the tropics. In temperate and polar zones there are fewer species overall but individual species can range over vast areas. For example, the wood frog (*Rana sylvatica*) is found across the entire boreal forest of North America and is the most northerly distributed amphibian or reptile, with a range far above the Arctic Circle (Fig. 7.1). Wood frogs expanded northwards following the retreat of the glaciers after the last ice age and phylogenetic analyses by Lee-Yaw *et al.* (2008) show that colonization of the north derived from a limited number of eastern and western refugia, with high-latitude refugia in the Appalachian highlands and modern-day Wisconsin having the biggest impact on northern populations. Most amphibian and reptile species in the northern USA and Canada show similar patterns of recolonization from southern refugia (Lee-Yaw *et al.*, 2008). Therefore, as a result of founder effects, colonizers should show high levels of genetic homogeneity in the north (Hewitt, 1996) and this could make them particularly susceptible to climate change. Impacts of global warming on northern species could occur in at least three ways:

1. Direct effects on individual cold-hardy species;
2. Effects on the food species and habitat (e.g. breeding or wintering sites) of cold-hardy species; and
3. Competition from less hardy species moving northward as the climate gets milder.

In the rest of this article we will focus on the first of these, beginning with a survey of the strategies used by amphibians and reptiles for survival in cold climates, and then examining the biochemical options that organisms have for adapting to new environmental challenges.

Winter presents multiple challenges to amphibians and reptiles. Cold temperatures greatly limit activity and food availability, and when environmental temperatures fall below 0°C, animals are at risk of lethal freezing. Most northern and alpine ectotherms choose to hide for the winter in thermally buffered sites. Some species go underground below the frostline, digging by themselves (e.g. toads) or exploiting natural tunnels or caves (e.g. salamanders, snakes). In the interlake region of Manitoba, for example, garter snakes migrate several kilometers to mass by the thousands in underground caverns in the limestone bedrock (Gregory and Stewart, 1975). These species do not appear to need highly specialized adaptations to endure winter cold, but they must be well-attuned to both photoperiod and thermoperiod cues to trigger their autumn movement to hibernacula and their emergence in the spring. Their key adaptation for winter survival may be the accumulation of sufficient body fuel reserves during summer/autumn feeding; enough is needed to fuel not only a 6–9 month period of cold inactivity, but also an intense period of spring breeding that frequently occurs before eating resumes.

Other ectothermic vertebrates hibernate underwater. By doing so, various frogs, turtles and snakes place themselves

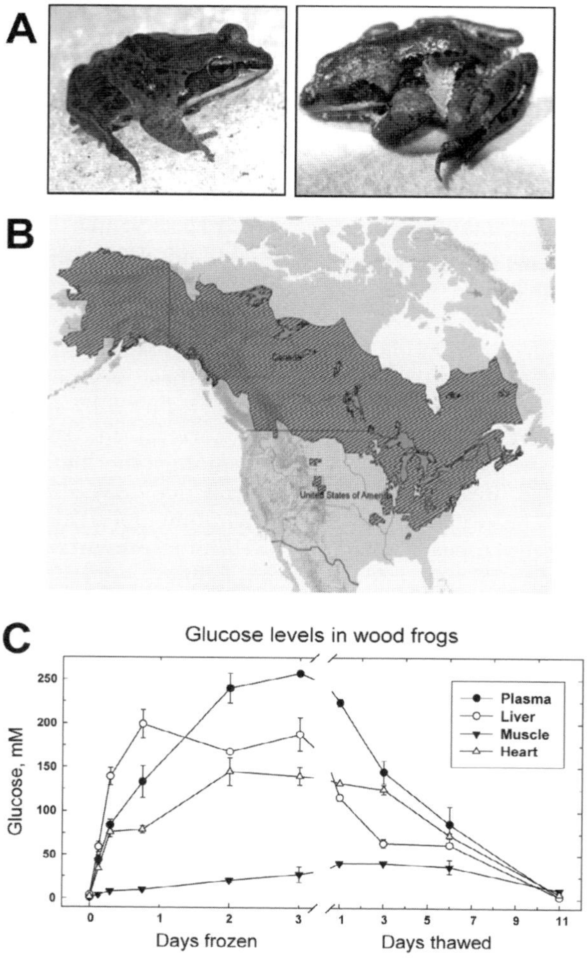

Fig. 7.1. (A) The wood frog, *Rana sylvatica* (syn. *Lithobates sylvaticus*), in unfrozen and frozen states. (B) Wood frog distribution in North America (IUCN, 2011). (C) Profiles of cryoprotectant glucose accumulation in blood and organs of wood frogs during freezing and glucose clearance after thawing. Data in (c) are replotted from Storey and Storey (1986).

advantageously in a near-constant 0°C–4°C environment although they are still vulnerable to winter-kill if the water freezes to the bottom. The main challenge of this winter strategy is oxygen availability, for two reasons:

1. These are lung-breathing animals; and
2. Organismal respiration (microbes, plants, animals) depletes oxygen in ice-locked bodies of water leading to hypoxia or even anoxia.

The problem of acquiring oxygen is solved for submerged frogs by simply switching from lung breathing to oxygen uptake across their skin (Tattersall and Ultsch, 2008). Some turtles also do this and others can absorb oxygen across other surfaces (cloaca, cloacal bursae, buccopharyngeal) (Jackson and Ultsch, 2010). Because oxygen content is highest in cold water and animal metabolism is lowest, these nonpulmonary solutions to oxygen uptake can be sufficient to sustain aerobic metabolism over the winter.

Other turtles have optimized anoxia tolerance to deal with long-term submergence. For example, painted turtles (*Chrysemys picta*) can survive in deoxygenated cold water for ~3 months (Jackson and Ultsch, 2010), and indeed, these and red-eared sliders (*Trachemys scripta elegans*) are widely used in biomedical studies to discover the molecular adaptations that allow survival of long-term oxygen deprivation by vertebrate brain and heart (Lutz and Milton, 2004; Storey, 2007; Bickler and Buck, 2007). The adaptations involved include acquisition of high levels of fermentable fuels in organs (massive stores of glycogen in liver), use of the shell as a site to store and buffer the products (lactate, H^+) of anaerobic glycolysis, and strong metabolic rate depression (Storey, 2007; Jackson and Ultsch, 2010). Indeed, the metabolic rate of a turtle submerged in cold deoxygenated water is only about 10% of the value in air at the same temperature (Herbert and Jackson, 1985), and just 0.6% of the metabolic rate in air at a common summer temperature of 20°C (Jackson and Ultsch, 2010). Metabolic rate depression is not just a feature of anoxia tolerance but is a major contributor to energy savings in many species confronted with environmental stress, including amphibians and reptiles aestivating in hot climates (Storey, 2002; Storey and Storey, 2010) and those frozen solid over the winter months (discussed below).

The third strategy for winter survival by terrestrial ectotherms is to endure subzero temperatures, by one of two ways (Storey and Storey, 1988). The first is to supercool, using strategies to maintain body fluids in a liquid state when Tb falls below its equilibrium freezing point (FP). The second is freeze tolerance, using strategies to manage and endure the conversion of as much as 50%–70% of total body water into extracellular ice. Freeze-tolerant animals endure the penetration of ice into all extracellular spaces (e.g. abdominal cavity, bladder, brain ventricles, eye lens, plasma) and survive the interruption of all vital functions including breathing, circulation, and nerve and muscle activity, regaining these in a coordinated manner upon thawing. The equilibrium FP of the body fluids of terrestrial ectotherms is typically only about −0.5°C. Frogs can rarely supercool below −2°C because their highly water-permeable skin allows easy nucleation of body water due to contact with environmental ice or the action of ice-nucleating bacteria on skin or in gut. Hence, several species of frogs that hibernate on the forest floor have developed freeze tolerance. The best studied is the wood frog, *R. sylvatica* (Fig. 7.1), but other North America species that are freeze tolerant include grey tree frogs, *Hyla versicolor* and *H. chrysoscelis*, spring peepers *Pseudacris crucifer* and chorus frogs, *P. triseriata* (Storey and Storey, 1992). Two freeze-tolerant amphibians are also known from Eurasia: moor frogs, *R. arvalis* (Voituron *et al.*, 2009) and Siberian salamanders, *Salamandrella keyserlingi* (Berman *et al.*, 1984). Among reptiles, ecologically-relevant freeze tolerance has been reported for the European common lizard, *Lacerta vivipara* (Voituron *et al.*, 2002) and for several turtle species. Both oviparous and viviparous strains of *L. vivipara* occur; they have equivalent supercooling capacities but only the viviparous strain can endure prolonged freezing. This correlates well with the more northerly distribution of the viviparous form (up to 69° N), compared with 47° N for the oviparous form, and supports the theory that cold climatic conditions are a key selective force favouring viviparity in reptiles (Voituron *et al.*, 2004). Box turtles (*Terrapene carolina, T. ornata*), at ~500 g in mass, are both the largest known freeze-tolerant animals and the only adult turtles that display this adaptation (Storey and Storey, 1992). However, hatchlings of several northern turtle species that spend their first winter on land are freeze tolerant, including painted turtles (*Chrysemys picta*), Blanding's turtles (*Emydoidea blandingii*), diamondback terrapins (*Malaclemys terrapin*) and ornate box turtles (*T. ornata*) (Storey and Storey, 1992; Costanzo *et al.*, 2008). Hatchlings of some other species rely instead on a capacity to supercool and resist ice nucleation; the map turtle, *Graptemys geographica*, is a good example (Baker *et al.*, 2003). Interestingly, both cold hardiness strategies have been reported for painted turtle populations from different geographic locales; some populations of *C. picta* show well-developed freeze tolerance whereas others show extensive supercooling (to −10°C or lower) (Storey, 2006; Costanzo *et al.*, 2008).

Freeze-tolerant vertebrates tend to be those that spend the winter on land at or near the soil surface. As compared with aquatic hibernation, the benefits of terrestrial hibernation can include predator avoidance over the winter months (important for neonatal turtles) (Gibbons and Nelson, 1978), the ability to start breeding very early in the spring, and possibly also energy savings due to an overall lower metabolic rate than is possible in aquatic sites. However, the low temperature limit for most freeze-tolerant vertebrates is only about −6°C (and often higher), so survival also depends on substantial insulation from layers of organic litter and snow to defend against air temperatures above the snowpack that may fall to much lower values. These species may be particularly vulnerable to global warming because milder winters will reduce the snowpack, lowering its insulation value and possibly exposing animals to lethal subzero temperatures. A reduced snowpack also means reduced meltwater in the spring to form the ephemeral ponds used for breeding by terrestrially-hibernating frogs, as well as reduced time available for tadpole development before ponds dry out. This might be offset by the effects of higher temperature on the rate of larval development, but optimal development of cold-hardy species might actually be geared to a cool temperature range.

The development of freeze tolerance was probably aided by pre-existing physiological and biochemical capacities that are broadly present in reptiles and amphibians. The hypoxia/anoxia tolerance displayed by many species would aid resistance to ischaemia caused by plasma freezing, and the general tolerance of amphibians for wide variation in body water content would help them endure the cell dehydration and volume reduction that occurs when water exits to join extracellular ice crystals. Metabolic adaptations that generally promote anoxia or dehydration tolerance would have been brought into play during the development of freeze tolerance. For example, the accumulation of high concentrations of urea in blood and tissues is a well-known colligative defence against water loss in aestivating anurans (Storey, 2002) and also occurs in freeze-tolerant frogs (Costanzo and Lee, 2005). Hence, it is not surprising that rudimentary freezing survival has been reported in a number of species (other than those above); several species can survive if the freeze duration is short (several hours maximum), temperatures are mild (e.g. −1°C to −2°C), ice accumulation is low (usually <20% of total body water), and ice is restricted to peripheral tissues (skin and skeletal muscles) (Storey and Storey, 1992). Such short-term tolerance may be useful for animals experiencing an unexpected overnight frost, but specific biochemical adaptations are needed to push freezing survival to days or weeks, the ecologically relevant requirement to survive through the winter.

7.4 Principles of Biochemical Adaptation

Adaptation allows organisms to sustain homeostasis as much as possible, preserving the underlying web of biochemical reactions that constitutes life but, when necessary, making fundamental adjustments to selected systems to cope with new realities. All animal adaptation can ultimately be traced to the molecular level, involving mechanisms such as:

1. Changes to gene sequences that dictate the structures and properties of proteins;
2. The evolution of new gene products that allow quantum leaps to be made in species capability (e.g. development of antifreeze proteins);
3. Changes to regulatory mechanisms that adjust type, amount, timing and signal responsiveness of gene expression, protein translation and protein/enzyme function; and

4. Changes to multiple other features of the cellular environment within which the 'business' of life takes place (e.g. levels of low molecular weight metabolites and ions, modification of buffering capacity, membrane lipid composition).

Our exploration of the biochemical adaptations that underlie the acquisition of cold hardiness and freeze tolerance in reptiles and amphibians illustrates the types of molecular tools and mechanisms that animals can bring to bear in adjusting to environmental stress. The remainder of this chapter will focus on this repertoire of molecular mechanisms.

7.5 Enzymatic and Metabolic Adaptation

7.5.1 Low molecular weight protectants

A fundamental principle of survival in almost all freeze-tolerant vertebrates and invertebrates (except reptile species) is the synthesis of high concentrations of low molecular weight cryoprotectants that act colligatively to limit the loss of water from cells and also protect/stabilize macromolecules. Freeze-tolerant frogs typically use glucose or glycerol, whereas cold-hardy insects rely on polyhydric alcohols (glycerol, sorbitol, etc.) (Storey and Storey, 1988). For example, Fig. 7.1C shows the high level of glucose accumulated in wood frog organs during freezing; glucose synthesis from liver glycogen is triggered within 5 min of ice nucleation on the skin. The synthesis of high concentrations of small protectants (sugars, polyols, amino acids, urea, etc.) is a widespread animal adaptation to water/osmotic stress in response to many stresses (dehydration, hypersalinity, freezing, etc.) (Yancey, 2005) and is a defence strategy that would be quite simple to implement on an evolutionary scale because it typically involves common metabolites and regulatory controls that are accomplished by quantitative rather than qualitative changes. For example, freeze-intolerant leopard frogs (*R. pipiens*) that hibernate underwater show a hyperglycaemic response to dehydration (a 24-fold increase in glucose in liver); this could be modified quantitatively in freeze-tolerant species to produce the increases of up to ~300-fold that are seen in *R. sylvatica* liver during extracellular freezing (Storey, 1997). Indeed, we have shown that the cryoprotectant synthesis response to freezing by wood frog liver is stimulated just as strongly by dehydrating the animals at 5°C.

7.5.2 Fuels and end products

Quantitative and qualitative changes to body stores of fuel reserves are features that should also be rapidly adaptable in response to environmental stress or climate change. Anoxia-tolerant and cold-hardy animals accumulate higher reserves of glycogen to meet their need for carbohydrate fuels than do intolerant species, whereas aestivating species and mammalian hibernators store massive fat reserves to support aerobic metabolism over many months of dormancy. If climate change alters the percentage of the year that animals can be active and foraging, then adjustments will need to be made to ensure that sufficient body fuel reserves are accumulated (or food is cached) to sustain viability over the nonactive season. The end products of energy metabolism are also a concern, especially if species are forced to use anaerobic metabolism for extended times. Again, pre-existing solutions can be modified quantitatively. This seems to be the strategy of anoxia-tolerant turtles that winter underwater. All turtles can accumulate lactate during prolonged diving, but species such as *C. picta* or *T. scripta* that winter underwater for months have taken this to the extreme with two main modifications:

1. Massive amounts of lactate are moved into shell and bone for storage; and
2. Calcium carbonate is dissolved from shell and bone to buffer the acid load associated with lactic acid production (Warren and Jackson, 2007).

7.5.3 Enzyme and protein regulation

Regulation of enzymes and functional proteins is critical to adaption (Hochachka and Somero, 1984; Storey, 2004a). Multiple levels of control can be involved. Coarse controls on the amounts of enzymes/proteins or on the types of isoforms expressed are achieved by altering gene expression (see below) or protein degradation. Fine controls are achieved by multiple methods including changes in:

1. Levels of substrates, products and allosteric effectors;
2. Post-translational modification of proteins;
3. Protein–protein binding interactions;
4. Subcellular localization; and
5. Enhanced stability or longevity of proteins via the action of protein chaperones and low molecular weight stabilizers (Storey, 2004a).

Post-translational modifications of enzymes are particularly useful for making stable changes to the activities and properties of enzymes and functional proteins in response to environmental stress, and would certainly be involved (at least as an early response) in animal mechanisms of dealing with climate change. Multiple forms of covalent modification exist that can make stable modifications to enzyme function in response to signals and stresses; these include phosphorylation, acetylation, methylation, ubiquitinylation, SUMOylation, GlcNAcylation and many others.

As a mechanism of stress-responsive differential regulation of enzymes, reversible protein phosphorylation (RPP) via the actions of protein kinases and protein phosphatases, is by far the best known. RPP is a central mechanism by which animals adjust the activities of enzymes and functional proteins in response to environmental stresses including cold, freezing, oxygen limitation and desiccation (Storey 2004a). Perhaps the earliest identification of RPP in a stress response was as the mechanism of cold-activation of glycogen phosphorylase to trigger cryoprotectant synthesis in insects (Ziegler *et al.*, 1979) and later in wood frogs (Storey and Storey, 1988). A huge number of cellular proteins are targets of RPP and the mechanism is now known to be widely used for coordinating the action of multiple cell functions in response to signals and stresses. Indeed, RPP is the primary mechanism that accomplishes metabolic rate depression, reprioritizing and turning down/off energy expensive cell functions (e.g. ATP-driven ion pumps, transcription, translation) under stress conditions, as well as rationing fuel supplies (Storey and Storey, 1990, 2007).

A recent study of creatine kinase (CK) from skeletal muscle of wood frogs provides an instructive example of the methodologies that can be used to determine whether an enzyme undergoes stress-responsive RPP and to discover the consequences of phosphorylation for enzyme function (Dieni and Storey, 2009). CK catalyses a reversible reaction that controls phosphagen pools and their interconversion with ATP (creatine + ATP ↔ creatine-P + ADP) and is particularly important in muscle where creatine-P hydrolysis instantly supplements ATP levels when contraction is initiated. Evidence for covalent modification is usually derived first from identification of one or more stable significant changes in the properties of an enzyme between control and stressed states. In this case, CK isolated from muscle of frogs frozen for 24 h at −3°C showed significant differences in its maximal activity (35% higher) and affinity (K_m) for creatine (29% lower) compared with CK from 5°C-acclimated control frogs (Fig. 7.2). An increase in activity could derive from a greater amount of CK protein, but immunoblotting found no change in total CK protein between control and frozen states. Strong evidence of probable RPP can then be derived from monitoring elution profiles of an enzyme off an ion exchange column, because phosphorylation alters the net charge on a protein and thereby shifts its elution profile. Figure 7.2A shows that frog muscle CK eluted in two peaks from DEAE Sephadex® and the percentage of activity in each peak changed in response to freezing; in control muscle 60% of CK was in peak II compared with only 35% in frozen animals.

Even stronger confirmation of RPP as the mechanism of stress-responsive stable modification of the enzyme can come from the use of *in vitro* incubations that promote the action of protein kinases versus protein phosphatases. Incubation of control CK under conditions that stimulated protein kinases led to an increase in CK activity and a decrease in K_m creatine (Fig. 7.2) comparable to the changes in properties of CK from frozen muscle. By contrast, stimulation of phosphatase activities in extracts from frozen muscle reduced CK maximal activity and increased K_m creatine to values that mimicked the control situation. Put together, the data show that phosphorylation increases CK activity and substrate affinity, that freezing stimulates increased CK phosphorylation, and implicates peak I as the phosphorylated enzyme form (Fig. 7.2). Further confirmation of stress-responsive changes in phosphorylation state can come from two procedures:

1. Incubations under kinase-stimulating conditions in the presence of ^{32}P-ATP followed by immunoprecipitation of radio-labelled enzyme (Dieni and Storey, 2009); or
2. The use of ProQ® Diamond phosphoprotein stain to detect the relative amounts of phosphate bound to the two enzyme forms (Bell and Storey, 2010).

Incubation studies can also be used to identify the probable protein kinase(s) and phosphatase(s) that act on an enzyme *in vivo* by adjusting the specific components of the *in vitro* incubation mixture. From such studies, we implicated AMP-activated protein kinase (AMPK), calcium-calmodulin dependent protein kinase (CaMK) and protein phosphatases 2B and 2C in the control of frog muscle CK (Fig. 7.2). Interestingly, other studies found that freezing triggered a 4.5-fold increase in AMPK activity in wood frog muscle (Fig. 7.2) (Rider *et al.*, 2006), suggesting that this could be the protein kinase that is involved *in vivo* in CK control. AMPK has an important role as the 'energy sensor' of cells (responding to rising AMP whenever ATP is depleted) and so it makes sense that it would also regulate CK that buffers ATP levels by mobilizing phosphagen stores.

7.5.4 Changes in protein types and levels

Stress-induced changes in the types and amounts of proteins in cells and tissues typically derive from changes in gene expression, although other factors are involved (Storey and Storey, 2004b). In recent years the methods available to search for gene expression responses to environmental stress have expanded greatly, and include multiple forms of DNA array screening (both homologous and heterologous), protein screening by 2D gel electrophoresis and mass spectrometry, and screening methods targeted at subsets of proteins (e.g. transcription factors) (Eddy and Storey, 2008; Storey and Storey, 2010). These are being used effectively in many different ways to highlight the changes in gene and protein expression patterns that are triggered by environmental stress. For example, we have used these approaches to identify genes that are freeze responsive in wood frogs, and freeze/anoxia responsive in hatchling painted turtles (Storey, 2004b, 2006).

Multiple options are open for achieving adaptive responses by proteins to stresses including climate change. Again, quantitative change may be the easiest to initiate and all that is needed in some cases. For example, the amount of glycogen phosphorylase is 12-fold higher in hepatocytes of wood frogs compared with leopard frogs, reflecting the need for rapid cryoprotectant glucose synthesis by the freeze-tolerant species (Mommsen and Storey, 1992). Enzymes involved in urea synthesis are also quickly up-regulated when anurans face desiccating or hypersaline conditions (Balinsky, 1981).

Generalized protein responses to stress are also used to preserve/protect existing cellular macromolecules by synthesizing specialized protective agents, both metabolites (discussed earlier) and proteins. The best known of the latter are the so-called heat-shock proteins (HSP) that act as molecular chaperones to refold proteins that are unfolded or malfolded under stress conditions. Although ubiquitous responses to heat stress in animals, they also respond to other stresses including hypoxia, osmotic stress, UV irradiation and heavy metals.

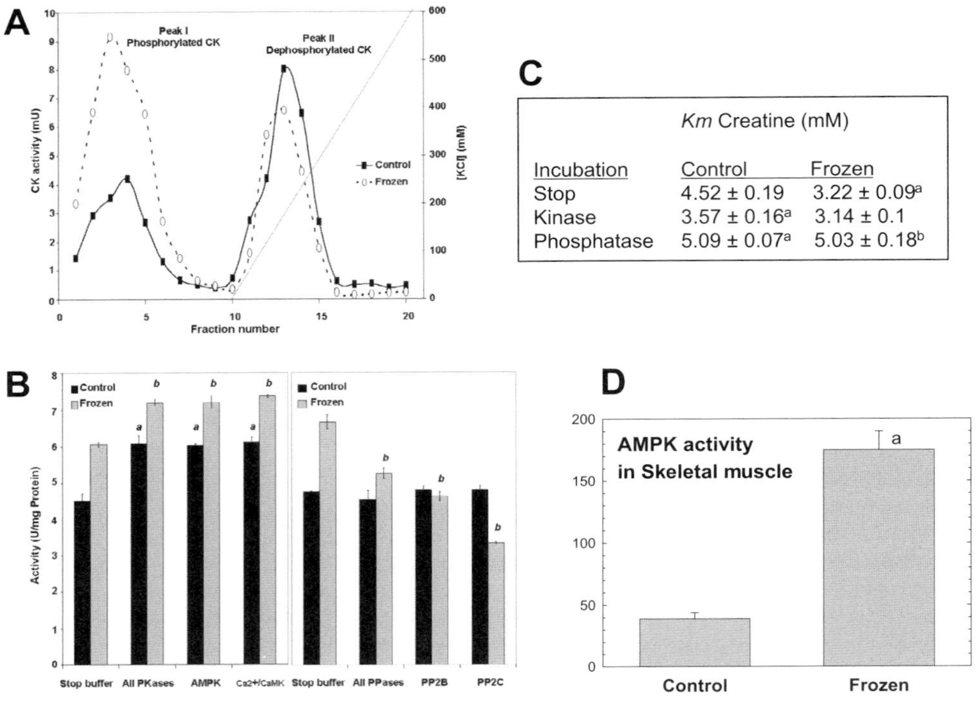

Incubation	*Km* Creatine (mM) Control	Frozen
Stop	4.52 ± 0.19	3.22 ± 0.09[a]
Kinase	3.57 ± 0.16[a]	3.14 ± 0.1
Phosphatase	5.09 ± 0.07[a]	5.03 ± 0.18[b]

Fig. 7.2. Regulation of skeletal muscle creatine kinase (CK) from freeze tolerant wood frogs by reversible protein phosphorylation. (A) Elution profiles of muscle CK from control (5°C acclimated) versus frozen (24 h at –3°C) frogs on DEAE Sephadex®. (B) Effect of *in vitro* incubations promoting protein kinase (total kinases, AMPK, CaMK) or protein phosphatase (total phosphatases, PP2B, PP2C) action on CK maximal activity as compared with the situation in STOP buffer where inhibitors of both kinases and phosphatases are present to preserve the phosphorylation state in the tissue. (C) Effect of *in vitro* incubations promoting protein kinases or phosphatases on the K_m creatine of muscle CK. (D) AMPK activity in skeletal muscle of frozen frogs. [a] Significantly different from the corresponding control (or [b] frozen) condition in STOP buffer (no additions), $P < 0.05$. Compiled from Dieni and Storey (2009) and Rider *et al.* (2006).

For example, anoxia exposure triggered coordinated increases (1.8–2.9 fold) in the levels of Hsp25, Hsp40, Hsp70, Hsp90 and Hsc70 (the constitutive chaperone) in skeletal muscle of turtles (*T. s. elegans*) (Krivoruchko and Storey, 2010a). Antioxidant defences are another protective strategy for dealing with environmental stress. Their importance is emphasized in two ways. First, stress-tolerant species typically have higher constitutive activities of antioxidant enzymes than do intolerant species. For example, activities of most antioxidant enzymes in liver of freeze tolerant wood frogs are two- to threefold higher than in leopard frogs (and are even higher in liver of anoxia-tolerant *T. s. elegans*) (Hermes-Lima *et al.*, 2001). Second, tolerant species can show rapid stress-mediated enhancement of antioxidant enzyme expression that typically occurs under the stress condition (while anoxic, frozen, etc.) to prepare animals for a burst of reactive oxygen species formation that accompanies the reintroduction of oxygen when animals exit the stress situation (Hermes-Lima *et al.*, 2001). Indeed, appropriate responses by these and other defensive mechanisms form a generalized cellular stress proteome that is found across phylogeny (Kültz, 2005) and gives all organisms basal mechanisms to help preserve viability in the face of climate change or other environmental stresses.

A final mode of protein adaptation that is available to animals to deal with persistent stress is the elaboration of completely new protein types. Because significant evolutionary time is needed to create these, this is not a strategy for dealing with rapid climate change but is a longer term adjustment that can re-optimize species in an altered environment. For example, wood frogs show strong freeze-stimulated expression of three novel genes/proteins that are not found in other animals, even closely related frogs, although their functions are not yet known (Storey, 2004b). The antifreeze proteins (AFP) of marine fish are excellent examples of some of the options for creating new protein types (Fletcher *et al.*, 2001). In evolutionary terms AFPs are quite young, the need for them having arisen only 10–14 million years ago when sea level glaciation recurred. Several types evolved rapidly, each with a different story. For example, Type II AFP have been shown to be derived from the carbohydrate recognition domains of C-type lectins, proteins that play a role in immunity by recognizing surface carbohydrates on pathogens. By contrast, antifreeze glycoproteins in Antarctic fish are composed of a repeating glycotripeptide that turned out to be derived from a gene segment crossing the first exon–intron boundary of the trypsinogen gene. Trypsinogen is a protein that is normally secreted into the gut lumen, one of the main sites where antifreeze action is needed because polar fish often consume ice crystals along with their food. Probably the antifreeze protein evolved by exploiting a minor pre-existing antifreeze action of the ancestral protein. The lesson here is that environmental stress can place strong pressures on species to quickly come up with solutions that allow them to readjust and preserve life when their environment changes.

7.6 Regulation of Gene Expression

Genes define the total biological repertoire that any organism can bring to bear in its effort to survive and adapt to environmental stress. A variety of mechanisms contribute to the regulation of gene expression, but we will focus here on selected categories that illustrate important concepts and mechanisms that are applicable to studying gene expression responses to environmental stress, including during climate change.

7.6.1 Epigenetic regulation of DNA transcription

Epigenetics is the study of heritable changes made via mechanisms other than changes in the underlying DNA sequence of genes. Primary mechanisms include post-translational modifications of histones (e.g. methylation, acetylation, phosphorylation and others) and methylation of cytosine residues in DNA itself (Zhang, 2008). Epigenetic mechanisms are best known in the regulation of cell differentiation during development and mostly affect somatic cells over the course of an individual's lifetime. For example, these are the reason why a differentiated liver cell only produces more liver cells when it divides, despite having the full DNA complement needed to make any cell type. However, epigenetic modifications that occur in sperm or egg can potentially be carried forward to the next generation, providing a means for environmental stresses that alter the DNA or chromatin structure of the parent generation to influence their offspring. Importantly, epigenetic variation may be generated at a much higher rate than equivalent variation in DNA nucleotide sequences, especially under rapidly changing environmental conditions when organisms are under pressure to produce alternative phenotypes (Angers *et al.*, 2010). Accumulating evidence indicates that the detection of epigenetic modifications of gene expression could be a significant and exciting new factor in our understanding of adaptation and evolution (Turner 2009); indeed, over 100 cases of inherited epigenetic variations in bacteria, protists, fungi, plants and animals are now known (Jablonka and Raz, 2009). A new study links epigenetic modification with climate change. Paun *et al.* (2010) report that three species of allotetraploid orchids that arose during or since the last glacial maximum show species-specific epigenetic patterns have had a direct impact on the ecology, distribution and evolution of these lineages. Epiloci were pinpointed that correlated with environmental variables (water availability, temperature) suggesting that stable epigenetic divergence led to persistent ecological differences, and set the stage for species-specific genetic patterns to accumulate in response to further selection and/or drift.

Recent studies in our laboratory have shown that epigenetic mechanisms also have a natural role to play in animal transitions into environmental stress-induced states of hypometabolism, and this suggests to us the possibility that epigenetic controls will be identified more and more often as significant regulators of animal responses to both short and prolonged environmental change. Our initial work focused on chromatin modification as a means of global inhibition of transcription during winter hibernation in ground squirrels (Morin and Storey, 2006). Histone proteins alter chromatin structure and gate access to DNA by the transcriptional machinery. Increased methylation of lysine residues on histones leads to a more closed chromatin structure whereas acetylation and phosphorylation open up the DNA–protein structure to allow the transcriptional machinery to bind (Jenuwein and Allis, 2001). Analysis of ground squirrel skeletal muscle during torpor showed significant histone modifications that would repress transcription during torpor by creating a more closed chromatin structure. Specifically, acetylated histone H3 (Lys 23) and phosphorylated histone H3 (Ser 10) contents during torpor were reduced by 25% and 40%, respectively, and this correlated with increased protein levels and activity of histone deacetylase (HDAC) (Morin and Storey, 2006). The same mechanism of transcriptional silencing was identified during anoxia-induced metabolic rate depression in turtles (*T. s. elegans*). Transcript and protein levels of five HDACs increased by 1.3–4.6- and 1.7–3.5-fold, respectively, in turtle skeletal muscle during 20 h anoxic submergence. Total HDAC activity also rose by 1.5-fold and levels of acetylated histone H3 decreased by 40%–60% (Krivoruchko and Storey, 2010b). Epigenetic regulation also contributes to chromatin remodelling for long-term gene silencing during aestivation of green-striped burrowing frogs, *Cyclorana alboguttata* (Hudson *et al.*, 2008). A comparison of mRNA abundance in cruralis muscle of control versus 6-month aestivated frogs found significantly increased transcript levels of transcriptional co-repressor SIN3A and DNA cytosine-5-methyltransferase 1 (by 1.7- and 3.5-fold, respectively), two genes whose protein products have established roles in gene silencing.

7.6.2 MicroRNA control of mRNA translation

Our understanding of the control of gene expression in cells has changed greatly in recent years with the discovery of the regulatory roles played by various classes of small RNA. In particular, microRNAs are proving to have major involvement in regulating mRNA transcripts outside the nucleus. MiRNA are small, non-coding transcripts (19–25 nucleotides long) that bind to target mRNA to regulate translation in normal, stressed and disease states (Filipowicz *et al.*, 2008; Bartel, 2009). A perfect sequence match between the miRNA and its target tends to direct mRNA into degradation pathways, whereas an imperfect match results in translational inhibition via storage in cytoplasmic P-bodies. In recent studies with both hibernating mammals and freeze-tolerant frogs, we have found that levels of key miRNA species change significantly when animals enter hypometabolism, suggesting that they have key roles to play in sequestering and preserving mRNA transcripts during stress-induced hypometabolism. In 13-lined ground squirrels, several microRNAs were differentially expressed in kidney, skeletal muscle and heart when animals entered torpor (Morin *et al.*, 2008). Furthermore, the amount of Dicer, one of two main enzymes involved in microRNA processing, increased during torpor. Significant changes in miR-16 and miR-21 levels in liver and skeletal muscle also occurred in response to freezing in wood frogs (Fig. 7.3) (Biggar *et al.*, 2009). Taken together, the two studies suggest that miRNA regulation of mRNA transcription will prove to be a principle of hypometabolism, with two consequences:

1. Contributing to global translational suppression by sequestering transcripts away from ribosomes; and
2. Preserving and storing transcripts over prolonged periods of hypometabolism so that mRNA is immediately available again for translation when animals arise from dormant states.

Furthermore, actions of specific miRNA types may be key to the inhibitory control of selected cell functions during hypometabolism. For example, the cell cycle is highly energy expensive and a target for strong suppression when organisms enter hypometabolism (Biggar and Storey, 2009). Several miRNA species are known to regulate cell cycle genes, for example miR-15a and miR-16 target the mRNA transcripts of proteins associated with the first gap phase including cyclin D1, cyclin E, cdc25a, checkpoint kinase 1 and E2F (Kaddar *et al.*, 2009). Indeed, the power of miRNA regulation was shown in studies where cells were transfected to express high levels of miR-16; this resulted in elevated numbers of quiescent cells and reduced numbers of cells in S, G2 and M phases (Linsley *et al.*, 2007). The results for wood frogs (elevated levels of miR-16 in liver, a proliferating tissue) suggest a natural role for miR-16 in cell cycle arrest in freeze-induced hypometabolism and, together with the results from mammalian hibernators, indicate that miRNA will prove to be a whole new level of regulatory control in animal response to environmental stress. Indeed, changes in miRNA patterns may potentially be developed into biomarkers of cells under stress that could be used to indicate individuals and populations that are under pressure from altered environments such as those due to climate change.

7.6.3 Tracing signal transduction pathways

The components and typical targets of multiple signal transduction pathways that link sensing of environmental change (generally by cell surface receptors) to downstream actions (e.g. changes in enzyme function, ribosomal translation, gene expression) are now well known, particularly in mammalian systems. These signalling pathways are highly conserved across the animal kingdom and detection of their activation can be another good diagnostic of an animal system under stress. Indeed, several examples of the activation of mitogen-activated protein kinase (MAPK) signalling cascades in response to anoxia, freezing, or osmotic stresses in amphibians and reptiles are now known (Cowan and Storey, 2003). Therefore, one way to initiate a search for the effects of environmental change on organisms is to evaluate one or more key components of signalling cascades. When positive responses to stress are found, a follow-up with analyses of both upstream and downstream members of the cascade, as well as assessment of known target proteins/enzymes, can lead to identification of the major metabolic actions that are triggered by the environmental signal. In a recent example of this approach, we analysed the responses of the extracellular signal-regulated kinase (ERK) signalling cascade (one member of the MAPK superfamily) to whole-body dehydration by African clawed frogs, *Xenopus laevis*. In their native environment, these frogs can experience seasonally arid conditions that require metabolic responses to ameliorate the effects of dehydration. Figure 7.4 shows the coordinated response by multiple pathway components that occurred in lung when frogs were challenged with medium and high levels of dehydration (Malik and Storey, 2009). All three tiers in the MAPK cascade responded positively to dehydration, with the amounts of phosphorylated active kinases increasing

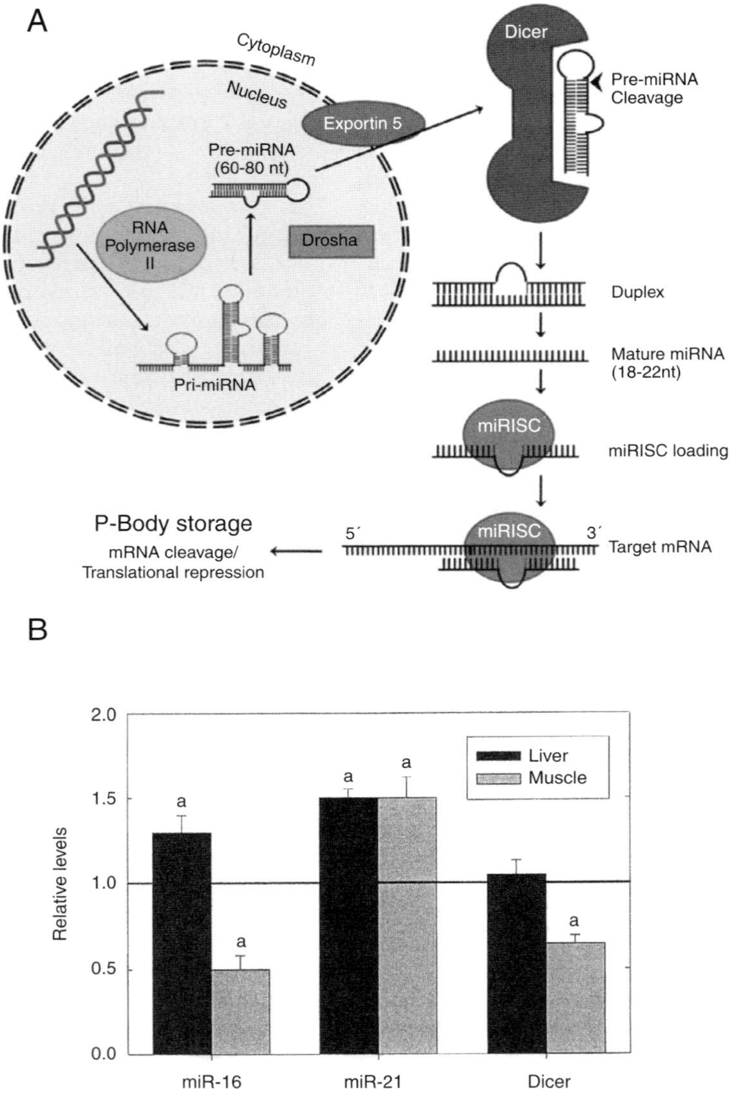

Fig. 7.3. (A) Synthesis of microRNA. Primary transcripts are transcribed by RNA polymerase II and processed by riboendonucleases (Drosha, Dicer) into single-stranded mature microRNAs. These then join microRNA-induced silencing complex (miRISC) and bind to mRNA transcripts at their 3′- UTR to repress translation. (B) Relative levels of miR-16, miR-21 and Dicer protein in liver and skeletal muscle of frozen wood frogs (24 h at –3°C) compared with 5°C acclimated controls (set to 1). Data are means ± SEM, n = 4. [a] – Significantly different from the control value, $P < 0.05$. Compiled from Biggar and Storey (2009) and Biggar *et al.* (2009).

for the initiating MAPK kinase kinases (c-Raf, MEKK), the MAPK kinase (MEK1/2), and finally the MAPK (ERK1/2). Two downstream targets of ERK2 also showed robust increases in amount of active phosphorylated protein: the p90 ribosomal S6 kinase (RSK^{Ser380}) and the transcription factor $STAT3^{Ser727}$ (not shown). RSK activation also led to strong phosphorylation of its target, the S6 ribosomal protein.

Furthermore, we found a strong conserved activation of the ERK cascade in most *X. laevis* tissues in response to dehydration (Malik and Storey, 2009). The ERK cascade was also powerfully activated by stress stimuli in heart of *Rana ridibunda* (Gaitanaki *et al.*, 2003). In *X. laevis*, activation of ERK signalling during dehydration was particularly prominent in lung, an organ that is highly susceptible to respiratory water loss. Although not yet experimentally tested, it is interesting to note that the expression of mucin genes in the mammalian respiratory tract is under ERK1/2 control (Choi *et al.*, 2009). This might suggest that ERK activation in lung of dehydrated *X. laevis* might mediate changes in the amount or composition of mucins in airway epithelia to contribute to limiting respiratory water loss.

7.6.4 Transcription factors and identification of stress-responsive genes

Transcription factors bind to DNA at specific sites in the promoter region of genes, and activate gene transcription. Typically each individual transcription factor regulates the expression of a select group of genes whose protein products are involved in a specific cell function. Hence, identification of the particular transcription factors that respond to a stress provide another good way of determining which cellular functions are important in the adaptive response to stress. The recent development of array screening technologies that can measure levels of activated transcription factors in cell nuclei has provided a powerful way to screen for those factors that are more active under stress conditions and, thereby, identify 'cassettes' of genes that are putatively up-regulated in a coordinated way in response to the imposed stress (Storey, 2008). Figure 7.5 shows the application of this method to analysing gene expression responses by turtle liver to 5 h anoxia exposure (Krivoruchko and Storey, 2010c). We chose to evaluate the possible role of the nuclear factor κB (NF-κB) transcription factor in anoxia-responsive gene expression due to its known link with antioxidant defence. NF-κB is a heterodimer (p50 and p65 subunits) that is maintained in an inactive state in the cytoplasm by binding to an inhibitory protein, IκB. Dissociation of IκB (after it is phosphorylated by an IκB kinase) allows the NF-κB dimer to migrate to the nucleus where it binds to specific sites in the promoter region of genes and activates their transcription. The data in Fig. 7.5 show five different ways in which an activation of NF-κB transcriptional activity can be evaluated:

1. The amount of phosphorylated IκB (rose twofold in anoxia);
2. mRNA transcript levels of p50 and p65 (increased 2.2–2.9-fold);
3. Levels of p50 and p65 protein (increased 1.4–2.3-fold);
4. Amount of p50 and p65 protein in the nucleus (increased 1.7–7.3-fold); and
5. p50/p65 binding to DNA (increased 1.9-fold).

Furthermore, all five of these agree with the conclusion that NF-κB is activated by anoxia in turtle liver. In practice, changes in the amount of phosphorylated IκB or in NF-κB binding to DNA are often the easiest to measure experimentally. This clear evidence of NF-κB activation then provides justification for analysing the anoxia-responsiveness of various genes under NF-κB control to identify the key targets that help to provide protection from stress. Of particular interest, NF-κB controls selected antioxidant proteins and the three that we tested were significantly elevated under anoxia: ferritin (an iron-binding protein) and both the Cu/Zn (cytoplasmic) and Mn (mitochondrial) forms of superoxide dismutase. This again highlights the importance of antioxidant defences for long-term viability under stress conditions. Macromolecules need to be protected under anoxia because the capacity to replace damaged proteins using ATP-expensive biosynthesis is strongly suppressed in the hypometabolic state. In addition, an expanded search of NF-κB targets known from other systems identified elevated levels of two anti-apoptotic proteins (Bcl-2 and

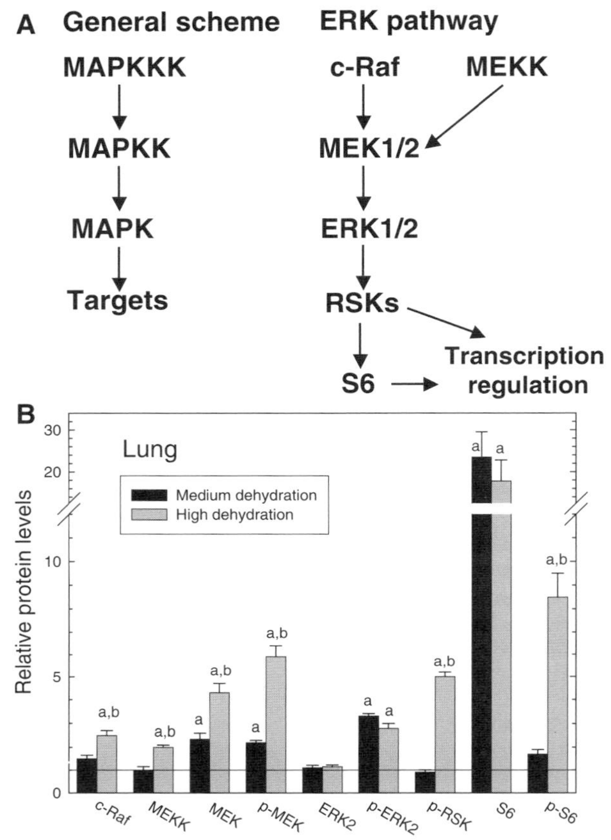

Fig. 7.4. Response of the ERK signalling pathway to whole-body dehydration in lung of African clawed frogs, *Xenopus laevis*. (A) Schematic showing the ERK pathway members analysed and their positions in the 3-tier signalling cascade that typifies the MAPK family. (B) Relative changes in protein or phosphoprotein (p-) levels, compared with controls (set to 1), in response to medium or high dehydration (16.6 ± 1.59% or 28.0 ± 1.6% of total body water lost, respectively) of frogs. Immunoblotting was used to analyse MAPK family members including total protein levels of MEKK, MEK1/2 and ERK2 and phosphorylated active forms: p-cRaf^{ser338}, p-MEK1/2$^{ser217/221}$ and p-ERK$^{thr202/tyr204}$. Downstream target proteins phosphorylated by ERK2 included p-p90 ribosomal S6 kinase (RSK)^{ser380}, total S6 ribosomal protein and p-S6$^{ser235/236}$. Data are means ± SEM, n=3-6 independent trials. [a] – Significantly different from the control value, or [b] – from the medium dehydration value, P < 0.05. Compiled from Malik and Storey (2009).

Bcl-xL) in liver of anoxic turtles. This is an important finding as it represents some of the very first evidence that inhibition of apoptosis is important for natural anoxia tolerance. Damage to cells by oxygen lack frequently triggers cell death but this would clearly be detrimental for species that can endure cycles of anoxia naturally. The increase in anti-apoptotic markers indicates that animals that are naturally stress tolerant can block apoptosis while they adjust or adapt to the stress condition (Krivoruchko and Storey, 2010d).

7.7 Concluding Remarks

Since the time when ectothermic vertebrates first appeared on land, they have been required to adapt to changes in a wide range

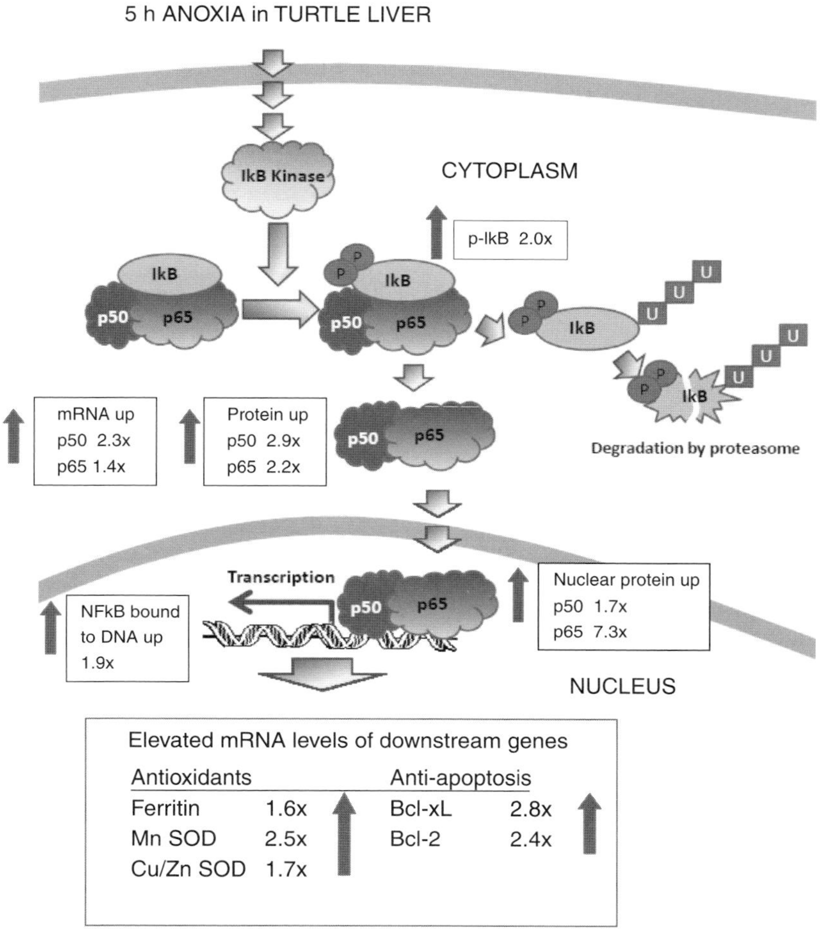

Fig. 7.5. Activation of the NF-κB signal transduction pathway in response to 5 h of anoxia exposure in liver of turtles, *T. s. elegans*. Under normal conditions, the NF-κB heterodimer (p50 and p65 proteins) is retained in the cytoplasm bound with its inhibitor protein, IκB. In response to a stimulus (e.g. anoxia), IκB kinase is activated and phosphorylates IκB which then dissociates and is ubiquitinated and degraded. NF-κB is then free to move to the nucleus and activate transcription of various target genes. Compiled from Krivoruchko and Storey (2010c,d).

of environmental parameters. These include many episodes of climate change during which some species have prospered whereas others have faced extirpation or extinction. Our present-day concern is with the rapid pace of climate change and with the fact that human activities are a major causative factor in the current episode of climate change, affecting not just ourselves but all organisms on earth. Ectothermic vertebrates will be affected by climate change in ways that include direct effects of heat/cold and changing patterns of water availability on survival and reproductive capacity, as well as secondary effects on the availability of food and shelter and competition with other species. All organisms have a variety of biochemical strategies that they can draw on to accomplish adaptation to new environmental realities, although some species (such as those that are used to highly stable or highly specialized environments) are inherently poorer at this than others. Cold-hardy reptiles and amphibians living at high latitudes already deal with wide variation in seasonal parameters and, overall, may be better able to implement biochemical adaptations to deal with a changing climate than comparable tropical species. The molecular mechanisms available for adaptation to environmental stress are many; here we have discussed examples such as proliferation of protectants (low molecular weight metabolites, chaperone proteins); enhanced antioxidant defences; quantitative changes in gene and protein expression; differential regulation of enzymes, transcription factors and signalling pathways; epigenetic and post-transcriptional control of genes; and the synthesis of novel proteins. A number of these also have the potential to be effective markers of organisms under stress and we have outlined some of the procedures that can be utilized to search for and identify stress-responsive biochemical adaptations. The potential for human intervention to directly enhance/alter the adaptive strategies of reptiles and amphibians in nature is low, but a thorough understanding of how animals adapt to stress gives us the tools to understand what the key stressors are (and perhaps how we can ameliorate them), what the critical adaptations are, how to search for and evaluate molecular adaptations, and what biochemical parameters are most effective markers of stress.

7.8 Acknowledgements

We thank the many members of our laboratory who have contributed to unravelling the mysteries of amphibian and reptile biochemical adaptation to environmental stress, including those whose work is featured here: C.A. Dieni, A.I. Malik, A. Krivoruchko and K.K. Biggar. Research in the Storey laboratory is supported by NSERC Canada and a Canada Research Chair to KBS.

References

Angers, B., Castonguay, E. and Massicotte R. (2010) Environmentally induced phenotypes and DNA methylation: how to deal with unpredictable conditions until the next generation and after. *Molecular Ecology* 19, 1283–1295.

Aubret, F. and Shine, R. (2010) Thermal plasticity in young snakes: how will climate change affect the thermoregulatory tactics of ectotherms? *Journal of Experimental Biology* 213, 242–248.

Baker, P.J., Costanzo, J.P., Iverson, J.B. and Lee, R.E. (2003) Adaptations to terrestrial overwintering of hatchling northern map turtles, *Graptemys geographica*. *Journal of Comparative Physiology B* 173, 643–651.

Balinsky, J.B. (1981) Adaptation of nitrogen metabolism to hyperosmotic environment in *Amphibia*. *Journal of Experimental Zoology* 215, 335–350.

Bartel, D. (2009) MicroRNAs: target recognition and regulatory functions. *Cell* 136, 215–233.

Bell, R.A.V. and Storey, K.B. (2010) Phosphorylation of liver glutamate dehydrogenase: role in mammalian hibernation. *Comparative Biochemistry and Physiology B* 157, 310–316.

Berman, D.I., Leirikh, A.N. and Mikhailova, E.I. (1984) Winter hibernation of the Siberian salamander *Hynobius keyserlingi*. *Journal of Evolutionary Biochemistry and Physiology* 3, 323–327 (Russian with English summary).

Bickler, P.E. and Buck, L.T. (2007) Hypoxia tolerance in reptiles, amphibians, and fishes: life with variable oxygen availability. *Annual Review of Physiology* 69, 145–170.

Biggar, K.K. and Storey, K.B. (2009) Perspectives in cell cycle regulation: Lessons from an anoxic vertebrate. *Current Genomics* 10, 573–584.

Biggar, K., Dubuc, A. and Storey, K.B. (2009) MicroRNA regulation below zero: Differential expression of miRNA-21 and miRNA-16 during freezing in wood frogs. *Cryobiology* 59, 317–321.

Bosch, J., Carrascal, L.M., Duran, L., Walker, S. and Fisher, M.C. (2007) Climate change and outbreaks of amphibian chytridiomycosis in a montane area of Central Spain; is there a link? *Proceedings of the Royal Society B: Biological Sciences* 274, 253–260.

Choi, H.J., Chung, Y.S., Kim, H.J., Moon, U.Y., Choi, Y.H., Van Seuningen, I., Baek, S.J., *et al.* (2009) Signal pathway of 17beta-estradiol-induced MUC5B expression in human airway epithelial cells. *American Journal of Respiratory Cell and Molecular Biology* 40, 168–178.

Costanzo, J.P. and Lee, R.E. (2005) Cryoprotection by urea in a terrestrially hibernating frog. *Journal of Experimental Biology* 208, 4079–4089.
Costanzo, J.P., Lee, R.E. and Ultsch, G.R. (2008) Physiological ecology of overwintering in hatchling turtles. *Journal of Experimental Zoology A* 309, 297–379.
Cowan, K.J. and Storey, K.B. (2003) Mitogen-activated protein kinases: new signaling pathways functioning in cellular responses to environmental stress. *Journal of Experimental Biology* 206, 1107–1115.
Dieni, C.A. and Storey, K.B. (2009) Creatine kinase regulation by reversible phosphorylation in frog muscle. *Comparative Biochemistry and Physiology B* 152, 405–412.
Dynesius, M. and Jansson, R. (2000) Evolutionary consequences of changes in species' geographical distributions driven by Milankovitch climate oscillations. *Proceedings of the National Academy of Sciences USA* 97, 9115–9120.
Eddy, S.F. and Storey, K.B. (2008) Comparative molecular physiological genomics: heterologous probing of cDNA arrays. *Methods in Molecular Biology* 410, 81–110.
Filipowicz, W., Bhattacharyya, S. and Sonenberg, N. (2008) Mechanisms of post-translational regulation by microRNAs: are the answers in sight? *Nature Reviews Genetics* 9, 102–114.
Fletcher, G.L., Hew, C.L. and Davies, P.L. (2001) Antifreeze proteins of teleost fish. *Annual Review of Physiology* 63, 359–390.
Gaitanaki, C. Konstantina, S., Chrysa, S. and Beis, I. (2003) Oxidative stress stimulates multiple MAPK signalling pathways and phosphorylation of the small HSP27 in the perfused amphibian heart. *Journal of Experimental Biology* 206, 2759–2769.
Gibbons, J.W. and Nelson, D.H. (1978) The evolutionary significance of delayed emergence from the nest by hatchling turtles. *Evolution* 32, 297–303.
Gregory, P.T. and Stewart, K.W. (1975) Long-distance dispersal and feeding strategy of the red-sided garter snake (*Thamnophis sirtalis parietalis*) in the Interlake of Manitoba. *Canadian Journal of Zoology* 53, 238–245.
Hayes, T.B., Falso, P., Gallipeau, S. and Stice, M. (2010) The cause of global amphibian declines: a developmental endocrinologist's perspective. *Journal of Experimental Biology* 213, 921–933.
Herbert, C.V. and Jackson, D.C. (1985) Temperature effects on the response to prolonged submergence in the turtle *Chrysemys picta bellii*. II. Metabolic rate, blood acid-base and ionic changes, and cardiovascular function in aerated and anoxic water. *Physiological Zoology* 58, 670–681.
Hermes-Lima, M., Storey, J.M. and Storey, K.B. (2001) Antioxidant defenses and animal adaptation to oxygen availability during environmental stress. In: Storey, K.B. and Storey, J.M. (eds) *Cell and Molecular Responses to Stress*. Vol. 2, Elsevier, Amsterdam, pp. 263–287.
Hewitt, G.M. (1996) Some genetic consequences of ice ages, and their role in divergence and speciation. *Biological Journal of the Linnean Society* 58, 247–276.
Hillman, S.S., Withers, P.C., Drewes, R.C. and Hillyard, S.D. (2009) *Ecological and Environmental Physiology of Amphibians*. Oxford University Press, Oxford.
Hochachka, P.W. and Somero, G.N. (1984) *Biochemical Adaptation.* Princeton University Press, Princeton, New Jersey.
Hudson, N.J., Lonhienne, T.G., Franklin, C.E., Harper, G.S. and Lehnert, S.A. (2008) Epigenetic silencers are enriched in dormant desert frog muscle. *Journal of Comparative Physiology B* 178, 729–734.
Hulin, V., Delmas, V., Girondot, M., Godfrey, M.H. and Guillon, J.-M. (2009) Temperature-dependent sex determination and global change: are some species at greater risk? *Oecologia* 160, 493–506.
IUCN (2011) The IUCN Red List of threatened species. www.iucnredlist.org/apps/redlist/details/58728/0, accessed 13 June 2011.
Jablonka, E. and Raz, G. (2009) Transgenerational epigenetic inheritance: prevalence, mechanisms, and implications for the study of heredity and evolution. *Quarterly Review of Biology* 84, 131–176.
Jackson, D.C. and Ultsch, G.R. (2010) Physiology of hibernation under the ice by turtles and frogs. *Journal of Experimental Zoology A* 313, 311–327.
Jenuwein, T. and Allis, C.D. (2001) Translating the histone code. *Science* 293, 1074–1080.
Kaddar, T., Rouault, J.P., Chien, W.W., Chebel, A., Gadoux, M., Salles, G., French, M., *et al.* (2009) Two new miR-16 targets: caprin-1 and HMGA1, proteins implicated in cell proliferation. *Biology of the Cell* 101, 511–524.
Kearney, M., Shine, R. and Porter, W.P. (2009) The potential for behavioral thermoregulation to buffer 'cold-blooded' animals against climate warming. *Proceedings of the National Academy of Sciences USA* 106, 3835–3840.
Krivoruchko, A. and Storey, K.B. (2010a) Regulation of the heat shock response under anoxia in the turtle, *Trachemys scripta elegans*. *Journal of Comparative Physiology B* 180, 403–414.

Krivoruchko, A. and Storey, K.B. (2010b) Epigenetics in anoxia tolerance: a role for histone deacetylases. *Molecular and Cellular Biochemistry* 342, 151–161.

Krivoruchko, A. and Storey, K.B. (2010c) Molecular mechanisms of turtle anoxia tolerance: A role for NF-κB. *Gene* 450, 63–69.

Krivoruchko, A. and Storey, K.B. (2010d) Forever young: mechanisms of anoxia tolerance in turtles and possible links to longevity. *Oxidative Medicine and Cellular Longevity* 3, 186–198.

Kültz, D. (2005) Molecular and evolutionary basis of the cellular stress response. *Annual Review of Physiology* 67, 225–257.

Lance, V.A. (2009) Is regulation of aromatase expression in reptiles the key to understanding temperature-dependent sex determination? *Journal of Experimental Zoology A* 311, 314–322.

Lee-Yaw, J.A., Irwin, J.T. and Green, D.M. (2008) Postglacial range expansion from northern refugia by the wood frog, *Rana sylvatica. Molecular Ecology* 17, 867–884.

Linsley, P., Schelter, J., Burchard, J., Kibukawa, M., Martin, M., Bartz, S., Johnson, J., *et al.* (2007) Transcripts targeted by the microRNA–16 family cooperatively regulate cell cycle progression. *Molecular and Cellular Biology* 27, 2240–2252.

Lutz, P.L. and Milton, S.L. (2004) Negotiating brain anoxia survival in the turtle. *Journal of Experimental Biology* 207, 3141–3147.

Malik, A.I. and Storey, K.B. (2009) Activation of extracellular signal-regulated kinases during dehydration in the African clawed frog, *Xenopus laevis. Journal of Experimental Biology* 212, 2595–2603.

Margesin, R., Neuner, G. and Storey, K.B. (2007) Cold-loving microbes, plants and animals – fundamental and applied aspects. *Naturwissenschaften* 94, 77–99.

McCaffery, R.M. and Maxell, B.A. (2010) Decreased winter severity increases viability of a montane frog population. *Proceedings of the National Academy of Sciences USA* 107, 8644–8649.

McCallum, M.L., McCallum, J.L. and Trauth, S.E. (2009) Predicted climate change may spark box turtle declines. *Amphibia-Reptilia* 30, 259–264.

McMenamin, S.K. and Hadly, E.A. (2010) Developmental dynamics of *Ambystoma tigrinum* in a changing landscape. *BMC Ecology* 10, 10.

Mitchell, N.J. and Janzen, F.J. (2010) Temperature-dependent sex determination and contemporary climate change. *Sex and Development* 4, 129–140.

Mitchell, N.J., Kearney, M.R., Nelson, N.J. and Porter, W.P. (2008) Predicting the fate of a living fossil: how will global warming affect sex determination and hatching phenology in tuatara? *Proceedings of the Royal Society B: Biological Sciences* 275, 2185–2193.

Mommsen, T.P. and Storey, K.B. (1992) Hormonal effects on glycogen metabolism in isolated hepatocytes of a freeze-tolerant frog. *General and Comparative Endocrinology* 87, 44–53.

Morin, P. and Storey, K.B. (2006) Evidence for a reduced transcriptional state during hibernation in ground squirrels. *Cryobiology* 53, 310–318.

Morin, P., Dubuc, A. and Storey, K.B. (2008) Differential expression of microRNA species in organs of hibernating ground squirrels: a role in translational suppression during torpor. *Biochimica et Biophysica Acta* 1779, 628–633.

Paun, O., Bateman, R.M., Fay, M.F., Hedren, M., Civeyrel, L. and Chase, M.W. (2010) Stable epigenetic effects impact adaptation in allopolyploid orchids (Dactylorhiza: Orchidaceae). *Molecular Biology and Evolution* 27, 2465-2473, doi:10.1093/molbev/msq150.

Pidwirny, M. (2006) *Fundamentals of Physical Geography*. 2nd ed., www.physicalgeography.net, accessed 13 June 2011.

Rider, M.H., Hussain, N., Horman, S., Dilworth, S.M. and Storey, K.B. (2006) Stress-induced activation of the AMP-activated protein kinase in the freeze-tolerant frog *Rana sylvatica. Cryobiology* 53, 297–309.

Storey, K.B. (1997) Organic solutes in freezing tolerance. *Comparative Biochemistry and Physiology A* 117, 319–326.

Storey, K.B. (2002) Life in the slow lane: molecular mechanisms of estivation. *Comparative Biochemistry and Physiology A* 133, 733–754.

Storey, K.B. (2004a) *Functional Metabolism: Regulation and Adaptation.* Wiley-Liss, Hoboken, New Jersey.

Storey, K.B. (2004b) Strategies for exploration of freeze responsive gene expression: advances in vertebrate freeze tolerance. *Cryobiology* 48, 134–145.

Storey, K.B. (2006) Reptile freeze tolerance: metabolism and gene expression. *Cryobiology* 52, 1–16.

Storey, K.B. (2007) Anoxia tolerance in turtles: metabolic regulation and gene expression. *Comparative Biochemistry and Physiology A* 147, 263–276.

Storey, K.B. (2008) Beyond gene chips: transcription factor profiling in freeze tolerance. In: Lovegrove, B.G. and McKechnie, A.E. (eds) *Hypometabolism in Animals: Hibernation, Torpor and Cryobiology*. University of KwaZulu-Natal, Pietermaritzburg, pp. 101–108.

Storey, K.B. and Storey, J.M. (1986) Freeze tolerant frogs: cryoprotectants and tissue metabolism during freeze/thaw cycles. *Canadian Journal of Zoology* 64, 49–56.

Storey, K.B. and Storey, J.M. (1988) Freeze tolerance in animals. *Physiological Reviews* 68, 27–84.

Storey, K.B. and Storey, J.M. (1990) Facultative metabolic rate depression: molecular regulation and biochemical adaptation in anaerobiosis, hibernation, and estivation. *Quarterly Review of Biology* 65, 145–174.

Storey, K.B. and Storey, J.M. (1992) Natural freeze tolerance in ectothermic vertebrates. *Annual Review of Physiology* 54, 619–637.

Storey, K.B. and Storey, J.M. (2004a) Physiology, biochemistry and molecular biology of vertebrate freeze tolerance: the wood frog. In: Benson, E., Fuller, B. and Lane, N. (eds) *Life in the Frozen State*. CRC Press, Boca Raton, Florida, pp. 243–274.

Storey, K.B. and Storey, J.M. (2004b) Metabolic rate depression in animals: transcriptional and translational controls. *Biological Reviews of the Cambridge Philosophical Society* 79, 207–233.

Storey, K.B. and Storey, J.M. (2007) Putting life on 'pause' – molecular regulation of hypometabolism. *Journal of Experimental Biology* 210, 1700–1714.

Storey, K.B. and Storey, J.M. (2010) Metabolic regulation and gene expression during aestivation. In: Navas, C.A. and Carvalho, J.E. (eds) *Aestivation: Molecular and Physiological Aspects.* Springer, Heidelberg, pp. 25–45.

Storey, K.B., Storey, J.M., Brooks, S.P.J., Churchill, T.A. and Brooks, R.J. (1988) Hatchling turtles survive freezing during winter hibernation. *Proceedings of the National Academy of Sciences USA* 85, 8350–8354.

Stuart, S.N., Chanson, J.S., Cox, N.A., Young, B.E., Rodrigues, A.S., Fischman, D.L. and Waller, R.W. (2004) Status and trends of amphibian declines and extinctions worldwide. *Science* 306,1783–1786.

Tattersall, G.J. and Ultsch, G.R. (2008) Physiological ecology of aquatic overwintering in ranid frogs. *Biological Reviews of the Cambridge Philosophical Society* 83,119–140.

Turner, B.M. (2009) Epigenetic responses to environmental change and their evolutionary implications. *Philosophical Transactions of the Royal Society B: Biological Sciences* 364, 3403–3418.

Voituron, Y., Storey, J.M., Grenot, C. and Storey, K.B. (2002) Freezing survival, body ice content and blood composition of the freeze tolerant European common lizard, *Lacerta vivipara. Journal of Comparative Physiology B* 172, 71–76.

Voituron, Y., Heulin, B. and Surget-Groba, Y. (2004) Comparison of the cold hardiness capacities of the oviparous and viviparous forms of *Lacerta vivipara. Journal of Experimental Zoology A* 301, 367–373.

Voituron, Y., Paaschburg, L., Holmstrup, M., Barré, H. and Ramløv, H. (2009) Survival and metabolism of *Rana arvalis* during freezing. *Journal of Comparative Physiology B* 179, 223–230.

Wake, D.B. (2007) Climate change implicated in amphibian and lizard declines. *Proceedings of the National Academy of Sciences USA* 104, 8201–8202.

Warren, D.E. and Jackson, D.C. (2007) Effects of temperature on anoxic submergence: skeletal buffering, lactate distribution, and glycogen utilization in the turtle, *Trachemys scripta. American Journal of Physiology* 293, R458–R467.

Yancey, P.H. (2005) Organic osmolytes as compatible, metabolic and counteracting cytoprotectants in high osmolarity and other stresses. *Journal of Experimental Biology* 208, 2819–2830.

Ziegler, R., Ashida, M., Fallon, A.M., Wimer, L.T., Silver Wyatt, S. and Wyatt, G.R. (1979) Regulation of glycogen phosphorylase in fat body of *Cecropia* silkmoth pupae. *Journal of Comparative Physiology* 131, 321–332.

Zhang, X. (2008) The epigenetic landscape of plants. *Science* 320, 489–492.

8 The Relationship between Climate Warming and Hibernation in Mammals

Craig L. Frank

8.1 Introduction

Mean surface temperatures increased by 0.6°C worldwide during the 20th century, and are expected to rise by up to 5.8°C during the 21st century. Autumn/winter temperatures in North America, for example, are predicted to rise by 4°C–8°C within 70 years (Watson, 2001). A review of studies dealing with the effects of recent climate warming on both plants and animals revealed that whereas 667 species have displayed phenological shifts and 893 have changed their ranges limits during the past 50 years, the authors could find studies dealing with only two mammal species (Parmesan and Yohe, 2003). The effects of recent climate warming on mammalian phenologies are thus poorly understood. Mammals that may be particularly sensitive to climate warming are those that utilize torpor (Humphries *et al.*, 2002). Mammals and birds are unique among animals in that they are homeothermic endotherms, maintaining a constant core body temperature (T_b) over a wide range of ambient temperatures (T_a) through a high metabolic rate (Willmer *et al.*, 2000). The prolonged periods of high metabolic heat production by mammals and birds requires high rates of food intake. Food availability in the wild often fluctuates, and consequently the energetic costs of maintaining a high T_b (32°C–42°C) via endothermy becomes prohibitively expensive in some situations. Consequently not all mammals and birds are permanently homeothermic, but instead enter periods of torpor (Geiser, 2004).

Torpor is a period when metabolic rate and T_b are greatly reduced. It involves the regulation of T_b at a new and substantially lower level, with a new minimum T_b being maintained. Mammalian and avian species that employ torpor are therefore classified as heterothermic endotherms (Geiser, 2004). Metabolic rates during torpor can be less than 5% of basal metabolic rate with a corresponding T_b of just 0.5°C–1.0°C above ambient temperature (Geiser and Ruf, 1995). However, the T_b of a torpid mammal cannot fall below T_a and therefore ambient temperatures dictate the degree of reduction in T_b and metabolic rate that is possible during torpor (Kayser, 1965). Torpor is known to occur in at least 90 species of mammals and has been found in 20 species of birds (Geiser, 2004). About 270 extant species of marsupials (infraclass *Metatheria*) exist at present (Feldhamer *et al.*, 1999), and torpor occurs in 33 (12%) of them, with species employing torpor being found in 6 of the 7 marsupial orders (Geiser, 1994). Placental mammals (infraclass *Eutheria*) are classified into 18 different orders (Feldhamer *et al.*, 1999), six of which contain species that utilize torpor (Geiser, 2004; Mzilikazi *et al.*, 2002). Torpor (hibernation) is also employed by at least one (*Tachyglossus aculeatus*) of the three extant monotreme species (Nicol and Anderson, 2007). Thus, torpor has appeared repeatedly, frequently, and independently throughout the course of mammalian evolution.

Mammals and birds generally employ one of two common patterns of torpor, depending upon species: prolonged torpor

© CAB International 2012. *Temperature Adaptation in a Changing Climate: Nature at Risk* (eds K.B. Storey and K.K. Tanino)

during hibernation, and daily torpor. Hibernation is seasonal, usually from late summer/autumn to the following spring. Hibernators do not remain torpid continuously throughout the hibernation season; instead, bouts of torpor last from days to weeks, interrupted by brief (< 36 h) periods of high metabolic rates and high T_b called arousal episodes. Hibernation is the most common pattern of torpor found in mammals. By contrast, daily torpor is the most common pattern among birds. Daily torpor lasts for only a matter of hours each day, usually interrupted by periods of diurnal foraging and feeding (Geiser and Ruf, 1995; Geiser, 2004). Hibernating mammals can be divided into two categories. Spontaneous hibernators enter torpor during the autumn/winter, regardless of the amount of energetic stress experienced. Facultative hibernators, however, enter torpor during the autumn/winter only when they are in a negative energy balance caused by thermal stress and/or food deprivation. The onset of torpor by facultative hibernators is closely tied to ambient temperature (T_a), whereas the onset of torpor by spontaneous hibernators is largely independent of T_a. Therefore, the onset of torpor by spontaneous hibernators typically occurs much sooner than for facultative hibernators under the same autumn/winter conditions, and the torpor bouts of spontaneous hibernators tend to both be longer and result in a greater T_b reduction than those of facultative hibernators at the same T_a (Harlow and Menkins, 1986; Harlow and Frank, 2001).

8.2 Predicted Effects of Climate Warming on Hibernation

The predicted effects of climate warming on the torpor of heterothermic mammals and birds have received little previous consideration. One major mechanism by which mammalian torpor may be affected by climate warming is by influencing the concentration of polyunsaturated fatty acids found in their natural diets. A polyunsaturated fatty acid (PUFA) has more than 1 carbon–carbon double bond per molecule, as opposed to a saturated fatty acid that contains no carbon–carbon double bonds, or a monounsaturated fatty acid containing only 1 such bond per molecule. Mammals can synthesize saturated and monounsaturated fatty acids, but they are incapable of producing PUFA. Most plant species, however, produce two types of PUFA: linoleic acid (18 carbon atoms, 2 double bonds) and α-linolenic acid (18 carbon atoms, 3 double bonds). When mammals consume PUFA these are incorporated into their cellular membranes and storage lipids (Gunstone, 1996).

Laboratory experiments with chipmunks (*Tamias amoenus*), two ground squirrel species (*Spermophilus saturatus* and *S. lateralis*), two species of prairie dogs (*Cynomys ludovicianus* and *C. leucurus*) and marmots (*Marmota flaviventris*) have revealed that dietary levels of linoleic acid during summer/autumn fattening influences their ability to hibernate (Geiser and Kenagy, 1987, 1993; Florant *et al.*, 1993; Thorp *et al.*, 1994; Harlow and Frank, 2001; Frank, 2002). Similar results were found in studies with mice and two species of marsupials (Geiser, 1991; Geiser *et al.*, 1992; Withers *et al.*, 1996). A total of 17 laboratory studies detailing the effects of dietary PUFA levels on mammalian hibernation and daily torpor have been conducted to date (Munro and Thomas, 2004). Thus, it appears that dietary PUFA content influences the torpor of all heterothermic mammalian species. For example, studies with *S. lateralis* show that hibernation ability is greatest when the linoleic acid (18:2) content of the diet is at least 33 mg/g, but less than 74 mg/g. Squirrels fed a 33–74 mg linoleic acid/g diet:

1. Were more likely to hibernate;
2. Spent less time fasting prior to the onset of torpor;
3. Had lower metabolic rates during torpor, and
4. Had longer torpor bouts than those maintained on diets containing either less or more linoleic acid (Frank, 1992, 2002; Frank and Storey, 1995, 1996; Frank *et al.*, 1998).

Similarly, the amount of α-linolenic acid (18:3) in the diet influences the torpor patterns of *S. lateralis* in a manner identical to that for linoleic acid (Frank *et al.*, 2004). A 3-year field study detailing both the natural diet PUFA contents and the hibernation patterns of free-ranging arctic ground squirrels (*Spermophilus parryii*) revealed that the total (18:2 + 18:3) PUFA contents of the autumn diet varied as much as threefold between individuals (Frank *et al.*, 2008). Our study also demonstrated that free-ranging *S. parryii* that consumed a moderate (33–74 mg/g) PUFA diet had:

1. Longer torpor bouts;
2. Fewer arousals from torpor;
3. Shorter arousal periods from torpor; and
4. More days in torpor

than those that consumed a high (>74 mg/g) PUFA diet during the autumn. The cellular/biochemical basis of mammalian torpor is not fully known (Geiser, 2004), thus it is not fully understood how dietary PUFA influences this process (Ruf and Arnold, 2008). It has been proposed that a certain minimum amount of PUFA must be incorporated into depot fats in order for them to be fluid and metabolizable at the low T_b range associated with torpor (Frank, 1991). There also is some evidence that a high linoleic acid diet protects the hearts of torpid mammals from cardiac arrhythmia (Ruf and Arnold, 2008). The PUFA in mammalian cells undergo autoxidation more readily and rapidly than either saturated or monounsaturated fatty acids. Autoxidation (also called lipid peroxidation) is a self-sustaining chain reaction between PUFA and reactive oxygen species (ROS), and it produces lipid peroxides that are highly toxic to cells (Gunstone, 1996). Exceedingly high (>74 mg/g) dietary PUFA contents during the autumn feeding period, consequently, have been shown to increase of the rate of lipid peroxidation that occurs in brown adipose tissue during torpor by two species of mammals (Frank and Storey, 1995; Harlow and Frank, 2001). Laboratory diet selection experiments with adult *S. lateralis* revealed that PUFA content influences food item choice (Frank, 1994), and that adults normally maintain a dietary PUFA content of 33 mg/g through their food item combinations. This presumably maximizes hibernation ability while also minimizing the degree of lipid peroxidation during torpor (Frank *et al.*, 1998). Free-ranging adult *S. lateralis*, however, are often forced to maintain a slightly greater (>33 mg/g) diet PUFA content due to limitations in the natural food items available (Frank *et al.*, 2004).

The PUFA contents of herbaceous plant tissues vary greatly with species, season and between different parts of the same plant (Florant *et al.*, 1990; Frank *et al.*, 1998). The lipids found in herbaceous plant tissues are typically mixtures of phospholipids, glycolipids, waxes, and in the case of seeds, triacylglycerols, all of which are fatty acid esters (Andrews and Ohlrogge, 1990). Waxes cannot be digested by mammals, whereas phospholipids, glycolipids and triacylglycerols are hydrolysed into their constituent fatty acids and absorbed (Stevens, 1988). Thus, PUFA bound to waxes cannot be utilized by mammals, whereas the PUFA that are parts of phospholipids, glycolipids and triacylglycerols are absorbed. A study of 19 different sunflower (*Helianthus*) species found that the seed oils (triacylglycerols) of those species inhabiting relatively higher (colder) latitudes had greater PUFA contents than seed oils of species from lower (warmer) latitudes (Linder, 2000). Some agricultural plant species have been shown to have a great deal of phenotypic plasticity with respect to PUFA content. Experiments with several crop species have revealed that proportion of PUFA in both leaf (Williams *et al.*, 1988) and seed (Nagao and Yamazaki, 1984; Martin *et al.*, 1986; Deng and Scarth, 1998) lipids increases as the ambient temperature during growth and development decreases. Thus, a warming associated with climate change should result in a decreased PUFA content of most plant species/parts found in temperate regions. Therefore, the dietary PUFA content available to herbivorous mammals that use either daily torpor or hibernation will also decrease as a consequence of climate warming. This predicted decrease in plant tissue PUFA levels will probably affect heterothermic mammals worldwide, since most of these species feed primarily on seeds and herbaceous plant parts during their fattening prior to torpor (Geiser, 2004).

Field studies with two species of ground squirrels indicate that the particular effects of decreasing plant tissue PUFA levels associated with climate change on heterothermic mammals would also be likely to vary with location. In the tundra of Alaska, nearly half of the free-ranging Arctic ground squirrels (*S. parryii*) examined had autumn diets with PUFA contents that exceeded the upper limit (74 mg/g) for maximal hibernation ability, which in turn adversely affected both their torpor patterns and propensity to persist at the same site (Frank *et al.*, 2008). Herbivorous ground squirrels in this region thus appear to have limitations in food plant availability that constrain their ability to maintain a dietary PUFA content during the late summer/autumn within the range (33–74 mg/g) that facilitates maximum hibernation ability. Hence, a decrease in the average PUFA content due to climate warming in the Alaskan tundra could most likely result in a greater proportion of *S. parryii* being able to maintain a dietary PUFA content within the 33–74 mg/g. Consequently, climate warming in this system may enhance the hibernation ability of free-ranging *S. parryii*, as well as the over-winter survival of this population. By contrast, field studies on free-ranging golden-mantled ground squirrels (*S. lateralis*) in the White Mountains of California revealed that nearly all adults in this population maintain a dietary PUFA content within the 33–74 mg/g range prior to the onset of hibernation (Frank *et al.*, 1998, 2004). Hence, a decrease in plant PUFA levels due to climate warming in the White Mountains may increase the proportion of *S. lateralis* that cannot raise their dietary PUFA levels above the minimum (33 mg/g) required for hibernation. Increasing ambient temperatures in this region may therefore lead to a decrease in the propensity of *S. lateralis* to hibernate, thereby reducing the over-winter survival rate of this population.

An increase in the average T_a during autumn/winter may also have direct negative effects on hibernation by heterothermic animals by increasing both the number of arousal episodes, and metabolic rates during torpor bouts, independently of diet composition. Arousal episodes from torpor normally account for 80–90% of all energy utilized during the entire hibernation season, although their physiological function is still unknown (Kayser, 1965; Karpovich *et al.*, 2009). Laboratory hibernation experiments with both bats and ground squirrels have revealed that over a T_a range which is above the minimum T_b defended during torpor, the duration of individual torpor bouts greatly increases as the T_a during hibernation decreases (Geiser and Kenagy, 1988; Dunbar and Tomasi, 2006; Karpovich *et al.*, 2009). Consequently, a higher average winter T_a resulting from climate warming should result in more arousal episodes during hibernation. Laboratory hibernation experiments with bats and ground squirrels have also shown that over a T_a range that is above the minimum T_b maintained during torpor, the metabolic rate during torpor greatly increases with T_a during hibernation (Geiser and Kenagy, 1988; Szewczak and Jackson, 1992; Buck and Barnes, 2000; Dunbar and Tomasi, 2006). Thus, increasing the mean winter T_a may also result in a reduced degree of metabolic rate depression during torpor. Therefore, the elevated autumn/winter temperatures associated with climate warming will reduce the energetic savings associated with hibernation. This in turn may decrease the over-winter survival rates of some populations of heterothermic mammals.

8.3 Field Studies on Hibernation by Eastern Chipmunks in South-eastern New York

My laboratory has been conducting field studies on the torpor patterns of free-ranging eastern chipmunks (*Tamias striatus*) in the forests of the Louis Calder Center (LCC) since 2000 in order to test the hypothesis that climate warming in temperate regions results in a decreased propensity of free-ranging facultative hibernators to enter torpor in these areas. Laboratory experiments with *T. striatus* revealed that this species is a facultative hibernator (French, 2000). Chipmunks store large amounts of seeds in their underground burrows over the autumn, and these items are ingested during periodic arousals from torpor during hibernation (Thomas, 1974; Wrazen and Wrazen, 1982). In New York, individuals retreat to their below ground nests during November and do not appear above ground until the following April (Allen, 1938). Animals have a mean life expectancy of 2.3 years in the northeast, with one-third to one-half of the adults in autumn populations being less than 2 years old (Tryon and Snyder, 1973).

The LCC is the biological field station of Fordham University located in Armonk, New York (41°7' N, 73°43' W), about 42 km north of New York City. The study forest is 45.7 ha of mixed deciduous trees dominated by red oak (*Quercus rubra* L.), black oak (*Q. velutina* Lam.), white oak (*Q. alba* L.), chestnut oak (*Q. prinus* L.), bitternut hickory (*Carya cordiformis* K. Koch), shagbark hickory (*C. ovata* K. Koch), red maple (*Acer rubrum* L.), sugar maple (*A. saccharum* Marshall) and American beech (*Fagus grandifolia* Ehrh.). Soils are shallow, sandy loam inceptisols belonging to the Hollis series (Zhu and Carriero, 1999). Measurements of air temperature (T_{air}) at this site have been conducted continuously since 1998 using two automated weather stations. Mean T_{air} values at the LCC from November to April are summarized in Figs. 8.1 and 8.2. The mean T_{air} during November 2006 was the highest recorded, and that for November 2001 was the second highest, both being 0.4°C–4.0°C warmer than those of other years (Fig. 8.1). Likewise, the same was true of December T_{air} in 2006 and 2001, with both being 0.8°C–10.5°C greater than December means for other years (Fig. 8.1). November–December 2001 and 2006 were thus the warmest November–December periods observed at the LCC during these 7 years (Fig. 8.1). Air temperature data published by the National Climatic Data Center indicate that November and December 2001 were the warmest in New York City since air temperature recordings began 98 years previously, with 2006 being the second warmest. Therefore, it appears that November–December 2001 and 2006 were the warmest November–December periods at the LCC during the previous 98 years as well. Mean air temperatures during February, March and April during 2002 and 2007, however, were similar to those observed at the LCC during other years (Fig. 8.2). This field study thus encompasses measurements on the torpor patterns of free-ranging *T. striatus* during two exceedingly warm autumn periods (November–December 2001 and 2006) that occurred when this species normally begins hibernation.

The over-winter survival and torpor patterns of 36 adult chipmunks at this site were determined using an automated radio telemetry system refined by my laboratory. Temperature-sensitive radio collars were placed around the necks of 4–9 *T. striatus* collected every October (2000–2006). Each collar (Holohil model PD-2THX) emits a signal on a unique radio frequency. The pulse rates of the collar signals depend on skin temperature (T_{skin}) of the animal wearing it, which is equivalent to body temperature (T_b) in small mammals. All chipmunks were released immediately after being fitted with a collar and radio signals were continuously monitored throughout the next 7–8 months using an automated receiver. Signals were recorded at 1–3 h intervals using a data collection computer (DCC) interfaced with a model R4000 radio receiver produced by Advanced Telemetry Systems (Isanti, MN). The chipmunks were divided into two different groups based on the mean November and December T_{air} at the LCC during the year in which they were captured. The warm November–December T_{air} group consisted of chipmunks that were monitored during the 1 November–30 April periods of 2001–2002 and 2006–2007, whereas the normal November–December T_{air} group comprised those chipmunks examined during the same periods of 2000–

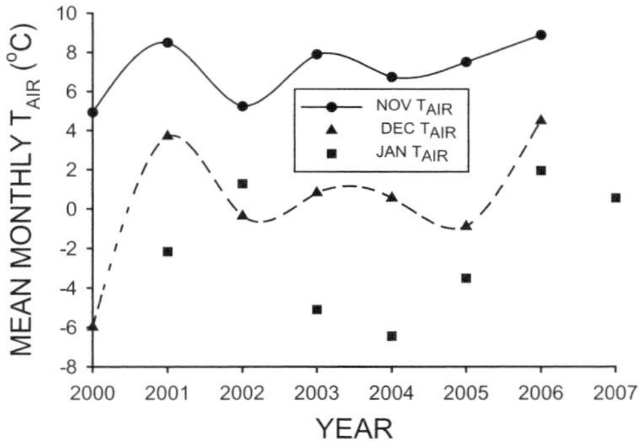

Fig. 8.1. Mean air temperatures over seven years at the Louis Calder Center (Armonk, New York; 41°7' N, 73°43' W) during November, December and January.

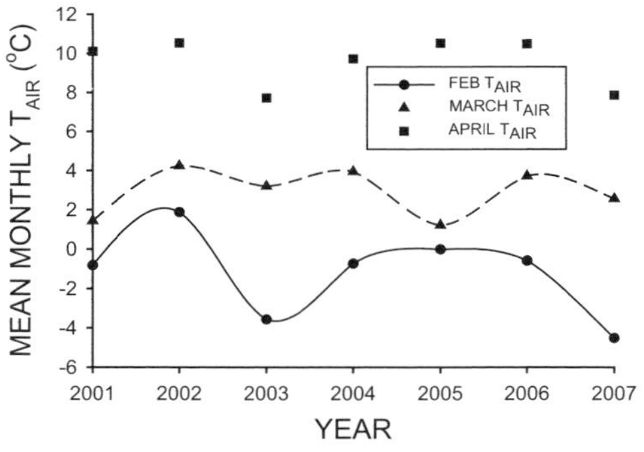

Fig. 8.2. Mean air temperatures over seven years at the Louis Calder Center during February, March and April.

2001, 2002–2003, 2003–2004, 2004–2005 and 2005–2006.

All 21 chipmunks in the normal November–December T_{air} group entered torpor during autumn/early winter, whereas only 6 of the 15 (40%) chipmunks in the warm November–December T_{air} category entered torpor during the entire study period, and this proportion of chipmunks hibernating was significantly lower than that of the normal November–December T_{air} group (FI = 17.35, P = 0.0001). Chipmunks hibernating after a normal November–December T_{air} spent more than 8 times more days in torpor (Table 8.1) during the entire winter than those that had experienced a warm November–December T_{air} period (t = 3.171, d.f. = 8, P = 0.013). Furthermore, the mean duration of a torpor bout for chipmunks after a normal November–December T_{air} period was more than 4 times longer (Table 8.1) than that of chipmunks that had experienced a warm November–December T_{air} period (t = 3.433, d.f. = 8, P = 0.009). The mean minimum T_b during torpor after a normal November–December period was 16.9°C lower (Table 8.1) than that of the chipmunks observed after a warm November–December period (t = –3.061, d.f. = 8, P = 0.016). Chipmunks also first entered torpor significantly earlier (t = –4.037, d.f. = 8.9, P = 0.03) during a November–December period with a normal mean T_{air} than during the same period in years (2001 and 2006) when the mean T_{air} was warm. The mean (± SE) Julian date of the first torpor observed in the normal November–December group was 294.5 ± 4.8 (N = 21), whereas that for the chipmunks in the warm November–December category was 332.0 ± 8.01 (N = 6).

Over-winter survival was estimated by continuous recordings of T_{skin} made at 1–3 h intervals for a total of eight adult free-ranging eastern chipmunks at the LCC during the 1 November–30 April periods of 2003–2004, 2004–2005 and 2005–2006. The autumn/winter air temperatures during

Table 8.1. Mean (± SE) time spent torpid, average torpor bout duration, and minimum T_b for free-ranging *Tamias striatus* during the autumn–spring periods of 2000–2001 to 2006–2007.

Autumn T_{air} group	Total time torpid (d)	Average torpor bout (d)	T_b minimum (°C)
Normal Nov.–Dec.	72.0 ± 22.0*	3.72 ± 1.01*	4.2 ± 0.7*
Warm Nov.–Dec.	8.95 ± 7.74	0.91 ± 0.14	21.1 ± 4.4

*Significantly different from the corresponding warm Nov.–Dec. mean at the P < 0.05 level; N = 4 for each normal Nov.–Dec. mean, N = 6 for each warm Nov.–Dec. mean.

these periods were similar (Figs. 8.1 and 8.2), and examples of the resulting torpor patterns are summarized in Figs. 8.3A and 8.3B. All of the chipmunks monitored during these years displayed consistent multi-day bouts of torpor throughout the autumn–winter period (Table 8.1), and seven of the eight survived the winter. A total of nine adult free-ranging *T. striatus* were fitted with radio collars during October 2006, and T_{skin} was measured and recorded from 1 November 2006 to 30 April 2007. Only two of these chipmunks had entered torpor by January 2007, and the T_{skin} patterns for these two individuals are summarized in Figs. 8.3C and 8.3D. Whereas during previous years all *T. striatus* at the LCC had numerous multi-day torpor bouts with a T_{skin} below 10°C during the January–April period (Figs. 8.3A and 8.3B), only one individual in the 2006–2007 study had multi-day torpor bouts (Fig. 8.3D) and this was the only individual that survived the hibernation period as well. The over-winter survival rate for 2006–2007 was therefore just one out of nine (11%). Thus, it appears that the abnormally high T_{air} during November 2006–January 2007 inhibited facultative torpor during that winter, which in turn greatly increased over-winter mortality.

8.4 Conclusions

The results of this 7-year study on the torpor patterns of free-ranging *T. striatus* clearly supports the hypothesis that climate warming in temperate regions results in a decreased propensity of free-ranging facultative hibernators to enter torpor in these areas. A 0.4°C–10.5°C increase in the mean T_{air} during the period when *T. striatus* normally retreats below ground (November–December) corresponded with:

1. An average 37.5 d delay in the first onset of torpor bouts;

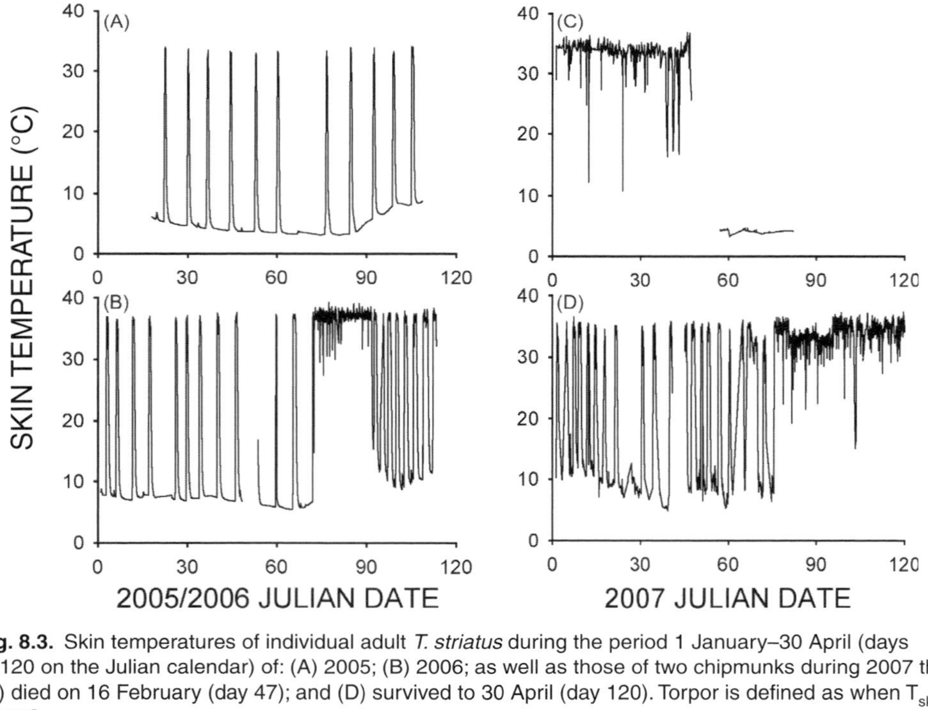

Fig. 8.3. Skin temperatures of individual adult *T. striatus* during the period 1 January–30 April (days 1–120 on the Julian calendar) of: (A) 2005; (B) 2006; as well as those of two chipmunks during 2007 that (C) died on 16 February (day 47); and (D) survived to 30 April (day 120). Torpor is defined as when T_{skin} < 30°C.

2. A 60% reduction in the proportion of individuals utilizing torpor at all; and
3. A corresponding 88% decrease in the total amount of time spent in torpor among those individuals that did utilize torpor during the subsequent winter.

Furthermore, the results of field studies conducted during the 1 November–30 April periods of 2003–2004 to 2006–2007 suggest that the reductions in torpor associated with warm autumn (November–December) periods greatly decreases over-winter survival. It is unclear, however, if the reduction in torpor observed after warm November–December periods was due to the direct effects of T_a on torpor bout length/ T_b minimum, the influences of T_a on the PUFA contents of food items (seeds) consumed during the autumn, or a combination of both types of effects. These findings also strongly suggest that the exceedingly warm autumn/winter periods associated with climate change will adversely affect species that have a facultative hibernation strategy, by reducing their propensity to enter torpor and thereby decreasing their over-winter survival rate. It also would be of interest to determine the effects of recent climate warming on the torpor patterns of free-ranging spontaneous hibernators and the associated effects on their over-winter survival rate.

Studies with marmots offer some insight into climate change effects on spontaneous hibernators. Yellow-bellied marmots (*Marmota flaviventris*) are spontaneous hibernators found in the alpine meadows of Colorado (Armitage, 1994; Wood and Armitage, 2003). A field study conducted on a *M. flaviventris* population from 1976 to 2000 revealed that the date in April at which the first marmot appeared above ground after hibernation at this site had advanced by 23 d by the end of this period, whereas the average April T_{air} at this location increased by 1.4°C during the same interval (Inouye *et al.*, 2000). Another field study on the same *M. flaviventris* population conducted from 1976 to 2008 revealed that the average body mass reached by adult females on 1 August of each year had increased by 339 g, and that during the last 7 years of the study, larger females had a greater over-winter survival rate than their smaller conspecifics (Ozgul *et al.*, 2010). Interpreting these two marmot studies together indicates that recent climate warming in this region has shortened the hibernation period of this particular *M. flaviventris* population, which in turn has increased the amount of body mass (depot fat) required to survive the winter. Climate warming will therefore probably decrease the over-winter survival rates of most spontaneous hibernator populations as well.

The results of two recent studies suggest that the degree to which climate warming will adversely affect heterothermic mammal populations may diminish, with greater plasticity in the torpor patterns characteristic of the particular species being considered. Eastern woodchucks (*Marmota monax*) are spontaneous hibernators found in eastern North America (Zervanos *et al.*, 2009). This species is found over a 2000 km latitudinal gradient extending from Maine to South Carolina. The hibernation period of northern (Maine) populations of *M. monax* averages 167 days, with torpor bouts averaging 185 h in length and a mean minimum T_b of 5.3°C. The southernmost populations of *M. monax*, in contrast, have a hibernation period that averages just 76 days, with torpor bouts that have a mean duration of 117 h, and a mean minimum T_b of 11.8°C (Zervanos *et al.*, 2010). Thus southern populations of eastern woodchucks have evolved the capacity to remain torpid for extended periods of time, even when hibernating at a much warmer T_a than the northern populations of this species. This indicates that the hibernation patterns of *M. monax* may not be affected by climate warming to the same extent as those of other heterothermic species. Heterothermic species that have already evolved the capacity to remain torpid at a relatively warmer T_a may thus be favoured by natural selection as the climate warms during the next 100 years. Further investigation of the influences of recent climate warming on the torpor patterns and subsequent over-winter survival of free-ranging heterothermic mammals will provide important insights into which populations/species may be most adversely affected by the climate warming predicted to occur during this century.

8.5 Acknowledgements

Many thanks to Kelly Stanton for her assistance with the radio telemetry. Additional assistance with data analysis was generously provided by Mike Stevens. This study was supported by two grants from the National Science Foundation awarded to C.L.F. (IOS-9986620 and IOS-0326330).

References

Allen, E.G. (1938) The habits and life history of the eastern chipmunk *Tamias striatus lysteri. New York State Museum Bulletin* 314, 7–122.

Andrews, J.E. and Ohlrogge, J. (1990) Fatty acid and lipid biosynthesis and degradation. In: Dennis, D.T. and Turpin, D.H. (eds) *Plant Physiology, Biochemistry, and Molecular Biology*. Longman Scientific and Technical, Essex, pp. 339–352.

Armitage, K.B. (1994) Unusual mortality in a yellow-bellied marmot population. In: Rumiantsev, V.Y. (ed.) *Actual Problems of Marmot Investigation.* A.B.F. Publishing House, Moscow, Russia, pp. 5–13.

Buck, C.L. and Barnes, B.M. (2000) Effects of ambient temperatures on metabolic rate, respiratory quotient, and torpor in an Arctic hibernator. *American Journal of Physiology* 279, R255–R263.

Deng, X. and Scarth, R. (1998) Temperature effects on fatty acid composition during development of low-linolenic oilseed rape (*Brassica napus L.*). *Journal of the American Oil Chemists' Society* 75, 759–766.

Dunbar, M.B. and Tomasi, T.E. (2006) Arousal patterns, metabolic rate, and an energy budget of eastern red bats (*Lasiurus borealis*) in winter. *Journal of Mammalogy* 87, 1096–1102.

Feldhamer, G.A., Drickhamer, L.C., Vessey, S.H. and Merritt, J.F. (1999) *Mammalogy: Adaptation, Diversity, and Ecology.* WCB McGraw-Hill, New York.

Florant, G.L., Nuttle, L.C., Mullinex, D.E. and Rintoul, D.A. (1990) Plasma and white adipose tissue lipid composition in marmots. *American Journal of Physiology* 258, R1123–1131.

Florant, G.L., Ameednuddin, H.S., and Rintoul, D.A. (1993) The effect of a low essential fatty acid diet on hibernation in marmots. *American Journal of Physiology* 264, R747–753.

Frank, C.L. (1991) Adaptations for hibernation in the depot fats of a ground squirrel (*Spermophilus beldingi*). *Canadian Journal of Zoology* 69, 2707–2711.

Frank, C.L. (1992) The influence of dietary fatty acids on hibernation by golden-mantled ground squirrels (*Spermophilus lateralis*). *Physiological Zoology* 65, 906–920.

Frank, C.L. (1994) Polyunsaturate content and diet selection by ground squirrels (*Spermophilus lateralis*). *Ecology* 75, 458–463.

Frank, C.L. (2002) The effects of short-term variations in diet fatty acid composition on mammalian torpor. *Journal of Mammalogy* 83, 1031–1019.

Frank, C.L. and Storey, K.B. (1995) The optimal depot fat composition for hibernation by golden-mantled ground squirrels (*Spermophilus lateralis*). *Journal of Comparative Physiology B*, 164, 536–542.

Frank, C.L. and Storey, K.B. (1996) The effect of total unsaturate content on hibernation. In: Geiser, F., Hulbert, A.J. and Nicol, S.J. (eds) *Adaptations to the Cold.* University of New England Press, Armidale, Australia, pp. 211–216.

Frank, C.L., Dierenfeld, E.S. and Storey, K.B. (1998) The relationship between lipid peroxidation, hibernation, and food selection in mammals. *American Zoologist*, 38, 341–349.

Frank, C.L., Hood, W.R. and Donnelly, M.C. (2004) The role of α–linolenic acid (18:3) in mammalian torpor. In: Barnes, B.M. and Carey, H.V. (eds) *Life in the Cold: Evolution, Mechanisms, Adaptation and Application.* Institute of Arctic Biology Press, Fairbanks, Alaska, pp. 71–80.

Frank, C.L., Karpovich, S. and Barnes, B.M. (2008) The relationship between natural variations in dietary fatty acid composition and the torpor patterns of a free-ranging arctic hibernator. *Physiological and Biochemical Zoology* 81, 486–495.

French, A.R. (2000) Interdependency of stored food and changes in body temperature during hibernation of the Eastern chipmunk, *Tamias striatus. Journal of Mammalogy* 81, 979–985.

Geiser, F. (1991) The effect of unsaturated and saturated dietary lipids on the patterns of daily torpor and the fatty acid composition of tissues and membranes of the deer mouse *Peromyscus maniculatus. Journal of Comparative Physiology B* 161, 590–597.

Geiser, F. (1994) Hibernation and daily torpor in marsupials: a review. *Australian Journal of Zoology* 42, 1–16.

Geiser, F. (2004) Metabolic rate and body temperature reduction during hibernation and daily torpor. *Annual Review of Physiology* 66, 239–74.

Geiser, F. and Kenagy, G.J. (1987) Polyunsaturated lipid diet lengthens torpor and reduces body temperature in a hibernator. *American Journal of Physiology* 252, R897–901.

Geiser, F. and Kenagy, G.J. (1988) Torpor duration in relation to temperature and metabolism in hibernating ground squirrels. *Physiological Zoology,* 61, 442–449.

Geiser, F. and Kenagy, G.J. (1993) Dietary fats and torpor patterns in hibernating ground squirrels. *Canadian Journal of Zoology* 74, 1182–1186.

Geiser, F. and Ruf, T. (1995) Hibernation vs. daily torpor in mammals and birds: physiological variables and classification of torpor patterns. *Physiological Zoology* 68, 935–966.

Geiser, F., Stahl, B. and Learmonth, R.P. (1992) The effect of dietary fatty acids on the pattern of torpor in a marsupial. *Physiological Zoology* 65, 1236–1245.

Gunstone, F.D. (1996) *Fatty Acid and Lipid Chemistry.* Blackie Academic and Professional, Glasgow.

Harlow, H.J. and Frank, C.L. (2001) The role of dietary fatty acids in the evolution of spontaneous and facultative hibernation patterns in prairie dogs. *Journal of Comparative Physiology B* 171, 77–84.

Harlow, H.J. and Menkens, G.E. (1986) A comparison of hibernation in the black–tailed prairie dog, white-tailed prairie dog, and Wyoming ground squirrel. *Canadian Journal of Zoology* 64, 793–796.

Humphries, M.M., Thomas, D.W. and Speakman, J.R. (2002) Climate mediated energetic constraints on the distribution of hibernating mammals. *Nature* 418, 313–316.

Inouye, D.W., Barr, B., Armitage, K.B. and Inouye, B.D. (2000) Climate change is affecting altitudinal migrants and hibernating species. *Proceedings of the National Academy of Sciences USA* 97, 1630–1633.

Karpovich, S.A., Tøien, Ø., Buck, C.L. and Barnes, B.M. (2009) Energetics of arousal episodes in hibernating Arctic ground squirrels. *Journal of Comparative Physiology B* 179, 691–700.

Kayser, C. (1965) Hibernation. In: Mayer, W. and VanGelder, R. (eds) *Physiological Mammalogy, Volume II.* Academic Press, New York, pp. 180–296.

Linder, C.R. (2000) Adaptive evolution of seed oils in plants: accounting for biogeographic distribution of saturated and unsaturated fatty acids in seed oils. *American Naturalist* 156, 442–458.

Martin, B.A., Wilson, R.F. and Rinne, R.W. (1986) Temperature effects upon the expression of a high oleic acid trait in soybean. *Journal of the American Oil Chemists Society* 63, 346–352.

Munro, D. and Thomas, D.W. (2004) The role of polyunsaturated fatty acids in the expression of torpor by mammals: a review. *Zoology,* 107, 29–48.

Mzilikazi, N., Lovegrove, B.G. and Ribble, D.O. (2002) Exogenous passive heating during torpor arousal in free-ranging rock elephant shrews, *Elephantus myurus. Oecologia* 133, 307–314.

Nagao, A. and Yamazaki, M. (1984) Effect of temperature during maturation on fatty acid composition of sunflower seed. *Agricultural and Biological Chemistry,* 48, 553–555.

Nicol, S.C. and Anderson, N.A. (2007). Cooling rates and body temperature regulation of hibernating echidnas (*Tachyglossus aculeatus*). *Journal of Experimental Biology* 210, 586–592.

Ozgul, A., Childs, D.Z., Oli, M.K., Armitage, K.B., Blumstein, D.T., Olson, L.E., Tuljapurkar, S., *et al.* (2010) Coupled dynamics of body mass and population growth in response to environmental change. *Nature* 466, 482–485.

Parmesan, C. and Yohe, G. (2003) A globally coherent fingerprint of climate change impacts across natural systems. *Nature* 421, 37–42.

Ruf, T. and Arnold, W. (2008) Effects of polyunsaturated fatty acids on hibernation and torpor: a review and hypothesis. *American Journal of Physiology* 294, R1044–1052.

Stevens, C.E. (1988) *Comparative Physiology of the Vertebrate Digestive System.* Cambridge University Press, New York.

Szewczak, J.M. and Jackson, D.C. (1992) Apneic oxygen uptake in the torpid bat, *Eptesicus fuscus. Journal of Experimental Biology* 173, 217–227.

Thomas, K.R. (1974) Burrow systems of the eastern chipmunk (*Tamias striatus pipilans* Lowry) in Louisiana. *Journal of Mammalogy* 55, 454–459.

Thorp, C.R., Kodanda, R.P. and Florant, G.L. (1994) Diet selection alters metabolic rate in the yellow-bellied marmot (*Marmota flaviventris*) during hibernation. *Physiological Zoology* 67, 1213–1229.

Tryon, C.A. and Snyder, D.P. (1973) Biology of the eastern chipmunk, *Tamias striatus*: life tables, age distributions, and trends in population numbers. *Journal of Mammalogy* 54, 145–168.

Watson, R.T. (2001) *Climate Change 2001: Synthesis Report.* Cambridge University Press, New York.

Williams, J.P., Khen, M.U., Mitchell, K. and Johnson

G. (1988) The effect of temperature on the level and biosynthesis of unsaturated fatty acids in diacylglycerols of *Brassica napus* leaves. *Plant Physiology* 87, 904–910.

Willmer, P., Stone, G. and Johnson, I. (2000) *Environmental Physiology of Animals.* Blackwell Science, Boston, Massachusetts.

Withers, K.J., Billingsley, D., Hirning, A., Young, P., McConnell, M. and Carlin, S. (1996) Torpor in Smithopsis macroura: effects of dietary fatty acids. In: Geiser, F., Hulbert, A.J. and Nicol, S.J. (eds) *Adaptations to the Cold.* University of New England Press. Armidale, Australia, pp. 217–222.

Woods, B.C. and Armitage, K.B. (2003) Effect of food supplementation on juvenile growth and survival in *Marmota flaviventris*. *Journal of Mammalogy* 84, 903–914.

Wrazen, J.A. and Wrazen, L.A. (1982) Hoarding, body mass dynamics, and torpor as components of the survival strategy of the eastern chipmunk. *Journal of Mammalogy* 63, 63–72.

Zervanos, S.M., Salsbury, C.M. and Brown, J.K. (2009) Maintenance of biological rhythms during hibernation in eastern woodchucks (Marmota monax). *Journal of Comparative Physiology* B 179, 411–418.

Zervanos, S.M., Maher, C.R., Waldvogel, J.A. and Florant, G.A. (2010). Latitudinal differences in the hibernation characteristics of woodchucks (Marmota monax). *Physiological and Biochemical Zoology* 83, 135–142.

Zhu, W.X. and Carriero, M.M. (1999) Chemo-autotrophic nitrification in acidic forest soils along an urban-to-rural transect. *Soil Biology and Biochemistry* 31, 1091–1100.

9 On Thin Ice: Marine Mammals and Climate Change

Michael Castellini

9.1 Introduction

Perhaps one of the most well-known public images of the impact of climate change on marine mammals is an iconic picture of a single polar bear apparently stranded on a small iceberg floating alone in the Arctic ocean (cover illustration: Huntington and Moore, 2008). The implications of the photograph are clear: the Arctic ice is melting and the animals that depend on it are in jeopardy. In fact, the diminishing ice scenario led to the listing of the polar bear as threatened under the United States Endangered Species Act in early 2009 (Fish and Wildlife Service, Department of the Interior, 2008). There is no doubt that the habitats of the polar bear and the other ice-associated marine mammals of both poles are changing. The annual summer minimum of the Arctic ice cover is decreasing at a high rate, both in surface area and thickness (Wang and Overland, 2009). For the ice-loving (pagophilic) marine mammals that depend on the ice for breeding, resting, nursing young or for protection from predators, the equation is simple: if the ice moves or significantly decreases, then the availability of that platform for these essential biological needs is compromised.

However, there is also the question of whether the predicted increase in oceanic water temperatures of several degrees Celsius in the next few decades would have a direct impact on the physiology of marine mammals themselves. Temperature regulation in marine mammals consists not only of protecting themselves against cold oceanic and sometimes ice-laden waters, but also of being able to release or dump heat when necessary. Would increasing ocean temperatures become a problem? Because they are homeothermic endotherms, marine mammals have evolved a suite of anatomical and physiological tools for closely regulating their body temperature throughout a vast range of ambient temperatures. For example, at about –40° (°F or °C), I have observed that the Antarctic Weddell seal usually spends its time in seawater (which is –1.8°C) and is reluctant to come out on the ice surface. Yet, on warm sunny days (5°C–10°C), the seals will haul out to rest on the ice and dump so much heat that they begin to melt into the sea ice surface (Fig. 9.1). On a colder day, the same seals might experience a blizzard and accumulate a layer of wind-driven snow on their bodies (Fig. 9.2). How is this possible? How can the same seal that can have snow build up on its body/fur surface then become warm enough to melt its way into solid ice under different conditions? The fur layer has not changed and neither has the mass nor thickness of the blubber layer. The answer is because they defend their core body temperature at around 37C° (as do most other mammals) and are able to increase circulation through the blubber layer to release excess heat to the skin surface on warm days, or shut down circulation to the skin surface to conserve heat on cold days. They are already adapted to handle extremely large ranges in external thermal challenges. Therefore, the major impact of changing climatic water temperatures on marine mammals will not come directly as an alteration in their thermoregulatory physiology, but more likely as 'bottom-up' alterations in food resources and, for polar marine mammals, the

© CAB International 2012. *Temperature Adaptation in a Changing Climate: Nature at Risk* (eds K.B. Storey and K.K. Tanino)

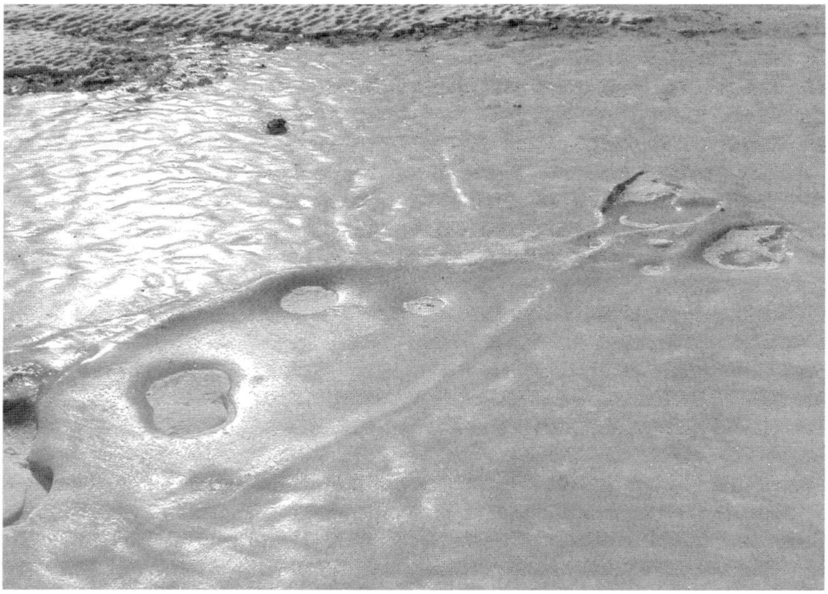

Fig. 9.1. A melt depression on the Antarctic sea ice made by a Weddell seal. The animal's skin surface was warm enough to melt into the ice a few centimetres.

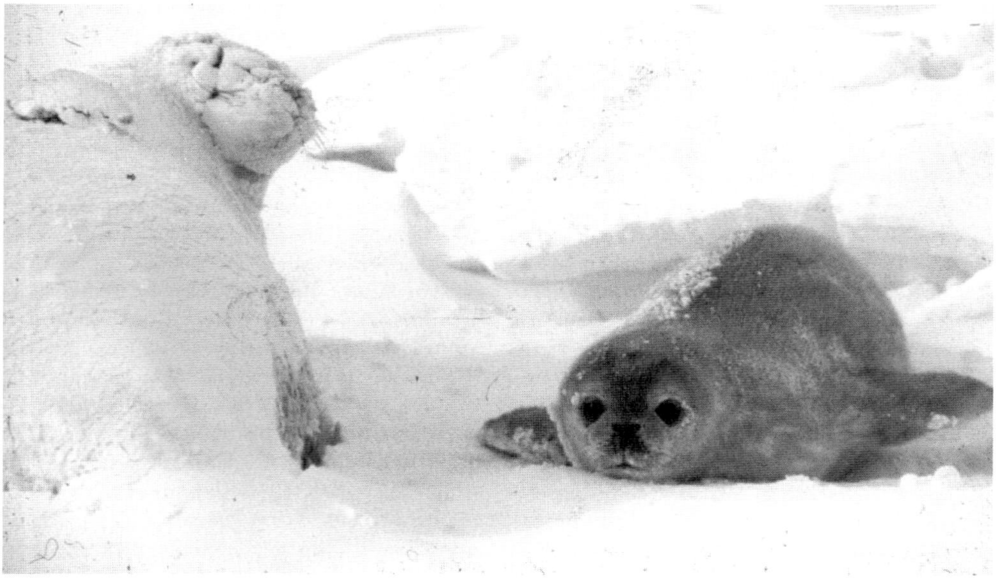

Fig. 9.2. A mother and pup Weddell seal after a blizzard in the Antarctic. The mother's skin and fur were cool enough so that the snow on her fur was not melted. The pup was protected from the blizzard by the mother.

reduction of sea ice as a stable platform for reproduction, resting, or refuge.

This chapter will cover two major aspects of climate change and marine mammals:

1. How changing ice conditions will alter the habitat of polar marine mammals, and the potential impacts of that change on their ecosystem relationships.
2. How marine mammals regulate body temperature and whether they would be directly impacted by climatic changes in oceanic water temperature. I have covered marine mammal thermoregulation in a previous review (Castellini, 2009) and discuss some of those concepts here in relation to the topic of climate change.

9.2 Impact of Climate Change on Ice-Associated Marine Mammal Habitat

9.2.1 Polar ice patterns

It is essential to begin this discussion with a basic review of the fundamental differences between the Antarctic southern ocean polar regions and those of the Arctic. The Antarctic is a continent, surrounded by the Southern Ocean. The continent is essentially entirely covered in freshwater ice to depths of over 3000 m near the centre. During the austral (southern) winter from mid-April to mid-October, the sea ice freezes around the continent out to distances beyond 500 km from the shoreline. This ice forms to an annual thickness of about 2 m and is termed 'fast ice' while it is held 'fast' to the shoreline. Conversely, starting in early November, the ice begins to melt back and break up until the cycle begins again in the austral autumn. The broken ice is termed 'pack ice' and in many cases along the Antarctic coastline, the ice melts and breaks up entirely up to the coast during the austral summer.

There are four species of phocid seals (true seals, as opposed to otariid sea lions) that inhabit this region, and they all use the sea ice for social and breeding behaviour during the spring, a safe platform for nursing their pups during the summer, and for resting and moulting during the early autumn (Bester and Hofmeyr, 2007; Boyd, 2009). In these same southern polar waters, there are many species of cetaceans (whales) that move through the region annually (Boyd, 2009), and several species of penguins that inhabit both land and annual sea ice regions (Stonehouse, 2007). An excellent collection of research findings and reviews on the Antarctic are found in the recent *Encyclopedia of the Antarctic* (Riffenburgh, 2007).

In the northern polar Arctic regions, the annual cycle is the opposite, with the annual sea ice minimum in October and the maximum in March. The most significant difference between the two polar regions is that the Arctic is an ocean surrounded by continents, and the Antarctic is a continent surrounded by oceans. Thus, the entire Arctic ocean freezes over in the winter, again to an average depth of about 2 m just as in the Antarctic, and breaks up during the northern summer. There are six species of true seals (plus the walrus) that are associated with the Arctic sea ice for the same reasons as their southern counterparts: social interactions, breeding, nursing, resting and moulting (Burns, 2009). In the north, while there are no ice-associated breeding birds such as penguins, there are birds that breed and forage near the ice edges. Also, and best known as representing 'climate change' to the public, the polar bear is linked to the ice and relies almost entirely on a diet of ice seals (Stirling, 2009). As with the southern ocean, there are also whales that move in and out of the Arctic waters and are associated with ice. Narwhals, beluga and bowhead whales are the best known of these (Burns, 2009).

Most of the geopolitical and economic focus on the impact of climate change on sea ice has been on the Arctic and not the Antarctic for several reasons:

1. Over 4 million people live above the Arctic circle and half a dozen nations border on the Arctic. By contrast, the Antarctic is not inhabited, no nations border it, and an international treaty prevents commercial or military operations inside the Treaty Zone.
2. The Arctic is an ocean and it is relatively

easy to observe annual and seasonal changes in sea ice patterns by air and satellite imagery. The Antarctic is a continent, and measuring changes in the thickness of continental ice sheets that are thousands of kilometres across and over 1 km deep is much more difficult.

3. The Arctic ocean will become important for oceanic intercontinental shipping if the ice begins to recede (Arctic Council, 2009) and therefore political, economic and military issues are coming to the forefront.

Remembering that the annual minimum in Arctic sea ice cover occurs in September–October, the National Snow and Ice Data Center (NSIDC, 2011) is one of many national and international organizations that have been following the change in the minimum coverage of the summer sea ice for several decades. Reliable satellite sea ice imagery and data collection began in 1979. Some of their data on the pattern of sea ice cover are shown in Fig. 9.3 and demonstrate the continued decline in the summer sea ice minimum (measured in October) for over 30 years. Regardless of the political, economic and social arguments about the causes or solutions to climate change, the implications of the data shown in Fig. 9.3 are clear: for sea ice-associated animals in the Arctic regions, their stable platform for breeding, resting and raising their pups or cubs is declining rapidly. Model predictions for when the summer sea ice in the Arctic ocean will disappear completely range from as early as 2020 to about 2050 (Wang and Overland, 2009). These data and predictions were central to the listing of the polar bear as endangered, due to loss of sea ice habitat.

The story is not as clear in the Antarctic because the marine mammals there are associated with the sea ice that forms around the continent each year. Each year it reaches the same minimum of essentially zero as it melts back to the continental edges during the summer. Therefore the metric of summer minimum cover alterations over time is not as applicable to the Antarctic. However, all four Antarctic seals and several penguin species have components of their life cycles that depend heavily on the pack ice phase of the annual ice cycle that occurs before the summer minimum. If there are major

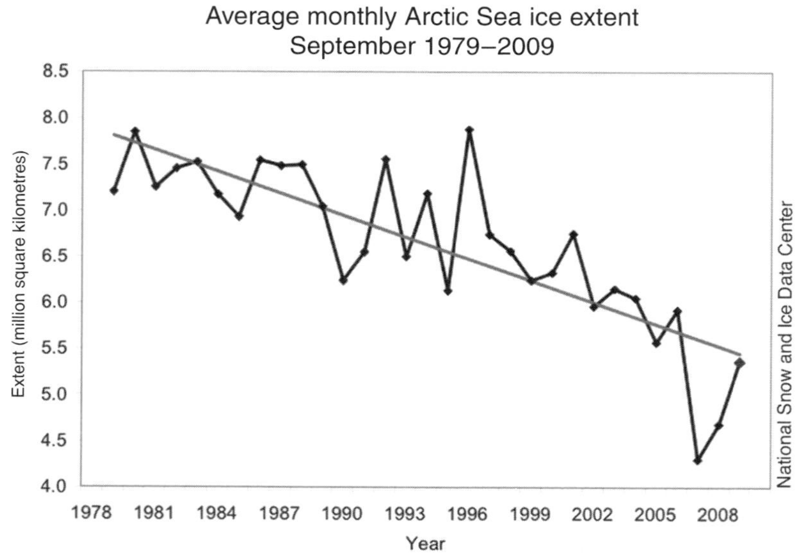

Fig. 9.3. Plot of annual Arctic sea ice minimum in October as collected from satellite imaging. Figure with permission from the NSIDC (2011).

changes in the pack ice dynamics or availability, then it is reasonable to assume that this will impact the life cycle of the seals and penguins that utilize the Antarctic pack ice.

Feeding patterns for ice-associated marine mammals

Beyond the obvious impact to seals and polar bears because of the loss of a stable sea ice platform during the summer, we must consider fundamental changes in the ecosystem relative to their feeding behaviour. Most of the research and modelling of the impact of climate change on marine mammals has been conducted using this ecosystem focus (Huntington and Moore, 2008; Marine Mammal Commission, 2009; Cooper *et al.*, 2009; Pook, 2009; Simpkins *et al.*, 2009). The concept is simple. If ocean temperature changes drive a shift in the distribution of prey items, then a 'bottom-up' scenario may become active. In this case, there is a fundamental shift in the distribution of the food web 'below' seals and whales. Their prey items may move away and be replaced by prey that have followed the warmer water. Alternately, the colder water species may move to deeper and colder waters and consequently move out of the diving depth abilities of marine mammals. In that case, marine mammal species that cannot dive deeply enough to find their regular prey may become compromised. Furthermore, adult marine mammals can dive deeply, but the younger and more immature pups and calves may not be able to reach their preferred prey. In all these scenarios, the core principle is constant. The changing climate alters the type, distribution of abundance of the prey, and the higher trophic level marine mammals are compromised.

Only in recent years have studies been designed to examine this particular aspect of climate change and marine mammals in the Arctic. They have shown what could be the first signs of prey differences in the diets of marine mammals over time (Cooper *et al.*, 2009; Smith *et al.* 2010). There have been many more studies that have shown distribution changes in plankton and fish species as warmer water species move northwards or polar species change their distribution (Hunt *et al.*, 2002; Ledhodey *et al.*, 2006; Wassmann, 2008; Wells *et al.*, 2008; Grebmeier *et al.*, 2010).

One of the complexities of these analyses is correcting for the plasticity of the diet requirements of marine mammals. Would it be a problem if prey species 'X' is replaced by species 'Y' in the normal hunting range of a population of marine mammals? Would seals simply switch their diet, or would they feed less and become compromised? Would whales change their diet if a species of plankton disappeared? These are notoriously difficult questions to answer and even in cases in the North Pacific where clear population declines of marine mammals have occurred, it is not obvious whether the fundamental issues are related to food, climate change or human interactions (Committee on the Alaska Groundfish Fishery and Steller Sea Lions, National Research Council, 2003). The fundamental point of the majority of studies on climate change and marine mammals is the same: the impact to marine mammals would come from ecosystem alterations or because of essential changes in human interactions, not because the basic physiology of marine mammals is compromised by oceanic water temperature changes.

Cross-species interactions

One particularly interesting theory that is somewhat midway between direct impacts on the physiology of marine mammals, and ecosystem processes that would occur because of changing ice conditions, has to do with cross-species interactions. In this scenario, marine mammals that do not usually interact with one another because of differences in how they inhabit the sea ice platforms, might find themselves competing for the same pelagic resources. In some cases, there could even be instances of cross-breeding between closely related species, and the creation of hybrid species (Kelly *et al.*, 2010). As the sea ice patterns change, it is also possible that species that rely on the movement of ice to disperse their breeding populations may become more isolated. The wide mix of possibilities of changing ice conditions would most likely affect breeding systems and therefore the genetic patterns of marine mammals. The impacts of these possibilities on the molecular genetics of marine mammals are covered extensively by O'Corry-Crowe (2008). Could this happen in the northern and southern oceans if the ice platforms that are central to breeding social behaviour are minimized? Again, testing data to support such a theory would be difficult, but it has been considered as one of the potential impacts of declining ice platforms.

9.2.2 Oceanic water temperatures

There are many more species of marine mammals associated with oceanic and non-ice conditions than the ice-associated species. What would the impact of changing open ocean water temperatures be on these species? Again, the primary impact would most likely come from bottom-up changes in prey distribution as plankton, fish, and other prey items shift their distributions. The direct impact of a change of a few degrees Celsius in temperature on the physiology of marine mammals is easy to thermoregulate with abilities already developed. For example, many species of seals and sea lions move back and forth between cold oceanic waters while feeding and warm beaches where they rest and have their pups. These large differences in temperature ranges are easily compensated for through their ability to conserve or dump heat through the blubber or fur layers. It is important to remember in these discussions that these mammals defend a core temperature of about 37°C although ambient temperature ranges from 4°C in the ocean to well over 30°C on many temperate or tropical beaches. Furthermore, the heat capacity of water is over 25 times that of air, so the actual ability to regulate heat flow across the body is both well developed and critical to marine mammals, regardless of whether they have part of their life cycle associated with sea ice.

Migrations and pup/calf biology

Despite the ability of adult marine mammals to thermoregulate over far wider ranges of temperature than would ever occur under climate change scenarios, there are two aspects of how oceanic water temperatures could impact the biology of these groups. Many species of marine mammals migrate seasonally every year, back and forth between breeding grounds (usually warmer) and feeding grounds (usually colder). If the temperature regimes of either the breeding areas or the feeding grounds change significantly, then it is possible that the migratory routes may have to shift to compensate. On first review, this might seem like a minor impact to marine mammals, but it could be significant for several reasons. For example, the adult male northern elephant seal migrates from its breeding/social grounds in central/southern California in the winter to its summer feeding grounds off the Aleutian trenches in the summer (LeBoeuf and Laws, 1994). If a temperature change altered the distribution of the prey for the male elephant seals, then presumably they would need to shift their annual migration patterns. How flexible are the patterns, and could the animals stay at sea longer if the prey moved further away (presuming the prey continue to move northwards to remain in cooler water)? Another example might be the grey whale that migrates down the western North American coastline to breed in the lagoons of Baja California and swims north to feed in the Bering sea region (Jones and Swartz, 2009). If that migration range becomes larger, if the prey distribution changes, or if the water temperatures needed by the calves to effectively thermoregulate is altered, then how will this impact the species? The answers to these questions are not known.

9.2.3 Summary of environmental changes

How would changing oceanic temperatures impact the ecosystems of marine mammals and how would that alter the biological success of these species? The salient points to consider include:

- Changes in prey distribution that would drive 'bottom-up' alterations in food availability.
- Changes in sea ice distribution that would alter breeding, social, feeding, pupping, and resting/moulting behaviours and success.
- Potential mixing of species that are usually separated by ice conditions that could impact feeding competition and breeding systems.
- Alterations in migratory behaviour.
- Minimal, if any, impact on the physiology/biochemistry of the marine mammals themselves due to environmental temperature changes.

9.3 Temperature Regulation in Marine Mammals

Given the caveat that the major impact of potential climate change to marine mammals is most likely to come about from alterations in their ecosystems, it is important to discuss the abilities of these species to thermoregulate. As noted above, most marine mammals are already capable of regulating their physiology over environmental temperature ranges far beyond what may come about because of climate change. They have developed these thermoregulatory abilities in order to move between warm and cool ocean basins for breeding/feeding and to exploit the ability to live both in the water and on land. They do this through a suite of physiological and biochemical adaptations, which are discussed below.

9.3.1 Insulation in marine mammals

A unifying characteristic of marine mammals is that they spend most, if not their entire lives, in an ocean that is significantly colder than their core temperature of 37°C. Based upon fundamental thermoregulation equations, this aquatic life represents a significant thermal challenge to these mammals. Marine mammals use either fur or blubber for insulation, and like all endotherms, balance their metabolic heat production with various pathways of heat loss. However, the uses of blubber or fur have their own biological costs. Whereas blubber is used for thermoregulation, it is also a primary source of metabolic fuel for a marine mammal during fasting periods and plays a role in buoyancy regulation and water balance. Blubber is a unique tissue in marine mammals and is not found outside of that group except for a similar tissue in polar bears and some penguins (Pabst *et al.*, 1999). Fur, however, is found in both terrestrial and marine mammals and the highest quality (density) fur is found in the sea otter. Because fur traps air in its hairs, it is a very good insulator as long as it is carefully maintained, groomed and kept dry on the layer next to the skin.

Marine mammals have no unusual heat-generating mechanisms or tissues compared to any other mammal. For example, while some large warm-bodied fishes have specialized heat-generating tissues behind their eyes, no such organs or tissues exist in marine mammals, except perhaps as brown fat in newborns of some seal species (Blix *et al.*, 1979). As noted before, marine mammals also have a typical mammalian body core temperature. In fact, upon close examination of the data, there appears to be nothing special about marine mammals that would distinguish them from terrestrial mammals when it comes to heat-generating mechanisms.

Given the particularly nondescript aspects of marine mammal heat generation, there must be something that is different about them because they live in a cold liquid environment that would be fatal to all terrestrial mammals. Given the fundamental balance equation of thermoregulation, this suggests that they must have adapted significant ways to alter the heat loss through reduced conduction and convection, and we find that they have done this through the use of blubber, fur and vascular adaptations.

Blubber

Blubber is often assumed to be an inert fat layer beneath the skin. However, it is actually a complex, active tissue that consists of a loose, spongy material where the matrix of the sponge is made up of collagen fibres and the volume is made of adipocytes (Pabst *et al.*, 1999). As the blubber layer increases or decreases, the collagen matrix remains the same and it is the movement of lipid in and out of that matrix that accounts for the change in blubber quality and characteristics. Blubber depth can range from just 1–2 mm in newborn pinniped pups to 50 cm thick in large whales. The key issue here is that blubber, by itself, is a good insulator because it can be up to 93% lipid with very little water content. Lipid has a conductance of only about one-third that of water and acts as a relatively good insulator. Furthermore, because blubber is deposited below the skin, the skin layer itself will be only marginally warmer than the surrounding water. In polar waters, for example, the skin of a whale or a seal would be just a degree or two above freezing, whereas the core temperature would remain at about 37°C.

Blubber should be thought of as a very dynamic tissue with multiple stressors and pressures on its biology. Because it is a critical tissue for several different processes in marine mammals, it cannot be modelled in a strictly thermal scenario. For example, during a time of fasting, the animal will utilize blubber heavily, which would be a problem if it was also being challenged with an increasing thermal demand. Hence, fasting periods associated with breeding occur in warmer months or in warmer water for most marine mammals. Rosen and Renouf (1997) and Ryg *et al.* (1988) have written about the relationships between the seasonal distribution of blubber and thermal problems in seals. We have recently noted (Castellini *et al.*, 2009) that even in the same Antarctic environment of pack ice, there is variation in the relative blubber content of the four species of seals that inhabit the pack ice zone. We attributed this difference not so much as to thermal challenges (because they were all in same thermal zone), but because of the different metabolic requirements that each species may possess for the lipids in the blubber. In fact, it appears that Weddell seals might be over-insulated in terms of blubber thickness, but put on excess blubber for energy supplies during the breeding cycle.

Fur

As with terrestrial mammals, fur in marine mammals functions by trapping dry air next to the skin and keeping water (or cold air for a land mammal) away from the skin surface. Thus, the temperature gradient is from the skin outwards with a warm skin surface and cold outer layers of fur (Boyd, 2000). The most-cited example of the use of fur by a marine mammal is that of the sea otter, and it provides an excellent example of how this animal lives in a cold environment. The sea otter is faced with a major thermal challenge, as it is a small mammal with a large surface area to volume ratio through which to lose heat. It utilizes a dense fur with a series of guard hairs and under furs to keep its skin warm. However, the cost of this luxurious fur coat is a tremendous amount of maintenance with up to 12% of daily energy expenditure being spent on grooming the coat (Williams *et al.*, 1992).

It is the reliance on a high quality fur in the sea otter and fur seals that makes these mammals particularly vulnerable to oil spills. Oil permeates the fur and destroys the air pockets that provide the thermal insulation for the animal. After the *Exxon Valdez* oil spill (EVOS) in Alaska, there was a massive clean-up operation on the hundreds of sea otters that were brought to rescue and rehabilitation centres. The goal was to clean the fur to restore its thermal insulation properties. However, cleaning the fur of man-made oils also cleans the fur of the natural oils (primarily squalene) that help make the fur water resistant. Therefore, small amounts of lipid had to be added and groomed back into the fur of the otters after they were cleaned of the heavy oil. For a general summary of the impact of the EVOS event on marine mammals, see Loughlin (1994), and for a detailed discussion on otters, see Williams and Davis (1995).

9.3.2 Vascular adaptations

It is in the area of vascular thermoregulation that marine mammals have evolved several unusual adaptations. The first of these is termed the *rete mirabile* which is Latin for a 'wonderful net'. This net, which is a countercurrent heat exchanger (Scholander and Schevill, 1955), involves an intertwined network of veins and arteries such that the cold blood returning from the extremities in the veins runs next to the warm blood going out to extremities in the arteries. Heat flows from the arteries to the nearby veins thus tending to conserve the heat in the interior and cool the arterial blood going out to the colder regions of the body. Marine mammals have exquisite control of blood flow in their bodies, not only for thermoregulation but also for diving. However, these two demands are themselves interrelated, and the control of one impacts the control of the other. For example, it would not benefit a diving seal to be closely controlling blood flow for oxygen conservation but then to override that control to dump or gain heat. In fact, Elsner and Gooden (1983) discussed some experiments with seals where the diving response inhibited thermoregulatory-driven circulatory adjustments. In another innovative study, divers were able to apply heat flow probes to the skin of dolphins while both divers and dolphins were underwater. The results showed that the animals tend to defer heat regulation and favour oxygen conservation vascular adjustments when both must coincide (Noren *et al.*, 1999; Williams *et al.*, 1999).

These circulatory retes are found in several locations in marine mammals (and in some cold-adapted birds), with the most-cited examples being in the flukes of whales and the flippers of pinnipeds (Tarasoff and Fisher, 1970; Kvadsheim and Folkow 1997; Meagher *et al.*, 2002). Interestingly, another fascinating rete is used to cool down the reproductive organs of dolphins and seals by bringing in cold blood from the extremities (Rommel *et al.*, 1995).

Another important vascular adjustment seen in marine mammals deals with those mammals that utilize thick blubber as an insulating material. Although having thick blubber is a good method for staying warm, it can cause serious problems when animals try to cool. In fact, the large whales have such a tremendous thermal mass and a low surface area to volume ratio that they may have a much more serious problem dumping heat than conserving it (Hokkanen, 1990). Whereas some fur-bearing marine mammals have been shown to use sweat glands as a method of dumping heat (Rotherham *et al.*, 2005), these would not be functional underwater. Of greater significance for thermoregulation is the fact that blubber is not simply an inert organic blanket surrounding the animal, but it is vascularized with a series of anastomoses, or blood flow shunts. These shunts control the amount of blood moving through the blubber and reaching the skin, thereby regulating the amount of heat lost to the environment. If a seal needs to dump heat, the anastomoses open and warm blood can reach the surface of the skin. Indeed, when Weddell seals dump excess heat in this manner, clouds of steam come off the animals as the blood reaches the surface of their skin. In some cases, the seals become so warm that they partially melt their way into the ice and leave perfect 'seal shadows' (Fig. 9.1). Conversely, when these circulatory shunts are closed, the same seals can be completely covered in snow with no signs of melting at any location except near the eyes and nose (Fig. 9.2).

Hence, the balance of blood flow throughout the bodies of marine mammals can be complex and is controlled by multiple demands: diving, exercise and heat regulation. Diving requires limited blood circulation, simultaneous underwater exercise requires increased circulation, and thermoregulation can require both. How these animals balance those conflicting demands is an area where much more work needs to be done. This can be seen with even simple manipulations of seals and sea lions. For example, when taking blood samples from the flippers of pinnipeds, the flippers must be warm or there is no blood flow out to the periphery. However, if anaesthesia or sedation is required in order to work with the animal, those procedures may also cause a series of vascular adjustments and can dump great amounts of heat quickly. Under these conditions, externally generated heat needs to be added to the animal to keep the core temperature up and blood flow open to the flippers.

9.3.3 Young animals and temperature development

For the ice-associated breeders such as the Arctic and Antarctic seals, there is an aspect of the stable platform of sea ice that is essential to their reproductive success. Many species of seals utilize blubber for thermal protection as adults, but use a specialized fur, called lanugo, as newborns. Lanugo, or pup fur, is a very effective insulator in the air and is usually long and very fluffy. On newborn pups, it functions as protection against the cold air during the time that they are on land or ice for nursing. Lanugo is useless in water and allows the skin to chill to essentially water temperature. A pup must shed its lanugo and develop a significant blubber layer before it can enter the water and be an effective swimmer and diver. Not all species of seal or sea lion pups are born with lanugo, but its purpose is well documented in many cases. Lavigne and Kovacs (1988) provide an excellent description of the first few days of life for harp seals as they adapt from the warm temperature inside the womb to the frigid cold of being born on the ice. McCafferty *et al.* (2005) discuss the thermoregulatory problems faced by grey seal pups while still at their nursing age and Dunkin *et al.* (2005) examine the thermal properties of blubber during development in dolphin calves.

Because seal pups are generally not adapted to be able to thermoregulate in cold water, they must remain on the ice surface and nurse for a time long enough to gain a significant layer of blubber. This nursing time ranges from an extremely short 4–5 days in hooded seals (pups gain ~5 kg/day by nursing; Lavigne and Kovacs, 1988) to a month or longer in other species. In all cases, however, the pups acquire significant amounts of blubber before they are weaned, and along with the blubber comes the ability to maintain their body temperature in cold water. The core concepts of blubber development, thermal characteristics, and the ability to control the circulation of blood through the blubber are the essential physiological and biochemical principles involved.

Another possibility to consider is whether young marine mammals have specialized abilities to generate heat to stay warm before their blubber or fur insulation is fully developed. The only specialized heat-generating tissue that has ever been found in marine mammals is brown fat in harp seal pups (Blix *et al.*, 1979). This tissue is thermogenically active via oxidation of lipids, but only for about the first 3 days after birth. This is an important source of heat for these young pups, but not unique since brown fat is found in many terrestrial mammals where it serves the same purpose.

9.3.4 Impact of diving

What are the essential elements of thermoregulation in marine mammals? In common with all endotherms, these mammals must obey the physics of heat balance when holding body temperature constant. The methods for producing heat (resting metabolism and exercise) must balance the windows for heat loss (primarily conduction and convection) (Whittow, 1987). Because marine mammals do not appear to have any special adaptations for producing excess heat, most of their ability to thermoregulate comes with their ability to control heat loss. Control of heat loss is regulated via biochemical, anatomical, physiological and behavioural mechanisms. However, as in all levels of adaptation to the environment, systems cannot be considered or modelled in isolation. The problem with balancing blood flow for thermoregulation while also controlling blood flow for diving is an excellent example of this. For example, as a consequence of marine mammals adjusting their physiology to maximize dive time by reducing metabolic rate and reducing blood flow to the periphery, their periphery begins to cool and it is possible to measure a drop of a few degrees in body temperature during such dives (Kooyman *et al.*, 1980; Hill *et al.*, 1987). The advantage is that a reduced body temperature then reduces metabolic rate via Q_{10} principles, which can extend dive time. However, the animal must not let this continue unchecked, because it must maintain minimum body temperatures for essential functions and be able to regain that temperature loss when at the surface after diving. This is not like the case of hibernating mammals that become inactive and dormant while cold. A marine mammal must be able to swim, hunt, digest and process information while diving. While small drops in body temperature are tolerable and even advantageous, the diving seal cannot let its body temperature drop too low.

9.4 Conclusions

The core concepts of the impact of potential climate change on marine mammals are simple in theory, but complex in nature.

1. The animals already have the biochemical, physiological and anatomical adaptations to withstand extremely large changes in temperature far beyond any potential change in water temperature due to climate change. Therefore, it is extremely unlikely that there would be a direct alteration in their biology due to environmental temperature changes.

2. The major issue that will alter marine mammal biology will come from 'bottom-up' ecosystem changes. As their prey changes distribution, either by moving to different regions or by moving to different depths in the water column, then feeding effort will need to change. How capable are marine mammals of switching prey if their preferred prey are not available? If prey moves to a deeper depth, will the animals be able to dive long enough to reach the prey?

3. For ice-associated marine mammals, significant changes in ice distribution will impact their breeding success, ability to moult, and have a safe area to rest, avoid predators, and raise their pups. Seal pups cannot be born in the water and do not have the ability to thermoregulate in cold water until they nurse long enough to develop a thick blubber layer. In public lectures about climate change and marine mammals, the question is often asked why disappearing ice would matter to a seal or a walrus: they can swim, so why does ice matter? My answer is always the same: this 'solution' would exist for exactly one generation and then there would be no pup survival.

4. Migratory patterns may have to change. This is another example of 'bottom up' impacts on the feeding component of migration and perhaps a calf/pup survival issue on the breeding side of migration. How capable are the animals of altering their migration patterns? Can they remain at sea for another few weeks, would they have to swim further and do they have enough energy reserves to do so?

In conclusion, there is no doubt that climate-induced changes on the water temperatures of the ocean will have an impact on marine mammals. This may not happen directly on the physiology and biochemistry of the animals themselves, but through ecosystem alterations, ice platform changes and subsequent behavioural modifications.

9.5 Acknowledgements

Sections of this review on thermoregulation in marine mammals are excerpts from Castellini (2009) with permission from Elsevier. Figs. 9.1 and 9.2 photos by M. Castellini and under Marine Mammal Protection Act, Antarctic Conservation Act, and UAF Institutional Animal Care and Use Permits to J.W. Testa and M. Castellini.

References

Arctic Council (2009) *Arctic Marine Shipping Assessment*. Protection of the Arctic Marine Environment Working Group, Arctic Council, Tromso, Norway.

Bester, M. and Hofmeyr, G. (2007) Seals: overview.

In: Riffenburgh, B (ed.) *Encyclopedia of the Antarctic*. Routledge, New York, pp. 877–880.

Blix, A.S., Grav, H.J. and Ronald, K. (1979) Some aspects of temperature regulation in newborn harp seal pups. *American Journal of Physiology* 236, R188–R197.

Boyd, I. (2000) Skin temperatures during free-ranging swimming and diving in Antarctic fur seals. *Journal of Experimental Biology* 203, 1907–1914.

Boyd, I. (2009) Antarctic marine mammals. In: Perrin, W.F., Wursig, B. and Thewissen, J.G.M. (eds) *Encyclopedia of Marine Mammals*. 2nd ed., Academic Press, Oxford, pp. 42–46.

Burns, J.J. (2009) Arctic marine mammals. In: Perrin, W.F., Wursig, B. and Thewissen, J.G.M. (eds) *Encyclopedia of Marine Mammals*. 2nd ed., Academic Press, Oxford, pp. 48–54.

Castellini, M.A. (2009) The thermoregulation of marine mammals. In: Perrin, W.F., Wursig, B. and Thewissen, J.G.M. (eds) *Encyclopedia of Marine Mammals*. 2nd ed., Academic Press, Oxford, pp. 1166–1171.

Castellini, M.A., Trumble, S.J., Mau, T.L., Yochem, P.K., Stewart, B.S., and Koski, M.A. (2009) Body and blubber measurements of Antarctic pack ice seals: Implications for the control of body condition and blubber regulation. *Physiological and Biochemical Zoology* 82, 113–120.

Committee on the Alaska Groundfish Fishery and Steller Sea Lions, National Research Council (2003) *Decline of the Steller Sea Lion in Alaskan Waters: Untangling Food Webs and Fishing Nets*. The National Academies Press, Washington, DC, pp. 216.

Cooper, M., Budge, S., Springer, A. and Sheffield, G. (2009) Resource partitioning by sympatric pagophilic seals in Alaska: monitoring effects of climate variation with fatty acids. *Polar Biology* 32, 1137–1145.

Dunkin, R.C., McLellan, W.A., Blum, J.E. and Pabst, D.A. (2005) The ontogenetic changes in the thermal properties of blubber from Atlantic bottlenose dolphin Tursiops truncatus. *Journal of Experimental Biology* 208, 1469–1480.

Elsner R. and Gooden, B. (1983) *Diving and Asphyxia*. Monographs of the Physiological Society 40, Cambridge University Press, Cambridge.

Fish and Wildlife Service, Department of the Interior (2008) Endangered and Threatened Wildlife and Plants; Special Rule for the Polar Bear. *Federal Register* 73, 76249–76269.

Grebmeier, J.M., Moore, S.E., Overland, J.E., Frey, K.E. and Gradinger, R. (2010) Biological response to recent Pacific Arctic sea ice retreats. *EOS* 91, 161–162.

Hill, R.D., Schneider, R.C., Liggins, G.C., Schuette, A.H., Elliott, R.L., Guppy, M., Hochachka, P.W., *et al.* (1987) Heart rate and body temperature during free diving of Weddell seals. *American Journal of Physiology* 253, R344–R351.

Hokkanen, J.E. (1990) Temperature regulation of marine mammals. *Journal of Theoretical Biology* 145, 465–485.

Hunt, G.L., P. Stabeno, G. Walters, E. Sinclair, R.D. Brodeau, J.M. Napp and Bond, N.A. (2002) Climate change and control of the southeastern Bering Sea pelagic ecosystem. *Deep Sea Research, Part II* 49, 5821–5853.

Huntington, H.P. and Moore, S.E. (2008) Assessing the impacts of climate change on Arctic marine mammal. *Ecological Applications* 18, S1–S2.

Jones, M.L. and Swartz, S.L. (2009). Gray whale. In: Perrin, W.F., Wursig, B. and Thewissen, J.G.M. (eds) *Encyclopedia of Marine Mammals*, 2nd ed., Academic Press, Oxford, pp. 503–511.

Kelly, B., Whiteley, A. and Tallmon, D. (2010) The Arctic melting pot. *Nature* 468, 891.

Kooyman, G.L., Wahrenbrock, E.A., Castellini, M.A., Davis, R.W. and Sinnett, E.E. (1980) Aerobic and anaerobic metabolism during voluntary diving in Weddell seals. Evidence of preferred pathways from blood biochemistry and behavior. *Journal of Comparative Physiology* 138, 335–346.

Kvadsheim, P.H. and Folkow, L.P. (1997) Blubber and flipper heat transfer in harp seals. *Acta Physiologica Scandinavica* 161, 385–395.

Lavigne, D.M. and Kovacs, K.M. (1988) *Harps and Hoods: Ice Breeding Seals of the Northwest Atlantic*. University of Waterloo Press, Waterloo, Canada.

LeBoeuf, J.B. and Laws, R.M. (eds) (1994) *Elephant Seals - Population Ecology, Behavior and Physiology*. University of California Press, Berkeley, California.

Lehodey, P., Alheit, J., Barange, M., Baumgartner, T., Beaugrand, G., Drinkwater, K., Fromentin, J.-M., *et al.* (2006) Climate variability, fish and fisheries. *Journal of Climate* 19, 5009–5030.

Loughlin, T.R. (ed.) (1994) *Marine Mammals and the Exxon Valdez*. Academic Press, San Diego, California.

Marine Mammal Commission (2009) *Climate Change and Marine Mammals: A Framework for Monitoring Arctic Marine Mammals*. Annual Report, Bethesda, Maryland, pp. 141–143.

McCafferty, D.J., Moss, S., Bennett, K. and Pomeroy, P.P. (2005) Factors influencing the radiative surface temperature of grey seal (Halichoerus grypus) pups during early and late lactation. *Journal of Comparative Physiology B* 175, 423–431.

Meagher, E.M., McLellan, W.A., Westgate, A.J., Wells, R.S., Frierson, D. and Pabst, D.A. (2002) The relationship between heat flow and vasculature in the dorsal fin of wild bottlenose dolphins Tursiops truncatus. *Journal of Experimental Biology* 205, 3475–3486.

Noren, D.P., Williams, T.M., Berry, P. and Butler, E. (1999). Thermoregulation during swimming and diving in bottlenose dolphins, Tursiops truncatus. *Journal of Comparative Physiology B* 169, 93–99.

NSIDC (2011) http://nsidc.org, accessed 23 June 2011.

O'Corry-Crowe, G. (2008) Climate change and the molecular ecology of Arctic marine mammals. *Ecological Applications* 18, S56–S76.

Pabst, D.A., Rommel, S.A. and McLellan, W.A. (1999) The functional morphology of marine mammals. In: Reynolds, J.E. and Rommel, S.A. (eds), *Biology of Marine Mammals*. Smithsonian Institution Press, Washington, DC, pp. 15–72.

Pook, M. (2009) Antarctic climate, weather and the health of Antarctic wildlife. In: Kerry, K.R. and Riddle, M.J. (eds) *Health of Antarctic Wildlife: A challenge for Science and Policy*. Springer-Verlag, Berlin pp. 195–209.

Riffenburgh, B. (ed.) (2007) *Encyclopedia of the Antarctic*. Routledge, New York.

Rommel, S.A., Early, G.A., Matassa, K.A., Pabst, D.A. and McLellan, W.A. (1995) Venous structures associated with thermoregulation of phocid seal reproductive organs. *Anatomical Record* 243, 390–402.

Rosen, D.A.S. and Renouf, D. (1997) Seasonal changes in blubber distribution in Atlantic harbor seals: Indications of thermodynamic considerations. *Marine Mammal Science* 13, 229–240.

Rotherham, L.S., van der Mewe, M., Bester, M.N. and Oosthuizen, W.H. (2005) Morphology and distribution of sweat glands in the Cape fur seal, Arctocephalus pusillus pusillus. *Australian Journal of Zoology* 53, 295–300.

Ryg, M., Smith, T.G. and Øritsland, N.A. (1988) Thermal significance of the topographical distribution of blubber in ringed seals (Phoca hispida). *Canadian Journal of Fisheries and Aquatic Science* 45, 985–992.

Scholander, P.E. and Schevill, W.E. (1955) Countercurrent vascular heat exchange in the fins of whales. *Journal of Applied Physiology* 8, 279–282.

Simpkins, M., Kovacs, K.M., Laidre, K. and Lowry, L. (2009) A framework for monitoring Arctic marine mammals. Findings of a workshop sponsored by the U.S. Marine Mammal Commission and U.S. Fish and Wildlife Service, Valencia, March 2007. CAFF International Secretariat, Conservation of Arctic Flora and Fauna, Circumpolar Biodiversity Monitoring Program Report 16, Borgir, Iceland.

Smith, P.A., Elliott, K.H., Gaston, A.J. and Gilchrist, H.G. (2010). Has early ice clearance increased predation on breeding birds by polar bears? *Polar Biology* 33, 1149–1153.

Stirling, I. (2009) Polar bears. In: Perrin, W.F., Wursig, B. and Thewissen, J.G.M. (eds) *Encyclopedia of Marine Mammals*. 2nd ed., Academic Press, Oxford, pp. 888–890.

Stonehouse, B. (2007) Penguins: overview. In: Riffenburgh, B. (ed.) *Encyclopedia of the Antarctic*. Routledge, New York, pp. 719–722.

Tarasoff, F.J. and Fisher, H.D. (1970) Anatomy of the hind flippers of two species of seals with reference to thermoregulation. *Canadian Journal of Zoology* 48, 821–829.

Wang, M. and Overland, J.E. (2009) A sea ice free summer Arctic within 30 years. *Geophysical Research Letters* 36, L0750.

Wassmann, P. (2008) Impacts of global warming on Arctic pelagic ecosystems and processes. In: Duarte, C.M. (ed.) *Impact of Global Warming on Polar Ecosystems*. FBBVA Press, Bilbao, pp. 111–138.

Wells, B.K., Field, J.C., Thayer, J.A., Grimes, C.B., Bograd, S.J., Sydeman, W.J., Schwing, F.B., *et al.* (2008) Untangling the relationships among climate, prey and top predators in an ocean ecosystem. *Marine Ecology Progress Series* 364, 15–29.

Whittow, G.C. (1987) Thermoregulatory adaptations in marine mammals: Interacting effects of exercise and body mass. A review. *Marine Mammal Science* 3, 220–241.

Williams, T.D., Allen, D.D., Groff, J.M. and Glass, R.L. (1992) An analysis of California sea otter pelage and integument. *Marine Mammal Science* 8, 1–18.

Williams, T.M. and Davis, R.W. (eds) (1995) *Emergency Care and Rehabilitation of Oiled Sea Otters: A Guide for Oil Spills Involving Fur-Bearing Marine Mammals*. University of Alaska Press, Fairbanks, Alaska.

Williams, T.M., Noren, D., Berry, P., Estes, J.A., Allison, C. and Kirkland, J. (1999) The diving physiology of bottlenose dolphins (Tursiops truncatus) – III. Thermoregulation at depth. *Journal of Experimental Biology* 202, 2763–2769.

10 Climate Change and Plant Diseases

Denis A. Gaudet, Anne-Marte Tronsmo and André Laroche

10.1 Introduction

Climate change resulting from greenhouse gas emissions associated with human activities is widely accepted among scientists as a fact. This change is expected to continue and even accelerate as present global CO_2 and methane emissions continue (House *et al.*, 2008; Matthews and Caldeira, 2008; Nordhaus, 2010). Models predict that global surface air temperatures will increase by 1.4°C–5.8°C by 2100 relative to 1990 (Houghton *et al.*, 2001). Continental regions and higher latitudes are projected to warm more than coastal regions and the tropics, night-time minimum temperatures are expected to increase more than daytime maximum temperatures, and winter temperatures at higher latitudes are expected to increase more than summer temperatures (Houghton *et al.*, 2001; Harvell *et al.*, 2002). Rising temperatures will have secondary impacts, in particular, on humidity and precipitation and some regions will experience higher rainfall and increased humidity while others have an increased risk of drought; however, globally, average rainfall, evaporation, and humidity are projected to increase (Houghton *et al.*, 2001). Because current terrestrial global food and fibre production is largely dependent on favourable climates worldwide, scientists, policy makers and the public at large have a vested interest in understanding the potential for climate change to drastically alter global plant-based biomass production.

Plant diseases have always been a threat to food production. Two frequently cited examples of large-scale human impacts of plant diseases are the Irish Potato Famine in the 1840s caused by *Phytophthora infestans*, whereby over 2 million persons died of starvation or were displaced; and the Great Bengal Famine of 1943 caused by drought and an epidemic by the fungus *Cochliobolus miyabeanus* on rice, when an estimated 2 million people died (Agrios, 2005; Strange and Scott, 2005). Currently, plant diseases cause annual losses of approximately 10% to global food and fibre supplies; this figure would be much higher without the use of pesticides commonly employed for their control (Strange and Scott, 2005). The impacts of plant diseases are greatest in those regions of the developing world where approximately 1 billion of the poorest people live and even seemingly small chronic losses of 5%–10% due to plant diseases can impact human populations, particularly when coupled with other environmental stresses such as drought or flooding. With predicted future population growth and the technological challenges in meeting the predicted food demands, the situation is likely to become even more critical.

Several treatises have been published on the impact of climate change on plant diseases (e.g. Coakley *et al.*, 1999; Anderson *et al.*, 2004; Boland *et al.*, 2004; Chakraborty and Newton, 2011). The present review emphasizes the effects of climate change on diseases in both natural and managed ecosystems, the drivers of pathogen variability, and strategies to mitigate future impacts of plant diseases. When considering the effects of climate change on plant disease, it is important to make a distinction between

© CAB International 2012. *Temperature Adaptation in a Changing Climate: Nature at Risk* (eds K.B. Storey and K.K. Tanino)

plants that are growing in managed systems that encompass agriculture, horticulture and some silviculture, and those in natural systems that include native forests, grasslands and other habitats. In managed systems, crops tend to possess a narrow genetic base and are grown in intensive monocultures (Burdon *et al.*, 2006). This situation is conducive for the development of disease epidemics that can rapidly destroy a crop. The Southern Corn Leaf Blight epidemic in the 1970s in the predominately Texas male sterile corn hybrids grown in the USA is such an example (Agrios, 2005). Natural systems, characterized by the long-lived nature of plants, their broader genetic base, and heterogeneity within the eco-system, are less prone to sudden disease epidemics (Burdon *et al.*, 2006). Important exceptions relate to the introduction of invasive plant pathogens into a natural system to which native plant species have little or no resistance. Dutch elm disease and chestnut blight are examples of invasive diseases that could eventually cause the extinction of native populations of American elm and American chestnut, respectively, in North America (Burdon *et al.*, 2006). Field crops and other annual or short-lived vegetation reflect changes in weather whereas trees, soil flora, and other long-lived flora reflect both weather and climatic changes (Hepting, 1963). Considering these distinctions and the past impact that plant diseases have had in managed and/or natural systems, it is pertinent to ask the question 'Will climate change substantially alter the nature of plant–pathogen interactions in such a way that diseases become more damaging than has been observed until now?'

Of the major groups of plant pathogens, the most important are the fungi, bacteria and viruses. Fungi and bacteria develop best under conditions of high moisture and humidity but each pathogen has its particular optimum temperature range for development (Agrios, 2005). Viruses are spread primarily by insect and mite vectors (Agrios, 2005) and dry weather tends to favour development and spread of these vectors. Hence, numerous plant viruses and some fungal diseases that are vectored are expected to increase under dry conditions (Anderson *et al.*, 2004). An obvious impact of climate change is on global weather circulation patterns that are associated with increasingly frequent extreme meteorological events such as hurricanes, unusually high temperature, high rainfall and snowfall, and long periods of drought (Rosenzweig *et al.*, 2001; Anderson *et al.*, 2004). For example, during the dry years in the 1980s, several new and unreported cereal viruses were observed in cereals in western Canada and Montana, including barley yellow streak mosaic virus vectored by the brown wheat mite (Robertson and Carroll, 1988). These viruses have not since been observed. Changes in global weather circulation patterns could favour the regional and long-range transport of new invasive species and their vectors (Coakley, 1988). Sugarcane rust (*Puccinia melanocephala)* was probably introduced into the Dominican Republic from Cameroon in West Africa through the transport of uredospores by cyclone winds in early 1978, and subsequently spread northwards into the southern USA and southwards into Venezuela due to prevailing wind currents later that year (Purdy *et al.*, 1985). A similar scenario was proposed for the introduction of the coffee rust pathogen (*Hemileia vastatrix)* uredospores from Angola to Brazil in 1970 (Bowden *et al.*, 1971). Regionally, Dutch elm disease was spread rapidly in the early 1950s throughout the New England States and eastern Canada by a series of hurricanes that disseminated the insect vector (Hepting, 1963).

It is inevitable that any long-term change in climate will impact the distribution and severity of plant diseases. The concept of the disease triangle (Fig. 10.1), taught in introductory plant pathology courses, describes the three factors that dictate whether a disease will be expressed. Highest disease levels are observed when a combination of susceptible host, aggressive pathogen and optimal environmental conditions occur (Agrios, 2005). Additional effects of natural enemies of plant pathogens such as parasites and herbivores, and on microbial interactions that occur in the phylloplane or in the soil may also impact disease expression (Berryman, 1982). Harvel *et al.* (2002) have predicted that climate warming will affect the incidence and severity of plant, animal and human diseases by:

1. Increasing pathogen development rates, transmission and number of generations per year due to extended and warmer growing seasons;
2. Reducing overwintering-related limitations on pathogen life cycle; and
3. Modifying host susceptibility to infection.

Elevated atmospheric CO_2 levels will have a direct effect on promoting plant growth, thereby increasing the crop canopy density and encouraging the development of pathogens (Pritchard *et al.*, 1999). Elevated CO_2 can also directly stimulate pathogen growth (Melloy *et al.*, 2010) and increase the aggressiveness and fecundity of certain plant pathogens (Chakraborty and Datta, 2003). It is probable that some host–pathogen interactions will be less severe, regionally, due to change in environmental conditions required for infection while new, potentially high impact disease interactions will result in more severe diseases.

Garrett *et al.* (2006) outlined the impacts of climate change on host–parasite interactions at the microclimate, local, regional and global levels. Needless to say, we are still in the earliest stages in gaining an understanding of the myriad of interactions between host, parasite and the changing environment. Based on our current understanding of host–parasite interactions and epidemiology, is it possible to predict whether climate change will substantively alter the nature of plant–pathogen interactions in such a way that diseases may become more damaging in managed and/or natural systems than has been observed until now? In the following discussion, we will focus on several specific plant diseases and their hosts in both managed and natural systems and explore how changing climates could potentially affect, or have already affected, the incidence and severity of disease in both managed and natural systems. We also examine the approaches and tools currently available for combating plant diseases and evaluate their future potential in mitigating losses predicted to increase as a result of climate change.

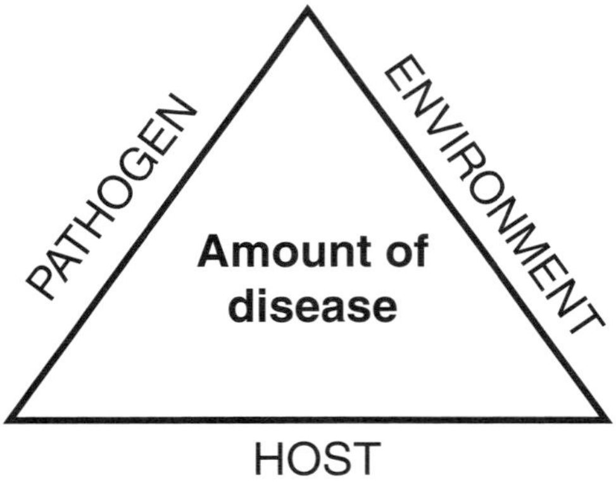

Fig. 10.1. The disease triangle whereby the host, the pathogen and the environment interact to produce disease.

10.2 Plant Diseases in Managed Systems

10.2.1 Cereal rusts

Cereal rusts constitute some of the most serious threats to global food supply (Saari and Wilcoxson, 1974; Lenné, 2000). Among the most important are the wheat rusts that include stem rust (*Puccinia graminis* f. sp. *tritici*), leaf rust (*P. triticina*) and stripe rust (*P. striiformis*) (Brown and Hovmøller, 2002). The most common method for controlling rusts has been through the incorporation of resistance genes into new varieties by breeding. Resistance genes, however, can be rendered ineffective by virulence shifts in the pathogen population. The high reproductive capacity and the novel genomic organization of rusts that is adapted to high mutation rates among virulence-related loci are important factors that continually drive the evolution of new virulent races; recombination of these new mutations with other groupings of avirulence/virulence loci through the pathogen's sexual or parasexual cycles are a constant threat to the effectiveness of resistance in wheat varieties worldwide (Chen, 2005; Hovmøller and Justesen, 2007; Jin *et al.*, 2009). For example, prior to the 1950s, stem rust caused devastating epidemics in North America but was eventually reduced to almost nonsignificant levels by 1990 by deploying combinations of race-specific and adult plant resistance in wheat varieties (Roelfs *et al.*, 1992). However, Ug99, a new and highly virulent race of stem rust that evolved in central Africa and was initially identified in Uganda in 1998, has now spread into Asia and is currently threatening global food security (Singh *et al.*, 2008). Year-round cropping of wheat in Africa permits sporulation throughout the year (Singh *et al.*, 2008), contributing to the high reproductive capacity that leads to pathogen evolution.

In North America, rusts normally overwinter in northern Mexico and in the southern USA on spring and winter wheat and are blown northwards along the Mississippi Valley on prevailing wind currents to later developing winter wheat and newly seeded spring wheat crops into northern USA and southern Canada (Nagarajan and Singh, 1990; Agrios, 2005). This is known as the Puccinia pathway. In Europe, the eastern European Puccinia pathway originates in Turkey and Romania and extends to Scandinavia, whereas the western pathway originates in Morocco and Spain and extends to Scandinavia (Nagarajan and Singh, 1990). Other Puccinia pathways exist for Asia and Australasia (Nagarajan and Singh, 1990).

A predicted outcome of climate change is the extension of the growing season in northern regions (Houghton *et al.*, 2001; Harvell *et al.*, 2002). For example, averaged across all agricultural regions in Canada, the first fall frost in 2040–2069 would be delayed by 16 d while the last spring frost (≤−2°C) would be advanced by 15 d (Rochette *et al.*, 2004). The projected change in temperature is 2.5°C–5°C in Norway, with an increase in growing season of 1–3 months during the present century (Hanssen-Bauer *et al.*, 2009). Collectively, these conditions would permit an earlier overall movement of rust disease along the different Puccinia pathways during the spring on winter and spring wheat, and extend the development period on winter wheat in the autumn. Epidemics that develop early in the crop's development have a much greater impact on yield compared to epidemics occurring later (Gaunt, 1995). The over-wintering of rusts on their hosts due to milder winters will permit the initiation of pathogen development even earlier in the spring, potentially resulting in more severe epidemics.

The evolution of the stripe rust pathogen in the Pacific Northwest (PNW) USA serves as an interesting case study of the potential for climate change to alter the nature of host–parasite interactions (Coakley, 1979). *P. striiformis* was not considered a serious pathogen in North America prior to the 1960s (Coakley, 1979). Stripe rust is normally a cool season pathogen with optimum temperature for germination of the spores ranging from 10°C–12°C with a minimum temperature for germination just above 0°C, and the optimum temperature for development of rust in plants ranging from 13°C–16°C (Newton and Johnson, 1936). The fungus does not survive temperatures below −10°C in an infected winter wheat plant and therefore does not normally overwinter on the Great Plains of North America, but will survive in some regions of PNW USA where ambient temperatures in winter do not reach critical temperatures (Chen, 2005). However, between 1961 and 1974, this pathogen caused severe epidemics in wheat and, consequently, significant yield losses in susceptible cultivars. Interestingly, January and February temperatures from 1961–1974 averaged 1.2°C higher and correlated with increased snowfall during this period, compared to the temperatures and snowfall during the previous three decades when the pathogen caused little or no damage (Coakley, 1979). Thus, relatively small differences in winter temperatures have permitted more frequent overwintering of the pathogen in the PNW USA that have resulted in dramatic increases in disease severity. Today, stripe rust is a highly aggressive and variable pathogen of wheat and triticale, quickly evolving new races that overcome resistance genes deployed in varieties worldwide (Chen, 2005; Milus *et al.*, 2009). Additionally, the fungus has evolved new, high temperature races that are adapted for growth at 18°C in North America (Milus *et al.*, 2006) and this has led to a dramatic expansion of its geographic range (Chen, 2005; Milus *et al.*, 2009; Chen *et al.*, 2010). Stripe rust could serve as an interesting model for pathogen evolution in response to climate change; increases in levels of disease could drive increases in pathogen variability globally, leading to more virulent and aggressive strains of the pathogen, the second component of the disease triangle described above. A predicted consequence of the favourable environmental conditions coupled with the rapid evolution of new, virulent races, according to the disease triangle model, would be a loss of durability of resistance genes that are used to control plant diseases.

10.2.2 Snow moulds

In regions that receive a persistent snow cover during winter, snow mould fungi can cause considerable damage to overwintering crops. These plant pathogens distributed among different species have different climatic requirements and geographic distributions, but a common characteristic among snow mould fungi is their requirement for a persistent winter snow cover during the winter and their ability to grow at temperatures down to and even below 0°C. The duration of snow cover will affect the severity of damage to the plants caused by snow moulds, and the longer the period of snow cover, the more severe the damage will generally be (Gaudet *et al.*, 1989). The numerous species of snow moulds include the extreme psycrophilic fungi *Sclerotinia borealis* and the low temperature basidiomycete (LTB) that have optimum temperature for growth rate on artificial growth media from 5°C to 10°C and can grow at temperatures as low as –8°C to –10°C (Årsvoll, 1977; Gaudet *et al.*, 1989). However, the majority grow best under snow at temperatures at or slightly above 0°C (Bruehl, 1982). The most widespread cause of biotic winter injury in winter cereals and grasses in the temperate and subarctic climates is pink snow mould caused by the fungus *Microdochium nivale*. Interestingly, this fungus has both a psycrophilic and a mesophyllic phase which permits its development both during the winter under snow, and during the warmer conditions of summer. During the summer growing season, it is the cause of *Fusarium* patch on turf grasses, and stem rot and leaf blotch as part of the fungal complex causing *Fusarium* head blight (FHB) in cereals (Tronsmo *et al.*, 2001).

Boland *et al.* (2004) and Roos *et al.* (2011) have suggested that climate change will result in a significant decrease in the prevalence and severity of snow moulds. This seems plausible because reduced snow cover duration brought upon by a warmer climate in the northern hemisphere will ultimately result in shorter incubation periods for the fungi. Indeed, under this scenario, snow moulds risk becoming extinct in some regions because of the lack of persistent snow. In another scenario described above, episodic extremes of weather including years of heavy snowfall and colder winters are predicted to become more common, according to current climate change models (Rosenzweig *et al.*, 2001; Anderson *et al.*, 2004). Some snow mould fungi can live saprophytically in the soil for years, and the inoculum of sclerotia-forming species can survive long periods in the soil in the absence of conducive environmental conditions (Hsiang *et al.*, 1999). Therefore, inoculum would be available to attack crops despite a long absence of snow, and less frequent but more severe overall attacks by snow moulds are likely to occur as a result of climate change. Moreover, the increased occurrence of weather extremes during the autumn, such as increased rain or drought, warmer temperatures and early frosts, will negatively impact the natural hardening of plants. In the autumn, plants that are adapted to northern climates respond to decreasing temperatures by undergoing physiological changes in their metabolism in a process commonly referred to as 'cold hardening'. This process is essential for development of maximal plant resistance against low temperatures and snow moulds in grasses and cereals (Årsvoll, 1977; Tronsmo, 1984a; Gaudet and Chen, 1988) and to other fungal plant pathogens (Tronsmo, 1984b; Gaudet *et al.*, 2010). Unhardened or insufficiently hardened plants are more susceptible to infection by snow moulds than fully cold-hardened plants (Tronsmo, 1984a; Gaudet and Chen, 1988; Tronsmo, 1994). Additionally, over-wintering crop plants such as grasses and cereals will be likely to face generally warmer autumns combined with increased precipitation, and consequently reduced irradiance. With warmer winter temperatures, plants will face winters with fluctuating temperatures whereby any levels of cold hardiness may be quickly lost. Under these conditions, it will be difficult to develop and maintain optimal hardening levels in plants, and hence optimal resistance to snow moulds.

Different taxa of snow moulds utilize different strategies for adaptation to cold environments (Bruehl and Cunfer, 1971; Hoshino *et al.*, 2009); therefore the effect of climatic change is expected to affect the various snow mould species differently. In Hokkaido, Japan, the distribution patterns of snow mould species have recently changed (Tamotsu Hoshino, Hokkaido, 2009, personal communication). In some regions, *Typhula ishikariensis* seems to have replaced *S. borealis* as the main snow mould species, indicating that prevailing environmental conditions have changed for sufficiently long periods to permit this shift in species distribution. However, some psychrotropic fungi such as *M. nivale* do not require snow cover for the mesophilic stage. In years that are not conducive to snow mould development this could permit accumulation of sufficient inoculum to replace other strictly psychrophilic fungi. Indeed, there are reports from Poland (Maria Wedzony, Krakow, 2010, personal communication) and the Netherlands indicating during recent years that injury due to *M. nivale* has been increasing. Its ability to grow at temperatures from sub-zero to 28°C may be a contributing factor in the apparent increase of this fungus in its area of adaptation and its spread to 'warmer' areas where it has not been recorded before, including golf courses in Hawaii, and in California with abnormal high rainfall and cooler growing season temperatures (Wong, 2007).

10.2.3 Fusarium diseases

A group of plant pathogens closely related to *M. nivale* are the *Fusarium* species. The *Fusarium* species that cause crown rot and *Fusarium* head blight (FHB) and *M. nivale* have very similar life cycles, except that *M. nivale* is better adapted to pathogenic growth at low temperatures. Other *Fusarium* spp. such as *F. avenaceum* and *F. poae* are also reported to cause snow mould on grasses (Årsvoll, 1975). FHB is caused by one or more different *Fusarium* spp. and the disease is a major concern internationally since the causal fungi produce mycotoxins that contaminate the grains. The problem has increased in magnitude over the last 25 years, and at the same time there has also been a remarkable shift in the *Fusarium* species (Parry *et al.*, 1995; Windels, 2000; Hofgaard *et al.*, 2010). In particular, *F. graminearum*, a species that readily produces the sexual stage, is transmitted both by conidia and ascospores and can be dispersed over long distances. The species has become dominant in both America and Europe during the last decade (Waalwijk *et al.*, 2003; Schmale *et al.*, 2006). The reasons for the increased severity of *Fusarium* diseases are not conclusive but changes in crop rotation and reduced tillage have been implicated (Dill-Macky and Jones, 2000; Pereyra and Dill-Macky, 2008). Agricultural practices such as higher fertility levels and use of pesticides may also be contributing factors (Edwards, 2004; Henriksen and Elen, 2005). Prevalence and abundance of different FHB species varies in different environments and climate change is projected to influence both their occurrence and production of mycotoxins (Xu *et al.*, 2008; Xu and Nicholson, 2009; Chakraborty and Newton,

2011). Diseases such as FHB are of immense importance to world food security, and the potential role of climate change in the sudden emergence of this disease during the late 1980s in temperate regions of the northern hemisphere needs to be investigated further.

10.2.4 Mechanisms for environmental adaptation among fungi

During the 21st century, the effects of a projected increase of 2°C–5°C in average temperatures in the northern hemisphere on diseases such as *Fusarium* and *Microdochium* species are likely be similar to those described above for the rusts, and translate to the production of many additional generations of spores. Increased inoculum production facilitates the evolution of the parasites towards environmental adaptation and overcoming host resistance by facultative necrotrophs, as described above in obligate biotrophs (Zhan *et al.*, 2007; Garrett *et al.*, 2009). A sexual stage has not been identified for most of the *Fusarium* spp. Environmental stresses are likely to induce/promote sexual reproduction in these fungi (Grishkan *et al.*, 2003). Recently, the sexual stage for stripe rust (*P. striiformis*) was elucidated after nearly 100 years of research (Jin *et al.*, 2010). Was this a stress-induced manifestation of the sexual stage associated with climate change? Microorganisms that can reproduce either sexually or asexually seem to express the sexual stage preferentially when exposed to environmental stresses (Taylor *et al.*, 1999). A recent study by Schoustra *et al.* (2010) showed that genotypes of *Aspergillus nidulans* reverted to sexual reproduction in stress environments in which their fitness was lowest. Even by employing only the asexual stage, isolates of *Fusarium* species display a great variability in morphology, spore production, pathogenicity and temperature requirement, as well as in mycotoxin production (Summerell *et al.*, 2010). Similar variability was observed in *M. nivale* (Hofgaard *et al.*, 2006). Variability within species of microorganisms has been attributed to the presence of transposable elements, specifically retrotransposons that are stimulated by environmental stresses (Anaya and Roncero, 1996; Gregory *et al.*, 2009). In a recent review, Gregory *et al.* (2009) discuss adaptive mutations in oomycete and fungal plant pathogens as a route for enhancing genome plasticity and rapid response to environmental changes, as well as the contribution of post-translational mechanisms. Collectively, these studies suggest that fungi react to environmental stresses by boosting their capacity for sexual reproduction. High reproductive capacity, as most plant pathogens possess, leads to large numbers of mutations that accumulate in the population. Sexual reproduction under stress environments may lead to recombination of advantageous mutations (i.e. new virulence combinations, higher aggressiveness, increased resistance to fungicides) to produce genotypes that are newer and fitter. By augmenting the stress levels encountered by plant pathogens, climatic change may ultimately expand the opportunities for spread and exploitation of new habitats for some pathogens such as *F. graminarium*, *M. nivale* and *P. striiformis*. In such a scenario, highly opportunistic microorganisms will predominate with the resulting negative impacts on crop and forest species.

10.3 Plant Diseases in Natural Systems

Coakley (1999) predicted that plants at greatest risk under a changing environment are those in which the associated pathogen has been historically contained at low levels because of unfavourable climatic conditions. Under conducive conditions, pathogens will extend their geographic range and thereby increase their regional or global impact. *Dothistroma* needle blight, caused by the fungus *Dothistroma septosporum*, has not been a problem historically in pine stands in northern temperate forests where the pest is native. Recently, however, *Dothistroma* has been responsible for extensive defoliation and mortality in plantations of lodgepole pine in north-western British Columbia, Canada (Woods *et al.*, 2005). The increasing severity of the disease has been attributed to the northerly spread of the pathogen due to a change in precipitation patterns resulting in increasing frequency of warm summer rains (Woods *et al.*, 2005). This situation is aggravated by intensive plantings of lodgepole pine in these regions. Another pathogen that is predicted to dramatically extend its range as a result of climate change is *Phytophthora cinnamomi*, considered one of the most devastating forest pathogens worldwide on oak, chestnut, eucalyptus and a wide range of nursery species (Zentmyer, 1980). The pathogen's current geographic range in Europe is in Mediterranean regions and its northern movement is restricted by freezing temperatures that limit winter survival of the fungus (Bergot *et al.*, 2004). Using current climate prediction models, a spread of the pathogen from 100–300 km eastwards from the Atlantic coast is anticipated within one century (Bergot *et al.*, 2004).

Warmer average temperatures in combination with pine wood nematodes and bark beetles are causing a decline in Scots pines in the Swiss Alps (Rebetez and Dobbertin, 2004). This situation illustrates the wider issue of climate change on the long-term decline among forest species. As early as 1963, Hepting discussed how a changing climate could act as a secondary factor in creating conditions stressful for tree species, thereby encouraging the proliferation of pathogen species (Hepting, 1963). Known as the cohort senescence theory (Mueller-Dombois, 1992), forest decline can be attributed to a physiological stress, whether catastrophic such as fire or hurricane; edaphic such as poor nutrient supply; or physiological, for example extreme weather events such as drought. Trees unable to recover from these shocks lose vigour, which in turn prevents them from maintaining normal defence responses to pathogens or insect pests. Over time, this process will lead to large-scale tree dieback. Under continued climate change, the frequency and severity of the stressors are predicted to increase, posing a serious threat to the future health and productivity in the earth's forest ecosystems. Dieback of aspen in Canadian boreal forests (Hogg *et al.*, 2002), decline of cypress in the Patagonian Andes (Filip and Rosso, 1999), oak decline and pine shoot blight in Europe, and charcoal disease in Mediterranean oaks are examples of forest decline (La Porta *et al.*, 2008).

10.4 Mitigating Losses due to Plant Diseases: Cultural Control and Pesticides

Disease control in managed systems, particularly for soil-borne pathogens that may persist in soil for long periods even in the absence of susceptible hosts, is more complex and difficult than in natural systems. This is partially due to the low organic content and lack of biological diversity in soils under managed systems. Soil biota includes termites, earthworms, N-fixing bacteria, mycorrhizal and other fungi, and nematodes. The amount of organic matter in managed systems has, until the recent advent of minimal and no-till cultivation practices, become limiting and this has contributed to reduced fertility and water retention, and increased soil erosion (Matson *et al.*, 1997). The reduced fertility has been offset by high nutrient inputs but these, particularly high nitrogen, have tended to encourage development of both soil and leaf pathogens, including insect vectors and their viruses (Mattson, 1980; Matson *et al.*, 1997). Moreover, the particular form of N fertilizer applied to a crop can have significant impacts on pathogen dynamics and disease severity. For example, soil-borne pathogens such as *Rhizoctonia* and *Fusarium* appear to cause more damage when N is applied as ammonia than as nitrate (Huber and Watson, 1974). The substantial changes in quantity and quality of the chemical composition of organic inputs to the soil in agricultural systems are also significant factors in changing the competitive balance between different organisms under organic production. Restoring a healthy balance of organic matter and soil biota by reduced tillage practices in managed systems could be important in reducing the future impact of plant diseases brought about by climate change.

Fungicides and other pesticides have traditionally been widely employed to control plant diseases, particularly in high value crops (Agrios, 2005). Due to their relatively low cost, seed treatment fungicides are often routinely employed to control seedling- and seed-borne diseases in most crops. However, fungicides are by no means universally effective or practical against all plant pathogens. For example, soil-borne fungal pathogens and nematodes are only controlled using soil fumigants that are expensive and tend to eliminate both beneficial as well as harmful organisms. While early fungicides were broad spectrum and generally persistent, modern fungicides are specific, effective at low concentrations, and are generally non-persistent in the environment. Because of their highly specific nature, modern fungicides are prone to loss of efficacy due to the evolution of new tolerant strains of the pathogens that frequently occurs within a very few years of the product's release (Brent and Hollomon, 2007). Indeed, the dynamics governing pathogen evolution of resistance to fungicides are similar to those governing the evolution of new virulent strains (described above). Expanded fungicide usage in crop production is in turn likely to lead to reduced biodiversity, as well as an increased risk of pollution and non-target effects on human health (Filser *et al.*, 1995). However, fungicides will remain an important means of control for many crop diseases and will be employed when other control measures are insufficient.

10.4.1 Mitigating losses using plant genetic resistance

Climate change will necessitate rapid genetic adaptation of our crop species to new and evolving diseases, and numerous breeding tools are now available to accelerate the release of improved crop cultivars. Prior to discussing these, it is useful to review the current models on host–parasite interactions (for a detailed discussion, refer to Abramovitch *et al.* (2006), Jones and Dangl (2006), de Wit (2007) and Boller and He (2009)). According to the model, metabolic substances are produced by both adapted and non-adapted pathogens early in the host–pathogen interaction (referred to as pathogen or microbe-associated molecular patterns: PAMP or MAMP). In non-host plants, the PAMP are recognized by pathogen recognitions receptors (PRR) as molecules that are 'non-self', and these initiate PAMP-triggered immunity (PTI) or non-host immunity (Bent and Mackey, 2007). Conversely, true pathogens will release effectors that suppress the PTI and permit continued compatibility within the host. However, resistant host plants possess a backup detection system governed by specific R-genes for races or biotypes of a pathogen species. Pioneering work by Harold Flor during the 1930s and 1940s led to the gene-for-gene concept where he determined that for each resistance gene in the host (R-gene), there was a corresponding avirulence gene (Avr-gene) in the pathogen (Flor, 1973). We now know that Avr-genes in the pathogen encode specific effectors that interact with and are recognized by specific cognate receptors (encoded by R-genes) situated either intracellularly or in the plasma membranes of host plants, and that these induce effector-triggered immunity (ETI) in plants (Jones and Dangl, 2006). Mutations in Avr-genes encoding the specificity regions of microbial effectors would prevent a normal interaction with the receptor region of the corresponding R-gene in the host, rendering the effector unrecognizable by the host, resulting in a susceptible or compatible interaction. Such a mutation would lead to the creation of a new race and each Avr-gene could be a target for such a mutation. Although most of the R-genes and their cognate Avr-genes have been characterized in model plants such as *Arabidopsis* and tobacco, several race-specific R-genes to obligate biotrophs such as the rusts and powdery mildews in cereals have been characterized, and these conform to the above model of host–parasite interaction (Collinge *et al.*, 2010). As knowledge of the R-genes and their corresponding Avr-genes expands, researchers will be better able to explain past virulence shifts and devise different strategies for deploying resistance genes.

The intense research related to the deciphering of the human genome has had tremendous benefits on research on plants and their plant pathogens. The release of the *Arabidopsis thaliana* genome sequence (The Arabidopsis Initiative, 2000) followed by the rice genome (Goff *et al.*, 2002) was the initial step in demonstrating that plant genomics could be applied to important plant species to study host–parasite interactions. Since then, genome sequences have been reported for maize (Schnable *et al.*, 2009), sorghum (Paterson *et al.*, 2009) and the new monocot model system *Brachypodium distachyon* (The International Brachypodium Initiative, 2010). These activities have paved the way for numerous additional genomics projects on crops, including bread wheat which possesses a huge genome in the order of 17 billion base pairs, 85% of which are repetitive elements (International Wheat Genome Sequencing Consortium; www.wheat-genome.org). So far, these research activities have provided sequences for reference genes and the elucidation of numerous different biosynthetic pathways. Currently, there are large-scale studies designed to modify plant organization fundamentally, such as the project to develop a photosynthetically efficient C_4 rice plant (the C_4 rice project, http://irri.org/our-partners/networks/c4-rice). Genomic research also continues to be very important for developing an understanding of pathogen dynamics. Sequencing projects for the following plant pathogens are under way or have been completed: *Botrytis cinerea*; *Fusarium* species including *F. graminearum*; *Blumeria graminis*; *Magnaporthe grisea*; *Phytophthora infestans*; *Puccinia* species including *Puccinia graminis*, *P. recondita* and *P. striiformis*; *Pyrenophora tritici-repentis*; *Sclerotinia sclerotiorum*; *Stagonospora nodorum*; *Ustilago maydis*; and *Verticillium dahlia* (The Broad Institute; www.broadinstitute.org//scientific-community/data). Putative Avr-genes have been identified and candidate Avr-genes can be predicted from different public databases based on sequence homology.

Identification of molecular markers linked to known resistance genes (known as molecular assisted selection or MAS) is extremely useful to breeding programmes as these can be applied to seed or seedlings, thereby reducing both the time and facilities needed for the identification and selection of resistant progenies. Numerous types of molecular markers are available including restriction fragment length polymorphisms (RFLPs), and PCR-based DNA markers such as microsatellites, random-amplified polyzmorphic DNA (RAPD), sequence characterized amplified regions (SCAR), sequence-tagged sites (STS), and inter-simple sequence repeat amplification (ISA), amplified fragment length polymorphic DNA (AFLPs), and amplicon length polymorphisms (ALP), (for a thorough discussion, see Mohan *et al.* (1997) and Feuillet and Keller (2005)). Molecular markers are often used to identify quantitative trait loci (QTL) when more than one gene can affect a given trait. Web sites are available, such as MASWheat (marker assisted selection in wheat; http://maswheat.ucdavis.edu/Index.htm) for particular species, facilitating the implementation of molecular screening. With the recent advances in sequencing technology and the identification of numerous single nucleotide polymorphism (SNP), novel assays which are more efficient to screen large populations have been developed (Fan *et al.*, 2006).

DNA-based techniques can be easily applied using F_2 and back-cross populations, near-isogenic lines (NIL), recombinant inbred lines (RIL) and doubled haploid lines (DH). Since it was first utilized in potatoes in 1952, doubled haploidy has become increasingly important as a breeding tool in many different crop species (Kasha and Maluszinski, 2003). Following generation of a haploid plantlet, spontaneous or chemically induced chromosome doubling yields doubled haploid plants which are homozygous for all alleles; this greatly facilitates the selection for desired traits (Kasha and Maluszinski, 2003). While the first doubled haploids produced among the small grain cereals were in barley (Symko, 1969), most cereals are amenable to doubled haploidy using anther and immature microspore culture, or by crossing to wild or unrelated species. As production of doubled haploids is often limited to specific genotypes and species, thus limiting its applications, a new method based on absence of a specific centromeric histone leading to the production of spontaneous diploid seeds in *Arabidopsis* has been described (Ravi and Chan, 2010). It is very likely that this novel approach will be applied to many different species, thereby greatly simplifying the production of DH lines. MAS applied to DH-based breeding programmes is particularly efficient because the markers can be applied to haploid tissues in culture; susceptible lines (cultures) are immediately discarded, thereby obviating the need to regenerate large numbers of the susceptible lines (Howes *et al.*, 1998).

In the short term, widespread cultivation of transgenic crops with improved resistance traits is unlikely due to international concerns over genetically modified (GM) crops entering the food chain. Introduction of transgenic cotton carrying different Bt-genes for resistance to insects in China has been successful on many fronts including wide adoption of the technology by farmers, effective protection of the crop against devastating insects, and a significant decrease in the amount of pesticide utilization and associated health-related problems among farmers (Yang *et al.*, 2005). Given the huge impact of diseases such as FHB on wheat, barley and other small grain cereals worldwide, and the potential for this pathogen to expand its range in a changing climate, it will be necessary to adopt new breeding strategies in order to ensure an abundant and safe world food supply. In the case of FHB, no specific R-genes conferring strong resistance have been identified although numerous QTL appear to have a combined impact. In this case, other strategies based on up-regulating or interrupting key genes in signalling pathways involved in the host–parasite interaction, or expressing toxins or other antimicrobial proteins that impair pathogen development, may provide resistance to multiple plant pathogens and ultimately be more durable and effective (Shah, 1997; Gurr and Rushton, 2005; Collinge *et al.*, 2010).

10.5 Summary and Conclusions

In this chapter, we have attempted to address the question 'Will climate change substantially alter the nature of plant–pathogen interactions in such a way that diseases become more damaging than has been observed until now?' in both managed and natural systems. We believe that the answer is 'yes', and that a higher overall impact of plant diseases will result from the gradual warming of the earth, particularly among those pathogens that are highly opportunistic and/or are able to rapidly modify their genetic make-up to overcome host resistance or pesticides, or to adapt to changing environments. This situation will be further aggravated by more frequent extreme weather events that move pathogens or races across large distances, or that stress plants to increase their susceptibility to pathogens. However, will climate change necessarily have a catastrophic effect on our food production systems and environment? In managed systems, many tools will be utilized to reduce the future negative impacts of climate change on crop production losses caused by plant diseases, including conventional breeding and its molecular-based variations, pesticides and cultural and biological controls. We remain optimistic that technological advances will be able to keep pace with climate-accelerated changes in pathogen severity in our managed systems. However, these technological advances are predicated on developing a comprehensive understanding of the numerous host–parasite interactions that impact crop plants and the role of a changing environment on these interactions. Considering the myriad of existing and potential host–parasite interactions, a substantial and sustained effort will be necessary worldwide. In natural systems where neither a substantial nor a sustained effort is likely, the outlook is less optimistic. More frequent accidental introductions of non-endemic plant pathogens, or radical extensions of the current geographic ranges of endemic plant pathogens, could lead to catastrophic losses ranging from severe displacement to the extinction of indigenous species.

References

Abramovitch, R.B., Anderson, J.C. and Martin, G.B. (2006) Bacterial elicitation and evasion of plant innate immunity. *Nature Reviews of Molecular and Cell Biology* 7, 601–611.

Agrios, G.N. (2005) *Plant Pathology*. Academic Press, San Diego, California.

Anaya, N. and Roncero, M.I.G. (1996) Stress-induced rearrangement of *Fusarium* retro-transposon sequences. *Molecular and General Genetics* 253, 89–94.

Anderson, P.K., Cunningham, A.A., Patel, N.G., Morales, F.J., Epstein, P.R. and Daszak, P. (2004) Emerging infectious diseases of plants: pathogen pollution, climate change and agrotechnology drivers. *Trends in Ecology & Evolution* 19, 535–544.

Årsvoll, K. (1975) Fungi causing winter damage on cultivated grasses in Norway. *Meldinger fra Norges Landgrukshøgskole* 54, 1–49.

Årsvoll, K. (1977) Effects of hardening, plant age, and development in *Phleum pratense* and *Festuca pratensis* on resistance to snow mould fungi. *Meldinger fra Norges Landgrukshøgskole* 56, 1–14.

Bent, A.F. and Mackey, D. (2007) Elicitors, effectors, and R genes: The new paradigm and a lifetime supply of questions. *Annual Review of Phytopathology* 45, 399–436.

Bergot, M., Cloppet, E., Pérarnaud, V., Déqué, M., Marçais, B. and Desprez-Loustau, M.-L. (2004) Simulation of potential range expansion of oak disease caused by *Phytophthora cinnamomi* under climate change. *Global Change Biology* 10, 1539–1552.

Berryman, A.A. (1982) Biological control, thresholds, and pest outbreaks. *Environmental Entomology* 11, 544–549.

Boland, G.J., Melzer, M.S., Hopkin, A., Higgins, V. and Nassuth, A. (2004) Climate change and plant diseases in Ontario. *Canadian Journal of Plant Pathology* 26, 335–350.

Boller, T. and He, S.Y. (2009) Innate immunity in plants: An arms race between pattern recognition receptors in plants and effectors in microbial pathogens. *Science* 324, 742–744.

Bowden, J., Gregory, P.H. and Johnson, C.G. (1971) Possible wind transport of coffee leaf rust across the Atlantic Ocean. *Nature* 229, 500–501.

Brent, K.J. and Hollomon, D.W. (2007) *Fungicide Resistance in Crop Pathogens: How can it be Managed.* Fungicide Resistance Action Committee, Croplife International, Brussels, Belgium, 57 pp.

Brown, J.K.M. and Hovmøller, M.S. (2002) Aerial dispersal of pathogens on the global and continental scales and its impact on plant disease. *Science* 297, 537–541.

Bruehl, G.W. (1982) Developing wheats resistant to snow mold. *Plant Disease* 66, 1190–1193.

Bruehl, G.W. and Cunfer, B. (1971) Physiologic and environmental factors that affect the severity of snow mold of wheat. *Phytopathology* 61, 792–799.

Burdon, J.J., Thrall, P.H. and Ericson, L. (2006) The current and future dynamics of disease in plant communities. *Annual Review of Phytopathology* 44, 19–39.

Chakraborty, S. and Datta, S. (2003) How will plant pathogens adapt to host plant resistance at elevated CO_2 under a changing climate? *New Phytologist* 159, 733–742.

Chakraborty, S. and Newton, A.C. (2011) Climate change, plant diseases and food security: an overview. *Plant Pathology* 60, 2–14.

Chen, X. (2005) Epidemiology and control of stripe rust on wheat. *Canadian Journal of Plant Pathology* 27, 314–337.

Chen, X., Penman, L., Anmin, W. and Cheng, P. (2010) Virulence races of *Puccinia striiformis* f. sp. *tritici* in 2006 and 2007 and development of wheat stripe rust and distributions, dynamics, and evolutionary relationships of races from 2000 to 2007 in the United States. *Canadian Journal of Plant Pathology* 32, 315–333.

Coakley, S.M. (1979) Climate variability in the Pacific Northwest and its effect on stripe rust disease of winter wheat. *Climatic Change* 2, 33–51.

Coakley, S.M. (1988) Variation in climate and prediction of disease in plants. *Annual Review of Phytopathology* 26, 163–181.

Coakley, S.M., Scherm, H. and Chakraborty, S. (1999) Climate change and plant disease management. *Annual Review of Phytopathology* 37, 399–426.

Collinge, D.B., Jørgensen, H.J.L., Lund, O.S. and Lyngkjær, M.F. (2010) Engineering pathogen resistance in crop plants: Current trends and future prospects. *Annual Review of Phytopathology* 48, 269–291.

de Wit, P. (2007) How plants recognize pathogens and defend themselves. *Cellular and Molecular Life Sciences* 64, 2726–2732.

Dill-Macky, R. and Jones, R.K. (2000) The effect of previous crop residues and tillage on Fusarium head blight of wheat. *Plant Disease* 84, 71–76.

Edwards, S.G. (2004) Influence of agricultural practices on fusarium infection of cereals and subsequent contamination of grain by trichothecene mycotoxins. *Toxicology Letters* 153, 29–35.

Fan, J.-B., Chee, M.S. and Gunderson, K.L. (2006) Highly parallel genomic assays. *Nature Review Genetics* 7, 632–644.

Feuillet, C. and Keller, B. (2005) Molecular markers for disease resistance: The example wheat. In: Lörz, H. and Wenzel, G. (eds) *Molecular Marker Systems in Plant Breeding and Crop Improvement.* Springer, Berlin, Germany, pp. 353–370.

Filip, G.M. and Rosso, P.H. (1999) Cypress mortality (mal del cipres) in the Patagonian Andes: comparisons with similar forest diseases and declines in North America. *European Journal of Forest Pathology* 29, 89–96.

Filser, J., Fromm, H., Nagel, R. and Winter, K. (1995) Effects of previous intensive agricultural management on microorganisms and the biodiversity of soil fauna. *Plant and Soil* 170, 123–129.

Flor, H.H. (1973) Current status of the gene-for-gene concept. *Annual Review of Phytopathology* 9, 275–296.

Garrett, K.A., Dendy, S.P., Frank, E.E., Rouse, M.N. and Travers, S.E. (2006) Climate change effects on plant disease: genomes to ecosystems. *Annual Review of Phytopathology* 44, 489–509.

Garrett, K.A., Nita, M., De Wolf, E.D., Gomez, L. and Sparks, A.H. (2009) Plant pathogens as indicators of climate change. In: Letcher, T.M. (ed.) *Climate Change: Observed Impact on Planet Earth.* Elsevier, Amsterdam, the Netherlands, pp. 425–437.

Gaudet, D.A. and Chen, T.H.H. (1988) The effect of freezing resistance and low temperature stress on development of cottony snow mold (*Coprinus psychromorbidus*) in winter wheat. *Canadian Journal of Botany* 66, 1610–1615.

Gaudet, D.A., Bhalla, M.K., Clayton, G.C. and Chen, T.H.H. (1989) Effect of cottony snow mold and low temperatures on winter wheat survival in central and northern Alberta. *Canadian Journal of Plant Pathology* 11, 291–296.

Gaudet, D.A., Wang, Y., Frick, M., Puchalski, B., Penniket, C., Ouellet, T., Robert, L., Singh, J. and Laroche, A. (2010) Low temperature induced defence gene expression in winter wheat in relation to resistance to snow moulds and other wheat diseases. *Plant Science* 180, 99–110.

Gaunt, R.E. (1995) The relationship between plant disease severity and yield. *Annual Review of Phytopathology* 33, 119–144.

Goff, S.A., Ricke, D., Lan, T.-H., Presting, G., Wang, R., Dunn, M., Glazebrook, J., *et al.* (2002) A draft sequence of the rice genome (*Oryza sativa* L. ssp. *japonica*). *Science* 296, 92–100.

Gregory, P.J., Johnson, S.N., Newton, A.C. and Ingram, J.S.I. (2009) Integrating pests and pathogens into the climate change/food security debate. *Journal of Experimental Botany* 60, 2827–2838.

Grishkan, I., Korol, A.B., Nevo, E. and Wasser, S.P. (2003) Ecological stress and sex evolution in soil microfungi. *Proceedings of the Royal Society Series B: Biological Sciences* 270, 13–18.

Gurr, S.J. and Rushton, P.J. (2005) Engineering plants with increased disease resistance: what are we going to express? *Trends in Biotechnology* 23, 275–282.

Hanssen-Bauer, I., Drange, H., Førland, E.J., Roald, L.A., Børsheim, K.Y., Hisdal, H., Lawrence, D., *et al.* (2009) *Klima i Norge 2100 bakgrunnsmateriale til NOU klimatilpassing* [Climate in Norway 2100 background material to the NOU climate adaption]. Norsk klimasenter, Oslo.

Harvell, C.D., Mitchell, C.E., Ward, J.R., Altizer, S., Dobson, A.P., Ostfeld, R.S. and Samuel, M.D. (2002) Climate warming and disease risks for terrestrial and marine biota. *Science* 296, 2158–2162.

Henriksen, B. and Elen, O. (2005) Natural Fusarium grain infection level in wheat, barley and oat after early application of fungicides and herbicides. *Journal of Phytopathology* 153, 214–220.

Hepting, G.H. (1963) Climate and forest diseases. *Annual Review of Phytopathology* 1, 31–50.

Hofgaard, I.S., Wanner, L.A., Hageskal, G., Henriksen, B., Klemsdal, S.S. and Tronsmo, A.M. (2006) Isolates of *Microdochium nivale* and *M. majus* differentiated by pathogenicity on perennial ryegrass (*Lolium perenne* L.) and *in vitro* growth at low temperature. *Journal of Phytopathology* 154, 267–274.

Hofgaard, I.S., Udnes-Aamot, H., Klemsdal, S.S., Elen, O., Jestoy, M. and Brodal, G. (2010) Occurrence of *Fusarium* spp. and mycotoxins in Norwegian wheat and oat. *Bioforsk Fokus* 5, 9.

Hogg, E.H., Brant, J.P. and Kochtubajda, B. (2002) Growth and dieback of aspen forests in northwestern Alberta, Canada, in relation to climate and insects. *Canadian Journal of Forest Research* 32, 823–832.

Hoshino, T., Xiao, N. and Tkachenko, O. (2009) Cold adaptation in the phytopathogenic fungi causing snow molds. *Mycoscience* 50, 26–38.

Houghton, J.T., Ding, Y., Griggs, D.J., Noguer, M., Van Der Linden, P.J., Dai, X., Maskel, K. and Johnson, C.A. (2001) *Climate Change 2001: The Scientific Basis. Contribution of Working Group I in The Third Assessment Report of the Intergovernmental Panel on Climate Change*, Cambridge University Press, Cambridge, 881 pp.

House, J.I., Huntingford, C., Knorr, W., Cornell, S.E., Cox, P.M., Harris, G.R., Jones, C.D., Lowe, J.A. and Prentice, C. (2008) What do recent advances in quantifying climate and carbon cycle uncertainties mean for climate policy? *Environmental Research Letters* 3, 044002, doi: 10.1088/1748-9326/3/4/044002.

Hovmøller, M.S. and Justesen, A.F. (2007) Appearance of atypical *Puccinia striiformis* f. sp. *tritici* phenotypes in north-western Europe. *Australian Journal of Agricultural Research* 58, 518–524.

Howes, N.K., Woods, S.M. and Townley-Smith, T.F. (1998) Simulations and practical problems of applying multiple marker assisted selection and doubled haploids to wheat breeding programs. *Euphytica* 100, 225–230.

Hsiang, T., Matsumoto, N. and Millett, S.M. (1999) Biology and management of Typhula snow molds of turfgrass. *Plant Disease* 83, 788–798.

Huber, D.M. and Watson, R.D. (1974) Nitrogen form and plant disease. *Annual Review of Phytopathology* 12, 139–165.

Jin, Y., Szabo, L.J., Rouse, M.N., Fetch, T., Pretorius, Z.A., Wanyera, R. and Njau, P. (2009) Detection of virulence to resistance gene *Sr36* within the TTKS race lineage of *Puccinia graminis* f. sp. *tritici*. *Plant Disease* 93, 367–370.

Jin, Y., Szabo, L.J. and Carson, M. (2010) Century-old mystery of *Puccinia striiformis* life history solved with the identification of *Berberis* as an alternate host. *Phytopathology* 100, 432–435.

Jones, J.D.G. and Dangl, J.L. (2006) The plant immune system. *Nature* 444, 323–329.

Kasha, K.J. and Maluszinski, M. (2003) Production of doubled haploids in crop plants. An introduction. In: Maluszinski, M., Kasha, K.J., Forster, B.P. and Szarejko, I. (eds) *Doubled Haploid Production in Crop Plants: A Manual.* Kluwer Academic Publishers, Dordrecht, The Netherlands, pp. 1–4.

La Porta, N., Capretti, P., Thomsen, I.M., Kasanen, R., Hietala, A.M. and Von Weissenburg, K. (2008) Forest pathogens with higher damage potential due to climate change in Europe. *Canadian Journal of Plant Pathology* 30, 177–195.

Lenné, J. (2000) Pests and poverty: the continuing need for crop protection research. *Outlook on Agriculture* 29, 235–250.

Matson, P.A., Parton, W.J., Power, A.G. and Swift, M.J. (1997) Agricultural intensification and ecosystem properties. *Science* 277, 504–509.

Matthews, H.D. and Caldeira, K. (2008) Stabilizing climate requires near-zero emissions. *Geophysical Research Letters* 35, L04705, doi:10.1029/2007GL032388.

Mattson, W.J. (1980) Herbivory in relation to plant nitrogen content. *Annual Review of Ecology and Systematics* 11, 119–161.

Melloy, P., Hollaway, G., Luck, J., Norton, R., Aitken, E. and Chakraborty, S. (2010) Production and fitness of *Fusarium pseudograminearum* inoculum at elevated carbon dioxide in FACE. *Global Change Biology* 16, 3363–3373.

Milus, E.A., Seyran, E. and McNew, R. (2006) Aggressiveness of *Puccinia striiformis* f. sp. *tritici* isolates in the South-Central United States. *Plant Disease* 90, 847–852.

Milus, E.A., Kristensen, K. and Hovmøller, M.S. (2009) Evidence for increased aggressiveness in a recent widespread strain of *Puccinia striiformis* f. sp. *tritici* causing stripe rust of wheat. *Phytopathology* 99, 89–94.

Mohan, M., Nair, S., Bhagwat, A., Krishna, T.G., Yano, M., Bhatia, C.R. and Sasaki, T. (1997) Genome mapping, molecular markers and marker-assisted selection in crop plants. *Molecular Breeding* 3, 87–103.

Mueller-Dombois, D. (1992) Potential effects of the increase in carbon dioxide and climate change on the dynamics of vegetation. *Water, Air, and Soil Pollution* 64, 61–79.

Nagarajan, S. and Singh, D.V. (1990) Long-distance dispersion of rust pathogens. *Annual Review of Phytopathology* 28, 139–153.

Newton, M. and Johnson, T. (1936) Stripe rust, *Puccinia glumarum*, in Canada. *Canadian Journal of Research* 14, 89–108.

Nordhaus, W.D. (2010) Economic aspects of global warming in a post-Copenhagen environment. *Proceedings of the National Academy of Sciences USA* 107, 11721–11726.

Parry, D.W., Jenkinson, P. and McLeod, L. (1995) Fusarium ear blight (scab) in small grain cereals - a review. *Plant Pathology* 44, 207–238.

Paterson, A.H., Bowers, J.E., Bruggmann, R., Dubchak, I., Grimwood, J., Gundlach, H., Haberer, G., *et al.* (2009) The *Sorghum bicolor* genome and the diversification of grasses. *Nature* 457, 551–556.

Pereyra, S.A. and Dill-Macky, R. (2008) Colonization of the residues of diverse plant species by *Gibberella zeae* and their contribution to Fusarium head blight inoculum. *Plant Disease* 92, 800–807.

Pritchard, S.G., Rogers, H.H., Prior, S.A. and Peterson, C.M. (1999) Elevated CO_2 and plant structure: a review. *Global Change Biology* 5, 807–837.

Purdy, L.H., Krupa, S.V. and Dean, J.L. (1985) Introduction of sugarcane rust into the Americas and its spread into Florida. *Plant Disease* 69, 689–693.

Ravi, M. and Chan, S.W.L. (2010) Haploid plants produced by centromere-mediated genome elimination. *Nature* 464, 615–618.

Rebetez, M. and Dobbertin, M. (2004) Climate change may already threaten Scots pine stands in the Swiss Alps. *Theoretical and Applied Climatology* 79, 1–9.

Robertson, N.L. and Carroll, T.W. (1988) Virus-like particles and a spider mite intimately associated with a new disease of barley. *Science* 240, 1188–1190.

Rochette, P., Bélanger, G., Castonguay, Y., Bootsma, A. and Mongrain, D. (2004) Climate change and winter damage to fruit trees in eastern Canada. *Canadian Journal of Plant Science* 84, 1113–1125.

Roelfs, A.P., Singh, R.P. and Saari, E.E. (1992) *Rust Diseases of Wheat: Concepts and Methods of Disease Management*. CIMMYT, Mexico, DF.

Roos, J., Hopkins, R., Kvarnheden, A. and Dixelius, C. (2011) The impact of global warming on plant diseases and insect vectors in Sweden. *European Journal of Plant Pathology* 129, 9–19.

Rosenzweig, C., Iglesias, A., Yang, X.B., Epstein, P.R. and Chivian, E. (2001) Climate change and extreme weather events; implications for food production, plant diseases, and pests. *Global Change and Human Health* 2, 90–104.

Saari, E.E. and Wilcoxson, R.D. (1974) Plant disease situation of high-yielding dwarf wheats in Asia and Africa. *Annual Review of Phytopathology* 12, 49–68.

Schmale, D.G., Leslie, J.F., Zeller, K.A., Saleh, A.A., Shields, E.J. and Bergstrom, G.C. (2006) Genetic structure of atmospheric populations of *Gibberella zeae*. *Phytopathology* 96, 1021–1026.

Schnable, P.S., Ware, D., Fulton, R.S., Stein, J.C., Wei, F., Pasternak, S., Liang, C., *et al.* (2009) The B73 maize genome: Complexity, diversity, and dynamics. *Science* 326, 1112–1115.

Schoustra, S., Rundle, H.D., Dali, R. and Kassen, R. (2010) Fitness-associated sexual reproduction in a filamentous fungus. *Current Biology* 20, 1350–1355.

Shah, D.M. (1997) Genetic engineering for fungal and bacterial diseases. *Current Opinion in Biotechnology* 8, 208–214.

Singh, R.P., Hodson, D.P., Huerta-Espino, J., Jin, Y., Njau, P., Wanyera, R., Herrera-Foessel, S.A. and Ward, R.W. (2008) Will stem rust destroy the world's wheat crop? *Advances in Agronomy* 98, 271–309.

Strange, R.N. and Scott, P.R. (2005) Plant disease: A threat to global food security. *Annual Review of Phytopathology* 43, 83–116.

Summerell, B., Laurence, M., Liew, E. and Leslie, J. (2010) Biogeography and phylogeography of *Fusarium*: a review. *Fungal Diversity* 44, 3–13.

Symko, S. (1969) Haploid barley from crosses of *Hordeum bulbosum* (2 x) X *Hordeum vulgare* (2x). *Canadian Journal of Genetics and Cytology* 11, 602–608.

Taylor, J.W., Jacobson, D.J. and Fisher, M.C. (1999) The evolution of asexual fungi: Reproduction, speciation and classification. *Annual Review of Phytopathology* 37, 197–246.

The Arabidopsis Initiative (2000) Analysis of the genome sequence of the flowering plant *Arabidopsis thaliana*. *Nature* 408, 796–815.

The International Brachypodium Initiative (2010) Genome sequencing and analysis of the model grass *Brachypodium distachyon*. *Nature* 463, 763–768.

Tronsmo, A.M. (1984a) Predisposing effects of low temperature on resistance to winter stress factors in grasses. *Acta Agricultura Scandinavica* 34, 210–220.

Tronsmo, A.M. (1984b) Resistance to the rust fungus *Puccinia poae-nemoralis* in *Poa pratensis* is induced by low-temperature hardening. *Canadian Journal of Botany* 62, 2891–2892.

Tronsmo, A.M. (1994) Effect of different cold hardening regimes on resistance to freezing and snow mould infection in timothy varieties of different origin. In: Dørffling, K., Brettschneider, B., Tantau, H. and Pithan, K. (eds) *Crop Adaptation to Cool Climates*. Cost 814 Workshop, European Commission, Brussels, pp. 83–89.

Tronsmo, A.M., Hsiang, T., Okuyama, H. and Nakajima, T. (2001) Low temperature diseases caused by *Microdochium nivale*. In: Iriki, N., Gaudet, D.A., Tronsmo, A.M., Matsumoto, N., Yoshida, M. and Nishimune, A. (eds) *Low Temperature Plant Microbe Interactions Under Snow*. Hokkaido National Experiment Station, Sapporo, Japan, pp. 75–86.

Waalwijk, C., Kastelein, P., de Vries, I., Kerényi, Z.,

van der Lee, T., Hesselink, T., Köhl, J., *et al.* (2003) Major changes in *Fusarium spp.* in wheat in the Netherlands. *European Journal of Plant Pathology* 109, 743–754.

Windels, C. (2000) Economic and social impacts of Fusarium head blight: changing farms and rural communities in the northern great plains. *Phytopathology* 90, 17–21.

Wong, F.P. (2007) The 2006 California disease round up. USGA Green Section Meeting. http://faculty.ucr.edu/~frankw/Downloads/070108%20USGA.ppt, accessed 1 March 2011.

Woods, A., Coates, K.D. and Hamann, A. (2005) Is an unprecedented Dothistroma needle blight epidemic related to climate change? *Bioscience* 55, 761–769.

Xu, X. and Nicholson, P. (2009) Community ecology of fungal pathogens causing wheat head blight. *Annual Review of Phytopathology* 47, 83–103.

Xu, X.M., Nicholson, P., Thomsett, M.A., Simpson, D., Cooke, B.M., Doohan, F.M., Brennan, J., *et al.* (2008) Relationship between the fungal complex causing Fusarium head blight of wheat and environmental conditions. *Phytopathology* 98, 69–78.

Yang, P., Li, K., Shi, S., Xia, J., Guo, R., Li, S. and Wang, L. (2005) Impacts of transgenic Bt cotton and integrated pest management education on smallholder cotton farmers. *International Journal of Pest Management* 51, 231–244.

Zentmyer, G.A. (1980) Phytophthora cinnamomi and the diseases it causes. Monograph, American Phytopathological Society, no. 10, pp. 96.

Zhan, J., Mundt, C.C. and McDonald, B.A. (2007) Sexual reproduction facilitates the adaptation of parasites to antagonistic host environments: Evidence from empirical study in the wheat-*Mycosphaerella graminicola* system. *International Journal for Parasitology* 37, 861–870.

11 Trees and Boreal Forests

J.E. Olsen and Y.K. Lee

11.1 Introduction

Global warming has resulted in about a 0.8°C increase in annual mean global temperature since 1850 (IPCC, 2007). Within this period the atmospheric CO_2 concentration has increased from 280 to about 380 ppm. On the basis of a forecast of an even more rapid increase in CO_2 concentration in the near future, by the end of this century a further temperature increase between 1°C and 6°C is predicted, with particularly pronounced changes at northern latitudes (IPCC, 2007). In line with this, the temperature of the Arctic has increased by 1.6°C over the past 4 decades (McBean *et al.*, 2005). Also, since 1890, the temperature of the Atlantic water of the Gulf Stream entering the Arctic Ocean has increased by 2°C and has never been higher in 2000 years (Spielhagen *et al.*, 2011).

Due to their immense ecological and economic importance, the possible fates of forest tree species under global warming have attracted considerable attention. Since forests contain about 75% of the total terrestrial biomass, they are closely linked with atmospheric carbon budgets. The boreal forests constitute almost 30% of the world's forests and are the world's largest terrestrial biome (Way and Oren, 2010). A significant annual variation in temperature is already an inherent characteristic of boreal and temperate areas but even more temperature fluctuations in these regions are expected. Although tree species have been able to adjust to warming in the past, the rate of warming within the present century is expected to be higher than in earlier periods. Before 2100, a temperature increase up to 10°C in high latitude forest regions has been predicted (Christensen *et al.*, 2007). Such temperature changes are likely to challenge the adaptive capabilities of boreal trees, and alter their geographic distribution. However, it is evident that the impact on trees cannot be predicted on the basis of annual mean temperatures alone, since their growth and development are determined by regional and local climatic conditions through the different seasonal periods. Increasing temperatures will in turn affect a wide range of other important environmental factors.

Climate change will involve changes in seasonal and diurnal temperature patterns, precipitation, air humidity, cloud patterns, light intensity, storms, fires, snow cover and freeze–thaw cycles (Saxe *et al.*, 2001; IPCC, 2007). Generally, changed frequencies and severity of extreme conditions are expected. Alterations in biotic interactions involving pathogens, insect pests, and herbivores, and competition between species, will also occur. Many recent reviews have focused on the effects of increased temperature and CO_2 concentrations, and on changed biotic interactions, as well as alterations in water availability and nutrition on photosynthesis and growth of trees (Saxe *et al.*, 2001; Millard *et al.*, 2007; Kreuzwieser and Gessler, 2010; Langley and Megonigal, 2010; Lukac *et al.*, 2010; Millard and Grelet, 2010). Only some of the main issues on the effects of temperature in the growing season will be discussed here. Due to the great importance of growth–dormancy cycling for survival of boreal forest trees, the discussion will then focus on the effects of changed temperature and its interaction with other factors in this respect.

© CAB International 2012. *Temperature Adaptation in a Changing Climate: Nature at Risk* (eds K.B. Storey and K.K. Tanino)

11.2 Recent Warming-induced Changes in Phenology and Growth in Trees

The different events in the annual growth cycle, such as growth cessation, winter bud formation, leaf abscission, dormancy development, cold hardiness, dormancy release, bud break and resumption of growth in trees are controlled by the local climatic conditions including temperature and light parameters (Tanino *et al.*, 2010; Olsen, 2010). An effect of warming on woody plant phenology in recent decades has been demonstrated in a considerable number of studies. Earlier bud break, leafing, flowering, and fruiting in the spring or summer, and delayed autumn leaf coloration have been shown across Europe, Asia and North America (Chuine and Beaubien, 2001; Zhang *et al.*, 2004; Menzel *et al.*, 2006; Delbart *et al.*, 2008; Nordli et *al.*, 2008; Gordo and Sanz, 2010). Data on forest leaf coloration from 35° N to 70° N have indicated an average 5 day increase in growth season length per °C temperature increase from 1981 to 1991 (Zhang *et al.*, 2004). In urban areas where temperatures are commonly 1°C–3°C higher than in other terrestrial areas this was even more pronounced, with bud burst occurring 7 and 9 days earlier, and dormancy onset 2.5 and 4 days later in Europe and North America, respectively (Zhang *et al.*, 2004). Growth parameters such as elongation growth and radial growth have been affected in recent decades, both in deciduous and coniferous tree species, in some positively, in others negatively (Huang *et al.*, 2010). Increased mortality in water-limited forests due to temperature-driven drought stress has also been reported (Breshears *et al.*, 2005). In the high arctic, increased plant productivity in the past 30 years has been documented, with increased shrub abundance, particularly evergreen shrubs (Sturm *et al.*, 2001; Hudson and Henry, 2009). Changes in the alpine tree line in some regions have also been observed (Harsch *et al.*, 2009).

11.3 Adaptation, Migration or Extinction in Response to Continued Warming?

Survival and competitive successes depend on a precise timing of growth, dormancy and frost hardiness in synchrony with seasonal changes in temperature. So far, trees appear to have adapted to these changes with an acceptable fidelity. However, the long generation intervals, often more than 30 years, could make them less able to respond to rapid changes in temperature by evolutionary means (Rehfeldt *et al.*, 1999). Thus, the anticipated change in global climate may represent a significant challenge for rapid adaptation of the growth–dormancy cycle. Upon continued warming, extensive migration of plant species to higher latitudes and elevations is predicted. Although not uniform across sites, certain alpine treeline changes have been observed over the last century (Harsch *et al.*, 2009). Irrespective of migration, a crucial question is to what extent different species are able to cope with and adapt to the new conditions at the current or new location. Another important question is whether different species are able to move fast enough to avoid the climatic conditions to which they are not adapted. Since plants are sessile, movement is only possible through seed dispersal through wind, animals and human activities. Fragmented landscapes typical of areas with intensive human activity, natural barriers to seed dispersal, poor soil conditions, and lack of seed sources might reduce the migration possibilities and thus obstruct the capacity for adaptation of natural plant populations to rapid climatic change (Jump and Penuelas, 2005; Aitken *et al.*, 2008). In any case, upon migration, at least some northern species are likely to be ousted by southern species, and further northwards movement is limited by the Arctic Ocean.

Genetic data and records of fossil pollen indicate that migration of trees has previously occurred in response to large environmental changes such as under past glacial and postglacial periods (Davis and Shaw, 2001; Hamrick, 2004; Aitken *et al.*, 2008). These data, as well as development of locally adapted populations during the postglacial period, indicate that tree species have a significant capacity for range shifts. Migration rates in response to previous climatic changes, and predictions on migration potential of a number of boreal trees species, have been calculated to <100–200 m per year (Iverson *et al.*, 2004; McLachlan and Clark, 2004; Aitken *et al.*, 2008). Upon doubling of atmospheric CO_2 levels, a need for migration rates up to ten times higher to avoid temperature changes outside normal habitat ranges has been estimated (Malcolm *et al.*, 2002). In light of the estimates of earlier migration rates and predictions of future migration, such high migration rates appear quite unrealistic. Accordingly, extinction of species which are not able to adapt to new conditions or migrate sufficiently rapidly might be a possible scenario.

Although modelling is an important tool to predict responses of plants to climatic changes, few models on plant distribution appear to take non-climatic factors into consideration (Aitken *et al.*, 2008). A variety of such factors as biotic interactions, degree of range fragmentation, life history traits such as age to sexual maturity, and extent of seed production and dispersal can affect the migration capacity of different species. Also, interactions between different climatic factors including temperature, precipitation, and air and soil humidity, might be altered upon climatic change, and models do not commonly include this (Aitken *et al.*, 2008). Furthermore, emerging evidence, particularly from coniferous species such as Norway spruce (*Picea abies*) suggest that temperature under embryogenesis affects the climatic responses of the progeny (Johnsen *et al.*, 2005a,b; Kvaalen and Johnsen, 2008). This might be of significance in the climatic adaptation of forest trees, but models have not so far taken such a mechanism into consideration. This intriguing phenomenon and its possible implications in a climate change perspective will be discussed in more detail below.

In herbaceous tundra species increased vital rates including growth, survival and reproduction were mostly demonstrated upon warming from low to moderate temperatures (Doak and Morris, 2010). However, across a range of moderate temperatures some vitality parameters were positively affected, some were stable, and some negatively affected. Upon further warming from moderate to high temperatures a general decrease in vitality was observed. Thus, although a certain warming might be beneficial in the first place, these results suggest the existence of a tipping point with respect to a negative effect of temperature. This tipping point appears to differ for different physiological processes within a species. It is likely that this also applies to other plants including tree species. This emphasizes a need for detailed knowledge of the environmental and endogenous factors controlling different adaptive traits in different species to be able to predict responses to climatic change.

11.4 Temperature Effects in the Growing Season

11.4.1 Effect of temperature on photosynthesis-related processes

Temperature affects virtually all chemical and biochemical processes in plants. Photosynthesis is affected both by a temperature effect on the content of photosynthetic pigments and other processes. Thus, a positive effect on photosynthesis as a whole has been hypothesized upon a certain degree of warming, in boreal areas (Saxe *et al.*, 2001). Generally a maximum light-saturated photosynthesis occurs at 25°C–40°C. Above this photosynthesis decreases rapidly and damage to the photosynthesis apparatus eventually occurs. However, acclimation to relatively high temperatures during days or weeks can increase the temperature optimum of photosynthesis by up to 10°C (Battaglia *et al.*, 1996). The responses of photosynthetic processes to temperature changes are known to vary between species and environmental conditions, and these may be linked to different acclimation capacities (Berry and Björkman, 1980; Gray *et al.*, 1997; Őquist and Huner, 2003; Ensminger *et al.*, 2006). Some species are able to recover more rapidly from heat stress than others (Saxe *et al.*, 2001; Way and Oren, 2010). Adding to the complexity is the fact that photosynthetic rates are not only dependent on photosynthetic biochemical reactions but also on responses of stomata to a variety of conditions.

CO_2 concentrations, water supply, light, climate and temperature are well known to affect stomatal opening and closure (Damour *et al.*, 2010). Under drought, the hormone abscisic acid plays a central role in inducing stomatal closure. Under high air humidity in long photoperiods or high irradiances, development of less functional stomata has been shown in herbaceous plants and roses (Torre and Fjeld, 2001; Nejad and van Meeteren, 2005, 2007), probably linked to increased inactivation of abscisic acid (Okamoto *et al.*, 2009). This made plants more drought sensitive when the high humidity was followed by drought. In maritime pine (*Pinus pinaster*), a seasonal change in stomatal response to air humidity has been reported (Medlyn *et al.*, 2002). Differences among tree species in stomatal conductance in response to environmental parameters have also been documented (Bassow and Bazzaz, 1998). Changed air humidity might thus affect stress tolerance and growth of trees through an effect on stomata. Furthermore, under high air humidity, some plants might also be more vulnerable to overheating in high temperatures. High air humidity reduces transpiration, and leaves may then not be able to cool properly.

Due to its importance in compensating for deleterious effects of increasing temperature, thermal acclimation of photosynthesis in trees has been addressed in a number of investigations. A common garden study along a 900 km latitudinal transect in eastern North America observed a lack of photosynthetic temperature acclimation in four deciduous tree species, trembling aspen (*Populus tremuloides*), paper birch (*Betula papyrifera*), eastern cottonwood (*Populus deltoides*) and sweetgum (*Liquidambar styraciflua*) (Dillaway and Kruger, 2010). Similar situations have been reported in maritime pine, black spruce (*Picea mariana*) and *Eucalyptus regnans* (Medlyn *et al.*, 2002; Warren, 2008; Way and Sage, 2008). These and other studies might suggest that tree species from cool climates have a low capacity for photosynthetic acclimation to increasing temperatures (Way and Sage, 2008). Alternatively, it is possible that the relatively small photosynthetic response to temperature changes reflects other compensating acclimation mechanisms. Factors such as variations in phenological development or acclimation to changes in light conditions might have contributed to lack of temperature acclimation. In different deciduous boreal tree species, the correlation between leaf temperature and variation in photosynthesis during the main growth season from July to August was lower than the correlation with light levels (Bassow and Bazzaz, 1998). However, the effect of temperature increased in September when temperatures were generally lower. Thus, increased autumn temperatures might stimulate photosynthesis to a certain degree, at least in some species. Some differences between species were observed, with photosynthesis in red oak (*Quercus rubra*) appearing more temperature sensitive in autumn than yellow birch (*Betula alleghaniensis*), white birch (*Betula papyrifera*) and red maple (*Acer rubrum*).

A combination of low temperature and high light, which is common in the growing season in boreal areas, is generally known to induce stress resulting in photoinhibition. Warming might reduce this, and thus potentially increase the photosynthesis efficiency. However, a low temperature (5°C) exposure of jack pine (*Pinus banksiana*) grown at 20°C, decreased sensitivity to photoinhibition at low temperatures (Krol *et al.*, 1995). This was probably due to induction of protective pigment production and increased photosynthetic capacity through acclimation mechanisms. On the other hand, low night temperature was found to increase photoinhibition in *Abies lasiocarpa* seedlings (Germino and Smith,

1999). It is likely that temperature differences between day and night will change upon climatic change, particularly in northern areas. Increased night temperature might thus affect at least some plants positively, due to reduced photoinhibition. However, different species and even ecotypes within species may differ in their sensitivity to photoinhibition. Northern ecotypes of dogwood (*Cornus sericea*) appeared less sensitive to photoinhibition under low temperature than southern ones (Tanino *et al.*, 2010).

Boreal trees are in general C3 plants. Under warm and dry conditions which result in stomata closure and thus decreased CO_2 concentrations in the chloroplasts, these plants exhibit significant photorespiration. This is due to an increase in the oxygenase activity of Rubisco at the expense of carboxylation. Continued warming and extensive drought stress is thus expected to increase photorespiration and decrease photosynthetic efficiency in C3 plants. Nevertheless, different species apparently differ in this respect. When temperatures increased from 25°C to 35°C, photorespiration was enhanced more than photosynthesis in loblolly pine (*Pinus taeda*) (Samuelson and Teskey, 1991). In black locust (*Robinia pseudoacacia*) both processes were simultaneously stimulated when temperature was rising (Mebrathu *et al.*, 1991).

In response to a temperature increase, dark respiration increases, with an exponential increase with temperature over the short term (Saxe *et al.*, 2001). Respiration also varies with total biomass and exposure to different stress conditions. In light of this, and since respiration is a means to produce the cellular energy currency ATP, it is thought to play a substantial role in determining growth. The long-term respiration must necessarily be limited by the supply of substrates, which is linked to photosynthesis. Reduced productivity in response to substantial warming has been predicted due to a more rapid increase in respiration than photosynthesis (Saxe *et al.*, 2001). However, it appears that respiration can acclimate to increased temperatures (Way and Sage, 2008). Whereas photosynthesis in black spruce showed a modest ability to acclimate to temperature, respiration showed a significant acclimation (Way and Sage, 2008). Temperature acclimation of respiration rates suggests that long-term respiration is not only regulated by reaction kinetics (Saxe *et al.*, 2001).

Taken together, although a variety of experimental evidence indicate an effect of temperature on photosynthetic processes, many tree species may not adjust photosynthesis to maximize carbon fixation in response to warming (Dillaway and Kruger, 2010). This would have important implications for the responses to climate change. However, generalization on increased photosynthesis upon warming is not straightforward, due to the inherent complexities of control of the different processes and acclimation mechanisms involved. Furthermore, increased photosynthesis does not directly translate into increased growth (Saxe *et al.*, 2001). Nevertheless, temperature is known to affect growth-related physiological parameters such as resource allocation and morphology (Dillaway and Kruger, 2010). Morphology in turn is a determinant for the size of the photosynthetic area of a plant.

11.4.2 Effect of temperature on morphology

In addition to light parameters, temperature is a well-known morphogenetic signal, and morphogenetic responses to temperature differ in light and dark (Went, 1944). Plant hormones are important signalling compounds in temperature modulation of morphology. Enhanced elongation in response to increased temperature during the light phase is linked to increased content of active gibberellin (GA) (Stavang *et al.*, 2009). In *Arabidopsis thaliana* this was shown to result from a rapid transcriptional down-regulation of a specific GA inactivating gene (a *GA-2oxidase*). In addition, auxin biosynthesis was stimulated by transcriptional up-regulation of key IAA biosynthetic genes. Auxin and GA apparently act together with brassinosteroid to promote elongation growth (Stavang *et al.*, 2009). In contrast, a low growing temperature in light reduces the shoot elongation rate and leaf area substantially more than a low night temperature (Myster and Moe, 1995; Stavang *et al.*, 2005, 2007). This is linked to a rapid decrease in GA content due to increased GA inactivation, and reduced auxin content upon a temperature reduction in light but not in darkness (Thingnæs *et al.*, 2003; Stavang *et al.*, 2005, 2007, 2010).

Taken together, plant hormones play an important role in short-term temperature adaptation resulting in altered growth rates and morphology. Most studies have been done with herbaceous plants, but a higher leaf elongation rate was also observed in *Salix viminalis* during the day than at night (McDonald *et al.*, 1992). On the other hand, young seedlings of Norway spruce showed limited differences only in height and dry matter accumulation under different combinations of day and night temperatures (Fløistad and Patil, 2002). This, and a substantial volume of horticulture-related papers (Myster and Moe, 1995) indicate that morphogenetic responses to diurnal temperature variations differ, at least when stem elongation is concerned. However, Way and Oren (2010) re-analysed growth data from a large number of tree species, and showed that leaf area, leaf number and mass were linked to changes in day temperature. In downy birch (*Betula pubescens*) ecotypes leaf number and radial growth were most affected by day temperature, whereas internode length was more affected by night temperature (Håbjørg, 1972b).

Several studies have suggested a positive effect of GA on photosynthesis and biomass production (reviewed in Stavang *et al.*, 2010). Transgenic citrus with increased GA levels showed increased growth and photosynthesis (Huerta *et al.*, 2008). *GA-20-oxidase* overexpressing hybrid aspen (*Populus tremula x tremuloides*) had increased biomass production (Eriksson *et al.*, 2000). Conversely, in tomato (*Lycopersicon esculentum*) no difference in photosynthesis rate was observed between the wild type and a low GA mutant (Cramer *et al.*, 1995). In a GA-saturated GA-signalling mutant of pea (*Pisum sativum*), no effect of GA status was observed on photosynthesis or respiration compared to the wild type (Stavang *et al.*, 2010). Collectively, the relation between GA, photosynthesis rate and growth is still unclear. However, since leaf morphology is decisive for photosynthetic area, it might be speculated whether a limited degree of warming might result in a larger total photosynthesis due to increased leaf area.

11.4.3 Effect of temperature on tree growth

There is evidence that tree growth at high latitudes and altitudes is temperature limited and may thus benefit from a certain degree of warming, in contrast to tree species adapted to warmer areas (Way and Oren, 2010), but this might not be valid in all species. In black spruce, a major boreal tree in North America, dendrochronological data demonstrated a negative relationship between temperature in the growing season and annual ring width (Way and Sage, 2008). In a recent survey of data from 63 different investigations, Way and Oren (2010) concluded that growth including stem height, stem diameter and biomass was promoted more by increased temperatures in deciduous species than evergreen trees. No conclusive explanation of this could be provided. However, evergreens often occur in sites with relatively low water and nutrient availability. Also, stomatal conductance of deciduous trees is often more responsive to increased CO_2 levels than that in evergreen conifers, and it can be speculated if this might be also be valid for temperature. Differences in use of available carbon stores might also contribute to the differences in growth. Patterns of carbon allocation can change with temperature and different groups of trees may differ in this respect (Litton and Giardina, 2008). Way and Oren (2010) also showed that warming generally stimulated growth in boreal but not in tropical trees; this may be because boreal trees are commonly growing under suboptimal temperatures, in contrast to tropical trees. Tropical species may be close to a high temperature threshold, above which photosynthesis and growth are negatively affected. That many tree species or populations have their main distribution in areas with suboptimal growth temperatures (i.e. at high latitudes and altitudes) might partly explain why they are outcompeted in warmer areas by trees with inherently higher growth rates (Rehfeldt *et al.*, 1999; Way and Oren, 2010). Also, in colder areas there is a trade-off between growth potential and winter survival by attaining bud dormancy and cold tolerance in time before the onset of the harsh winter conditions.

In spite of warming, trees in boreal areas will still have to face freezing temperatures during the winter, and the inherent temperature instabilities in autumn and spring will probably be even more pronounced in the near future. An important question is how the wintering processes will be affected by these instabilities and higher autumn temperatures. Thus, to be able to predict responses of trees in these areas to warming, a thorough knowledge of the environmental and endogenous control of the adaptive mechanisms ensuring survival through the winter is also a prerequisite.

11.5 Dormancy and Winter Survival in a Climate Change Perspective

11.5.1 Warming-induced migration might challenge adaptation to light climate

In woody species there is a trade-off between exploiting the full growing season and minimizing frost damage through proper timing of growth cessation, winter bud formation, development of dormancy, and cold hardening in the autumn; and dormancy release, de-hardening, and bud break in the spring. Widely distributed species include latitudinal and altitudinal ecotypes adapted to the local climatic conditions at their site of origin. Clinal variation patterns in the different adaptive traits have been thoroughly documented in numerous tree species (Saxe *et al.*, 2001; Olsen, 2010; Tanino *et al.*, 2010). Photoperiod and light quality are central parameters controlling growth–dormancy cycling. Upon global warming, photoperiod and light quality conditions will not change as such at a particular site, although irradiance and UV light might be affected, due to alterations in the occurrence of clouds. However, upon extensive migration further north or to higher altitudes, plants will experience conditions differing from those to which they are adapted. The adaptive responses to light parameters and their temperature modulation, as well as possible implications in a climatic change scenario, will be discussed in the following section.

In line with an increasing photoperiod with increasing northern latitude during the growth season, a clinal variation in critical photoperiod for apical growth exists in woody species with a free growth pattern (Pauley and Perry, 1954; Håbjørg, 1972a,b; Heide, 1974; Junttila, 1980). Short days (SD) shorter than the critical length result in growth cessation, winter bud formation and dormancy development. Also, due to differences in sun angle, the relative proportions of red (R) and far-red (FR) light varies with latitude, and time of the day and year (Nilsen, 1985). An increasing demand for FR with increasing northern latitude or a combination of FR and R light to prevent apical growth cessation and bud set has been demonstrated in woody species with a free growth pattern (Junttila and Kaurin, 1985; Clapham *et al.*, 1998a,b; Tsegay *et al.*, 2005; Mølmann *et al.*, 2006). Needle growth in different ecotypes in Scots pine (*Pinus sylvestris*) was similarly affected by FR light (Clapham *et al.*, 2002). It is well established that photoperiod, or actually night length, is sensed by the phytochrome (PHY) system (Howe *et al.*, 1996, Olsen *et al.*, 1997b). Two types of phytochrome, PHYA and PHYB, are considered important in photoperiodic responses. In line with this, altering PHYA content in hybrid aspen by gene modification changed sensitivity to photoperiod (Olsen *et al.*, 1997b; Kozarewa *et al.*, 2010). Furthermore, allelic variation in a *PHYB* gene in latitudinal ecotypes in *Populus* has been suggested to be linked to natural variation in bud set (Ingvarsson *et al.*, 2006).

Blue (B) light is easily scattered in the

atmosphere, and during the growing season at high northern latitudes the low sun elevation for several hours at the end of the day results in the presence of substantial diffuse B light (Taulavuori *et al.*, 2010). At more southern latitudes the twilight period with diffuse B light is shorter. Thus, plants are exposed to more indirect B light at higher northern latitudes than further south. In accordance with this, the responses to B light differ between latitudinal populations of woody species. An extension of the day with B light delayed growth cessation in bay willow (*Salix pentandra*) and Norway spruce substantially, with a larger delay in more southern than more northern ecotypes (Junttila and Kaurin, 1985; Mølmann *et al.*, 2006). Furthermore, artificial removal of B light enhanced elongation significantly in Scots pine at 69° N, but not further south at 64°C (Taulavuori *et al.*, 2005).

It has become clear that UV-B light is not only detrimental to plants but also constitutes an important developmental signal (Jenkins, 2009). Due to the low sun angle at high latitudes, more of the UV-B light is absorbed by the atmosphere than further south, resulting in lower UV-B levels. In contrast, UV-B levels increase with increasing altitude due to a decreasing layer of atmosphere. Increasing cloud cover results in lower UV-B levels since more of the UV-B light is then absorbed. In *Populus*, reduced elongation growth, leaf expansion, and biomass production was observed in response to increased UV-B levels (Ren *et al.*, 2006; Xu *et al.*, 2010). A *Populus* species originating from an altitude of 3500 m (*P. kangdingensis*) accumulated more UV-B-absorbing compounds and enzymes involved in protection against oxidative damage than a species originating from 1500 m (*P. cathayana*) (Ren *et al.*, 2006). Thus the species originating from the higher altitude appeared most tolerant to UV-B radiation. Whether UV-B light also affects overwintering processes in boreal trees is not known.

Taken together, the relative significance of photoperiod and light quality differs in latitudinal populations. Northern and southern populations seem to exhibit a more light- and dark-dominated timekeeping system, respectively, which is similar to what has been described for flowering control (Clapham *et al.*, 1998b; Vince-Prue *et al.*, 2001; Mølmann *et al.*, 2006). Although not yet verified experimentally, this might be linked to a dominance of a PHYA and PHYB-based system in the northern and southern populations, respectively. In contrast to northern ecotypes, southern ecotypes respond to night breaks in a R–FR reversible manner (Clapham *et al.* 1998a, 2002), which is characteristic for PHYB responses. Northern ecotypes respond little to night breaks, and this is typical for PHYA responses, together with requirement for prolonged FR or R:FR for growth. Light receptors are known to provide input to the circadian clock. Recently, down-regulation of the clock-related genes *LATE ELONGATED HYPOCOTYL1* and *2* (*LHY1* and *LHY2*) and *TIMING OF CAB EXPRESSION 1* (*TOC1*) were shown to induce a shift in critical photoperiod (Ibanez *et al.*, 2010). Also, *PHYA* expression appears important in coordinating clock-regulated rhythms with light–dark changes (Olsen *et al.*, 1997b; Kozarewa *et al.*, 2010). Furthermore, a latitudinal cline in allelic variation in circadian clock genes has been demonstrated in fruit flies (*Drosophila melanogaster*) (Kyriacou *et al.*, 2007), and might also be hypothesized to be important in timekeeping in trees.

An important output of the circadian clock is the regulation of *CONSTANS* expression (*CO*). In *Populus* Böhlenius *et al.* (2006) showed that high levels of CO in leaves at the end of a long day (LD) stimulate expression of *FLOWERING LOCUS T* (*FT*). FT protein then acts as a growth stimulator. High *FT* expression in LD has also been observed in Sitka spruce (*Picea sitschensis*) (Holliday *et al.*, 2008). In Norway spruce, an *FT*-like gene which also shows similarities to the floral inhibitor gene *TERMINAL FLOWER 1* (*TFL1*) in *Arabidopsis*, exhibited strongly enhanced expression in SD, and thus seemed to act as a growth inhibitor (Gyllenstrand *et al.*, 2007; Asante *et al.*, 2011). Such a potential FT/TFL1-like growth inhibitor has up to now not been demonstrated in deciduous trees. Down-regulation of GA levels and probably also of auxin is central to induce growth cessation and bud set (Olsen *et al.*, 1995a,b, 1997a,b; Eriksson and Moritz, 2002; Mølmann *et al.*, 2003). ABA and ethylene appear important in the bud set process (Ruonala *et al.*, 2006; Ruttink *et al.*, 2007). The relationship between hormones and FT in growth promotion or an FT/TFL1-like growth inhibitor awaits elucidation.

Northwards migration in response to warming may perturb the time-keeping system, since more southerly ecotypes will experience longer photoperiods and different light quality conditions late in the growing season than they are adapted to. This period of active growth will be prolonged, and growth cessation and dormancy induction delayed. This will easily result in frost damage since apical growth cessation is a prerequisite for development of a deep frost tolerance in woody species (Weiser, 1970; Olsen *et al.*, 1997b). Cold hardening depends on exposure to short photoperiod and altered light qualities in its initial phase. Subsequently, low temperatures, above or just below 0°C, results in deep winter hardiness (Welling and Palva, 2006). Indeed, in common garden experiments with trees originating from different latitudes, and in forestry, frost damage has been widely observed after transfer of ecotypes from more southern to more northern locations (Pauley and Perry, 1954). Thus, changes in photoperiod and light quality conditions due to plant migration northwards may negatively affect traits important for cold acclimation and hence growth and survival. Changed occurrence of clouds and migration northwards and to higher altitudes will also result in altered UV-B conditions, and changes in at least some growth parameters are expected.

11.5.2 Temperature modulation of light climate responses

Although photoperiod has been considered the primary factor controlling growth cessation and winter bud formation, there is strong evidence that temperature modifies the responses. Understanding the effects of temperature in these processes is important for predicting the effects of warming on winter survival. As thoroughly reviewed by Tanino *et al.* (2010), different groups of responses to temperature can be distinguished across deciduous woody species and conifers (Fig. 11.1):

1. Low temperature under LD results in growth cessation and bud set in northern, but not southern ecotypes, and in photoperiodically insensitive types such as apple (*Malus pumila*) and pear (*Pyrus communis*).
2. Warm temperature under SD will result in accelerated growth cessation and bud set as well as increased depth of dormancy in photoperiodically sensitive species.
3. Night temperatures affect growth cessation, bud set, and dormancy more than day temperatures. Examples of the different response types are provided below.

Northern ecotypes of a variety of species such as downy birch, Norway spruce, bay willow, hybrid aspen and dogwood were

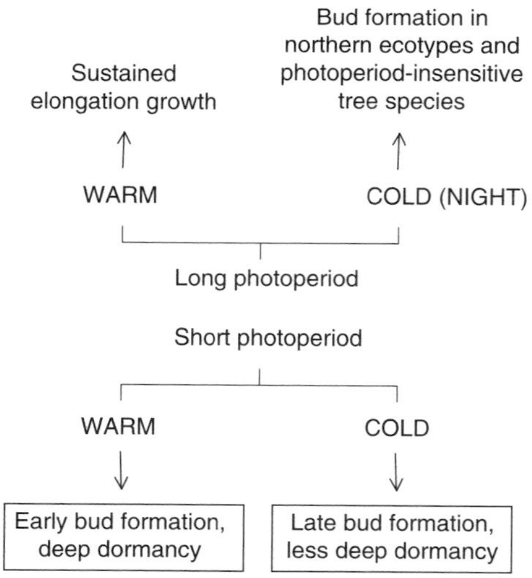

Fig. 11.1. Temperature modulation of dormancy induction in northern and southern ecotypes of photoperiod-responsive and photoperiod-insensitive trees.

shown to cease elongation growth and form buds in response to low temperature in LD (Håbjørg, 1972a,b; Heide, 1974; Junttila 1980; Mølmann *et al.*, 2005; Svendsen *et al.*, 2007). A low night temperature was generally very efficient. High altitudinal ecotypes of downy birch showed a similar response (Håbjørg, 1972b). High temperature under SD exposure resulted in a more rapid bud set and dormancy acquisition, as well as a deeper dormancy in species such as Norway maple (*Acer platanoides*), European white birch (*Betula pendula)*, downy birch, Norway spruce and hybrid poplar (Westergaard and Eriksen, 1997; Junttila *et al.*, 2002; Heide, 2003; Søgård *et al.*, 2008; Kalcsits *et al.*, 2009). Although relatively few studies have been performed on the effects of day versus night temperature, night temperature appears to affect growth cessation, bud set, and dormancy more than day temperature. In hybrid poplar, warm nights accelerated growth cessation and dormancy induction (Kalcsits *et al.*, 2009). Some species such as apple and pear utilize decreasing temperature, and not photoperiod, as their inwintering signal (Heide and Prestrud, 2005).

In conclusion, existence of two different response pathways in trees, one induced by low-temperature stress in northern and high altitude areas, and another induced by warm temperature and short photoperiod, ensures flexibility in the ability to maximize growth and reduces the risk of freezing injury (Tanino *et al.*, 2010). Generally, the effect of night temperature has not been included in models predicting growth under climate change. These observations might be of significance for the performance of trees upon migration northwards and to higher altitudes in response to warming. Thus, in contrast to what has been commonly claimed, trees might stop growing earlier under warmer autumn conditions. The risk of spring frost damage might then be lower than anticipated since warmer autumn results in deeper dormancy and probably later de-hardening.

The mechanisms of temperature sensing and signalling are poorly understood in general, but some factors of apparent importance will be discussed here. The photoequilibrium of the two PHY forms Pr and Pfr is sensitive to temperature (Borthwick *et al.*, 1952). Both dark reversion of Pfr to Pr, and Pfr destruction are enhanced by warm temperature (Butler and Lane, 1965; Schäfer and Schmidt, 1974). Thus, responses regulated by the Pr:Pfr ratio may be affected by temperature, including night temperature. PHY action has also been shown to be affected by temperature. In day length-insensitive *PHYA* overexpressing hybrid aspen, a low night temperature was shown to override the adverse effect of increased PHYA levels on bud set, and to a certain degree also cold hardiness (Mølmann *et al.*, 2005). A temperature effect on PHYB action and functional relationships between different PHY in control of petiole and stem elongation, flowering, and seed germination, have been demonstrated in *Arabidopsis* (Halliday *et al.*, 2003; Halliday and Whitelam, 2003; Heschel *et al.*, 2007; Thingnæs *et al.*, 2008). Thus, modulation of PHY action appears important in mediating a variety of temperature responses.

Expression of the transcription factors *PHYTOCHROME INTERACTING FACTOR 4* (*PIF4*) and *PIF3* increased under SD-induced growth cessation in *Populus* (Ruttink *et al.*, 2007). This might be a consequence of increased PIF4 accumulation when night length increases, since PIF4 is known to accumulate in the dark in *Arabidopsis* and be destabilized by PHYB in the early morning (Foreman *et al.*, 2011). Enhanced *PIF4* expression and post-translational regulation were also shown in *Arabidopsis* seedlings after a temperature increase (Koini *et al.*, 2009; Stavang *et al.*, 2009; Foreman *et al.*, 2011) Whereas high temperature and darkness both increase PIF4 abundance, low temperature and light both reduce it (Foreman *et al.*, 2011). Thus, a role of PIF4 in acceleration of growth cessation in trees in high temperature under SD might be hypothesized. On the other hand, if a high PIF4 abundance is needed for growth cessation, and a low temperature decreases *PIF4* expression/activity, low temperature-induced growth cessation and bud set in northern ecotypes cannot easily be explained in terms of PIF4 signalling. Another PIF family member, SPATULA (SPA), appears to be involved in cool temperature growth repression in *Arabidopsis*, but not in high temperature growth stimulation (Sidaway-Lee *et al.*, 2010). In common with other PIF, SPA acts downstream of PHY, but probably not PHYB. No information on SPA in trees is available yet.

It should be noted that temperature and light signalling in stem elongation control in trees might differ from signalling linked to seedling growth in *Arabidopsis*. Also, stem elongation in the growing season and growth cessation with formation of a terminal bud are not just opposite processes, since the latter involves a shift in the developmental programme. In any case, in *Arabidopsis*, a temperature effect on PHYB function in flowering control is linked to regulation of *FT* expression (Halliday *et al.*, 2003). So far, any relationships between FT and temperature under different photoperiods have not been shown in trees. It might be speculated that effects of day and night temperatures in growth cessation and bud set might be linked to modifications of GA levels, as shown in pea and *Arabidopsis* (Stavang *et al.*, 2005, 2007, 2009, 2010). However, the high night temperature acceleration of overwintering processes under SD, and the low night temperature-induced growth cessation and bud set in LD, are not easily explained in terms of thermoperiodic control of GA levels.

Oscillations in expression of circadian clock genes *TOC1* and *LHY* in chestnut (*Castanea sativa*) and hybrid aspen are disrupted under low temperatures (Ramos *et al.*, 2005; Ibanez *et al.*, 2010). Also, transgenic down-regulation of these clock genes shortened the critical day length, exerted a negative effect on cold hardiness, and delayed spring bud burst (Ibanez *et al.*, 2010). Thus the circadian clock appears important in adaptive responses to SD and low temperatures. However, in chestnut, reduced oscillations in clock gene expression was related to temperature, and was not an intrinsic characteristic of dormancy. In conclusion, our knowledge of temperature sensing and signalling, and how temperature modulates light and climate responses in trees is still fragmentary, and is certainly a highly interesting field for the future.

11.6 Temperature Under Embryogenesis Affects Adaptive Properties

Interestingly, a number of reports, particularly on Norway spruce, have indicated that environmental conditions during embryo development have a persistent effect on the climatic responses of the progeny (Johnsen *et al.*, 2005a,b; Kvaalen and Johnsen, 2008). This has been demonstrated both for zygotic and somatic embryogenesis *in vitro*. The timing of dehardening and bud burst in spring, apical growth cessation in summer, bud set, and cold acclimation in the autumn is advanced or delayed according to the temperature during female sexual reproduction (Johnsen *et al.*, 2005a,b; Kvaalen and Johnsen, 2008). Conditions colder than normal advance the timing while higher temperatures delay the onset of these processes. Exposure of clone materials of Norway spruce to 18°C, 23°C and 28°C during proliferation and embryo maturation resulted in subsequent differences in bud set timing corresponding to a 4°–6° latitudinal ecotype difference (Kvaalen and Johnsen, 2008). The warmer the temperature during embryo formation, the later plants formed terminal buds. Thus, due to a prolonged growth season, plants became longer than their cold embryogenesis counterparts. This indicates that a surprisingly large part of the variability between natural ecotypes with respect to bud phenology and frost hardiness can be explained by this memory of temperature under embryogenesis. It follows that the clinal variation patterns cannot only be due to natural directional selection, but rather a combination of this and memory effects of temperature and photoperiod under embryo development. There is little evidence of temperature-driven selection during the reproductive stages (Johnsen *et al.*, 2005a). The memory effect is suggested to be due to mechanisms acting under embryo development (Fig 11.2). This mechanism adjusts the timing of bud phenology traits in accordance with the temperature and photoperiodic conditions of the growing site of the mother tree.

Such a memory effect has also been demonstrated in progeny from white spruce (*P. glauca*), *P. glauca x P. engelmannii* crosses, Scots pine, *Larix* and shortleaf pine (*Pinus echinata*) (references in Yakovlev *et al.*, 2010b). The phenomenon has not yet been shown in deciduous trees and it is an open question whether it is valid for trees or plants in general. This might well be the case, since *Arabidopsis* seems to behave similarly. Progenies from a warm parental environment showed more rapid germination and growth as well as increased seed production compared to those from a cold parental environment (Blödner *et al.*, 2007). Hitherto the mechanisms behind the memory effect have been investigated to a limited extent only, but epigenetic mechanisms are probably involved (Yakovlev *et al.*, 2010a,b). Epigenetic mechanisms are involved in regulation of a variety of biological processes and include DNA methylation, histone modifications, and action of micro RNA (miRNA) (Fig. 11.2). DNA methylation and histone modifications are well known to affect gene activities, and miRNA are noncoding RNA important in gene regulation among others by mRNA cleavage (Bird, 2002; Carthew and Sontheimer, 2009). Yakovlev *et al.* (2010b) identified 44 miRNA in Norway spruce. Potential target genes were identified for four of these, including a *GAMYB* gene involved in GA signalling. Regulation of these genes by miRNA was hypothesized to be involved in or correlated with the epigenetic memory regulation. However, further evidence is needed in this respect. There is also evidence that temperature under embryogenesis induces changes in gene expression in the progeny (Yakovlev et al., 2010a,b). In this study eight candidate genes showed a differential response to cold or warm embryo development. Of these, two were transposon-regulated genes, three genes were unknown and three genes were related to the small RNA biogenesis pathway. Apart from this, little is known about the

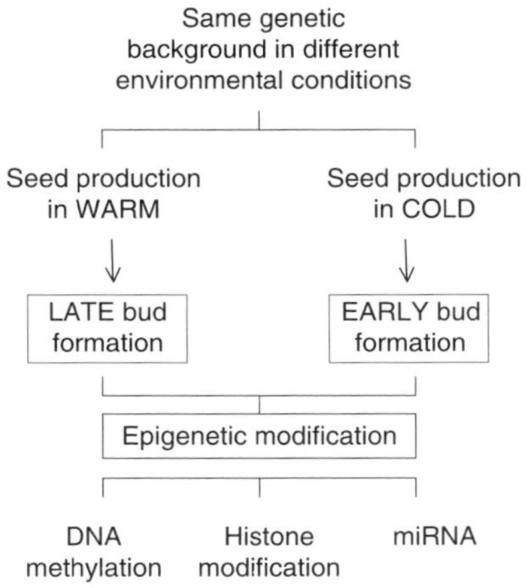

Fig. 11.2. Effect of climatic conditions under embryogenesis on the progeny and suggested epigenetic mechanisms.

background of the memory effect so far. Exciting new knowledge is certainly to come in the near future as the genome of Norway spruce is characterized, and the knowledge on epigenetic mechanisms in general is rapidly progressing.

The question arises whether the effect of climatic conditions on the progeny under embryogenesis will be able to counteract any negative effects of climate change. The delay of the inwintering processes after warm embryo development and enhancement after cool temperatures have important implications. The plasticity with respect to the critical conditions for growth cessation and bud dormancy development is apparently larger than anticipated earlier. Accordingly, negative effects due to changed light and temperature climate upon a gradual northwards migration (seed transfer) may not be as large as earlier thought (Kvaalen and Johnsen, 2008). It should be noted that our current knowledge on the epigenetic memory effect is still limited and it is not known whether this is universal among tree or plant species. There is a strong need to investigate the process in more depth, including in deciduous tree species. At any rate, this phenomenon challenges our existing perception of tree adaptation to climate.

11.7 Conclusions

A considerable amount of data has been collected on a progressing global warming and there is substantial evidence that this has already affected the phenology of trees. Earlier spring bud break and later autumn coloration has been demonstrated across Northern America and Europe. Trees in northern areas have so far mostly shown an acceptable ability to adapt to warmer conditions. Changes in alpine tree lines and species distribution in such regions have occurred in recent decades as well as in the past. The question is how trees and other plants will perform upon the predicted continued warming. It is evident that stress tolerance both with respect to temperature and other factors will be decisive in the fate of individual species. At high latitudes and altitudes growth appears temperature limited, at least for many tree species, and photosynthesis-related parameters and growth may benefit from a certain degree of warming in contrast to tree species adapted to warmer areas. Tropical species may already be close to a high temperature threshold, above which photosynthesis and growth are negatively affected.

Since adaptive dormancy-related traits and cold hardiness are dependent on the light climate, it might be predicted that extensive northwards migration might perturb the time-keeping mechanisms and thus lead to frost damage. This has been shown upon moving southern genotypes a large distance northwards. On the other hand, boreal and temperate trees appear to harbour an inherent flexibility in control of growth cessation and bud set. Acceleration of dormancy induction by warm temperature under SD and deeper dormancy than under cooler conditions might imply that trees could stop growing earlier under warmer autumn conditions. The risk of spring frost damage might then be lower than anticipated, since a warmer autumn may result in deeper dormancy and probably later de-hardening. The growth cessation and dormancy development under low (night) temperature in northern ecotypes under LD also contributes to minimizing the risk of frost damage. However, there is a need to investigate how climatic factors act in concert to regulate the different sub-processes involved in growth–dormancy cycling, as well as the general vital rates of trees. The memory of temperature under embryogenesis, as best described in Norway spruce, indicates that the plasticity of critical conditions with respect to overwintering processes is larger than thought earlier. Accordingly, the negative effects of gradual northwards migration may not be as large as anticipated. However, there is an urgent need to investigate this phenomenon in more detail and clarify whether it is universal among plants. If so, it might – at least partly – change our perception of climatic adaptation in trees and other plant species.

References

Aitken, S.N., Yeaman, S., Holliday, J., Wang, T. and Curtis-McLane, S. (2008) Adaptation, migration or extirpation: climate change outcomes for tree populations. *Evolutionary Applications* 1, 95–111.

Asante, D.K.A., Yakovlev, I.A., Fossdal, G.C., Holefors, A., Opseth, L., Olsen, J.E., Junttila, O., *et al.* (2011) Gene expression changes during short day induced terminal bud formation in Norway spruce. *Plant Cell Environment* 34, 332–346.

Bassow, S.L. and Bazzaz, F.A. (1998) How environmental conditions affect canopy leaf-level photosynthesis in four deciduous tree species. *Ecology* 79, 2660–2675.

Battaglia, M., Beadle, C. and Loughhead, S. (1996) Photosynthetic temperature responses of *Eucalyptus globulus* and *Eucalyptus nitens*. *Tree Physiology* 16, 81–89.

Berry, J. and Björkman, O. (1980) Photosynthetic response and adaptation to temperature in higher plants. *Annual Review of Plant Physiology* 31, 491–543.

Bird, A. (2002) DNA methylation patterns and epigenetic memory. *Genes and Development* 16, 6–21.

Blödner, C., Goebel, C., Feussner, I., Gatz, C. and Polle, A. (2007). Warm and cold parental reproductive environments affect seed properties, fitness and cold responsiveness in *Arabidopsis thaliana* progenies. *Plant Cell and Environment* 30, 165–175.

Böhlenius, H., Huang, T., Charbonnel-Campaa, L.,

Brunner, A., Jansson, S., Strauss, S. and Nilsson, O. (2006) CO/FT regulatory module controls timing of flowering and seasonal growth cessation in trees. *Science* 312, 1040–1043.

Borthwick, H.A., Hendricks, S.B., Parker, M.W., Toole, E.H. and Toole, V.K. (1952) A reversible photoreaction controlling seed germination. *Proceedings of the National Academy of Sciences, USA* 38, 662–666.

Breshears, D.D., Cobb, N.S., Rich, P.M., Price, K.P., Allen, C.D, Balice, R.G., Romme, W.H., *et al.* (2005) Regional vegetation die-off in response to global-change-type drought. *Proceedings of the National Academy of Sciences, USA* 102, 15144–15148.

Butler, W.L. and Lane, H.C. (1965) Dark transformation of phytochrome in vivo II. *Plant Physiology* 40, 13–17.

Carthew, R.W. and Sontheimer, E.J. (2009) Origins and mechanisms of miRNAs and siRNAs. *The Cell* 136, 642–655.

Christensen, J.H., Hewitson, B., Busuioc, A., Chen, A., Gao, X., Held, R., Jones, R., *et al.* (2007) *Regional Climate Projections, Climate Change, 2007: The Physical Science Basis.* Contribution of Working group I to the Fourth Assessment Report of the Intergovernmental Panel on Climate Change, Chapter 11, Cambridge University Press, Cambridge, pp. 847–940.

Chuine, I. and Beaubien, E. (2001) Phenology is a major determinant of tree species range. *Ecology Letters* 4, 500–510.

Clapham, D., Dormling, I., Ekberg, I., Eriksson, G., Qamaruddin, M. and Vince-Prue, D. (1998a) Latitudinal cline of requirement for far-red light for the photoperiodic control of bud set and extension growth in *Picea abies* (Norway spruce). *Physiologia Plantarum* 102, 71–78.

Clapham, D., Ekberg, I., Dormling, I., Eriksson, G., Qamaruddin, M. and Vince-Prue, D. (1998b) Dormancy: night timekeeping and day timekeeping for the photoperiodic control of bud set in Norway spruce. In: Lumsden, P.J. and Millar, A.J. (eds) *Biological Rhythms and Photoperiodism in Plants.* BIOS Scientific Publishers, Oxford, pp. 195–209.

Clapham, D., Ekberg, I., Eriksson, G., Norell, L. and Vince-Prue, D. (2002) Requirement for far-red light to maintain secondary needle extension growth in northern but not southern populations of *Pinus sylvestris* (Scots pine). *Physiologia Plantarum* 114, 207–212.

Cramer, M.D., Nagel, O.W., Lips, S.H. and Lambers, H. (1995) Reduction, assimilation and transport of N in normal and gibberellin-deficient tomato plants. *Physiologia Plantarum* 95, 347–354.

Damour, G., Simoneau, T., Cochard, H. and Urban, L. (2010) An overview of models of stomatal conductance at the leaf level. *Plant Cell and Environment* 33, 1419–1438.

Davis, M.B. and Shaw, R.G. (2001) Range shifts and adaptive responses to quaternary climate change. *Science* 292, 673–679.

Delbart, N., Picard, G., Toan, T.L., Kergoat, L., Quegan, S., Woodward, I., Dye D., *et al.* (2008) Spring phenology in boreal Eurasia over a nearly century time scale. *Global Change Biology* 14, 603–614.

Dillaway, D.N. and Kruger, E.L. (2010) Thermal acclimation of photosynthesis: a comparison of boreal and temperate tree species along a latitudinal transect. *Plant, Cell and Environment* 33, 888–899.

Doak, D.F. and Morris, W.E. (2010) Demographic compensation and tipping points in climate-induced range shifts. *Nature* 467, 959–962.

Ensminger, I., Busch, F. and Huner, N.P.A. (2006) Photostasis and cold acclimation: sensing low temperature through photosynthesis. *Physiologia Plantarum* 126, 28–44.

Eriksson, M.E. and Moritz, T. (2002) Daylength and spatial expression of a *gibberellin-20-oxidase* isolated from hybrid aspen (*Populus tremula* L x *P. tremuloides* Michx). *Planta* 214, 920–930.

Eriksson, M.E., Israelsson, M., Olsson, O. and Moritz, T. (2000) Increased gibberellin biosynthesis in transgenic trees promotes growth, biomass production and xylem fibre length. *Nature Biotechnology* 18, 784–788.

Fløistad, I.S. and Patil, G.G. (2002) Growth and terminal bud formation in *Picea abies* seedlings grown with alternating diurnal temperature and different light qualities. *Scandinavian Journal of Forest Research* 17, 15–27.

Foreman, J., Johansson, J., Hornitschek, P., Josse, E.M., Fankhauser, C. and Halliday, K.J. (2011) Light receptor action is critical for maintaining plant biomass at warm ambient temperatures. *Plant Journal* 65, 441–452, doi: 10.1111/j.1365–313X.2010.04434.x.

Germino, M.J. and Smith, W.K. (1999) Sky exposure, crown architecture, and low temperature photoinhibition in conifer seedlings at alpine treeline. *Plant Cell and Environment* 22, 407–415.

Gordo, O. and Sanz, J.J. (2010) Impact of climate change on plant phenology in Mediterranean ecosystems. *Global Change Biology* 16, 1082–1106.

Gray, G.R., Chauvin, L.P., Sarhan, F. and Huner, N.P.A. (1997) Cold acclimation and freezing tolerance. A complex interaction of light and temperature. *Plant Physiology* 114, 467–474.

Gyllenstrand, N., Clapham, D., Källman, T. and Lagercrantz, U. (2007) A Norway spruce

FLOWERING LOCUS T homolog is implicated in control of growth rhythm in conifers. *Plant Physiology* 144, 248–257.

Håbjørg A. (1972a) Effects of light quality, light intensity and night temperature on growth and development of three latitudinal populations of *Betula pubescens* Ehrh. *Meldinger fra Norges landbrukshøgskole* 51, 1–17.

Håbjørg, A. (1972b) Effects of photoperiod and temperature on growth and development of three latitudinal and altitudinal populations of *Betula pubescens* Ehrh. *Department of Dendrology and Nursery Management, Agricultural University of Norway Report* 44, 1–27.

Halliday, K.J. and Whitelam, G.C. (2003) Changes in photoperiod or temperature alters the functional relationships between phytochromes and reveal roles for phyD and phyE. *Plant Physiology* 131, 1913–1920.

Halliday, K.L., Salter, M.G., Thingnæs, E., Whitelam, G.C. (2003) Phytochrome control of flowering is temperature sensitive and correlates with expression of the floral integrator FT. *Plant Journal* 33, 1–11.

Hamrick, J.L. (2004) Response of forest trees to global environmental changes. *Forest Ecology and Management* 197, 323–335.

Harsch, M.A., Hulme, P.E., McGlone, M.S. and Duncan, R.P. (2009) Are treelines advancing? A global meta-analysis of treeline response to climate warming. *Ecology Letters* 12, 1040–1049.

Heide, O.M. (1974) Growth and dormancy in Norway Spruce ecotypes (*Picea abies*) I. Interaction of photoperiod and temperature. *Physiologia Plantarum* 30, 1–12.

Heide, O.M. (2003) High autumn temperature delays spring bud burst in boreal trees, counterbalancing the effect of climatic warming. *Tree Physiology* 23, 931–936.

Heide, O.M. and Prestrud, K. (2005) Low temperature, but not photoperiod, controls growth cessation and dormancy induction and release in apple and pear. *Tree Physiology* 25, 109–114.

Heschel, M.S., Selby, J., Butler, C., Whitelam, G.C., Sharrock, R.A. and Donohue, K. (2007) A new role for phytochromes in temperature-dependent germination. *New Phytologist* 174, 735–741.

Holliday, J.A., Ralph, S.G., White, R., Bohlmann, J., Aitken, S.N. (2008) Global monitoring of autumn gene expression within and among phenotypically divergent populations of Sitka spruce (*Picea sitchensis*). *New Phytologist* 178, 103–122.

Howe, G.T., Gardner, G., Hackett, W.P. and Furnier, G.R. (1996) Phytochrome control of short-day-induced bud set in black cottonwood. *Physiologia Plantarum* 97, 95–103.

Huang, J., Tardif, J.C., Bergeron, Y., Denneler, B., Berninger, F. and Girardins, M.P. (2010) Radial growth response of four dominant boreal tree species to climate along a latitudinal gradient in the eastern Canadian boreal forest. *Global Change Biology* 16, 711–731.

Hudson, J.M.G. and Henry, G.H.R. (2009) Increased plant biomass in a high arctic community from 1981 to 2008. *Ecology* 90, 2657–2663.

Huerta, L., Forment, J., Gadea, J., Fagoaga, C., Peña, L., Perez-Amador, M.A. and Garcia-Martinez, J.L. (2008) Gene expression analysis in citrus reveals the role of gibberellins on photosynthesis and stress. *Plant Cell and Environment* 31, 1620–1633.

Ibanez, C., Kozarewa I., Johansson, M., Őgren, E., Rohde, A. and Eriksson, M.E. (2010) Circadian clock components regulate entry and affect exit of seasonal dormancy as well as winter hardiness in *Populus* trees. *Plant Physiology* 153, 1823–1833.

Ingvarsson, P.K., Garcia, M.V., Luquez, V., Hall, D. and Jansson, S. (2006) Nucleotide polymorphism and phenotypic associations within and around the phytochrome B2 locus in European Aspen (*Populus tremula*, *Salicaceae*). *Genetics* 178, 2217–2226.

IPCC (2007) *Climate Change 2007: Synthesis Report.* www.ipcc.ch/pdf/assessment-report/ar4/syr/ar4_syr.pdf, accessed 17 June 2011.

Iverson, L.R, Schwartz, M.W. and Prasad, A.M. (2004) How fast and far might tree species migrate in the eastern United States due to climatic change? *Global Ecology and Biogeography* 13, 209–219.

Jenkins, G. (2009) Signal transduction in responses to UV-B radiation. *Annual Review of Plant Biology* 60, 407–431.

Johnsen, Ø., Dæhlen, O.G., Ostreng, G. and Skrøppa, T. (2005a) Daylength and temperature during seed production interactively affect adaptive performance of *Picea abies* progenies. *New Phytologist* 168, 589–596.

Johnsen, Ø., Fossdal C.G., Nagy, N., Mølmann, J., Dæhlen, O.G. and Skrøppa, T. (2005b) Climatic adaptation in *Picea abies* progenies is affected by the temperature during zygotic embryogenesis and seed maturation. *Plant Cell and Environment* 28, 1090–1102.

Jump, A.S. and Penuelas, J. (2005) Running to stand still: adaptation and the response of plants to rapid climatic change. *Ecology Letters* 8, 8010–8020.

Junttila, O. (1980) Effect of photoperiod and temperature on apical growth cessation in two ecotypes of *Salix* and *Betula*. *Physiologia Plantarum* 48, 347–352.

Junttila, O. and Kaurin, Å. (1985) Climatic control of apical growth cessation in latitudinal ecotypes of *Salix pentandra* L. In: Kaurin, Å., Junttila, O. and Nilsen, J. (eds) *Plant Production in the North*. Norwegian University Press, Oslo, pp. 83–91.

Junttila, O., Nilsen, J. and Igeland, B. (2002) Effect of temperature in the induction of bud dormancy in various ecotypes of *Betula pubescens* and *B. pendula*. *Scandinavian Journal of Forest Research* 18, 208–217.

Kalcsits, L., Silim, S. and Tanino, K. (2009) Warm temperature accelerates short-photoperiod induced growth cessation and dormancy induction in hybrid poplar (*Populus* x spp). *Trees* 23, 971–979.

Koini, M.A., Alvey, L., Allen, T., Tilley, C.A., Harberd, N.P., Whitelam, G. and Franklin, K. (2009) High temperature-mediated adaptations in plant architecture requires the bHLH transcription factor PIF4. *Current Biology* 19, 408–413.

Kozarewa, I., Ibanez, C., Johansson, M., Ogren, E., Mozley, D., Nylander, E., Chono, M., *et al.* (2010) Alteration of PHYA expression change circadian rhythms and timing of bud set in Populus. *Plant Molecular Biology* 73, 143–156.

Kreuzwieser, J. and Gessler, A. (2010) Global climate change and tree nutrition: influence of water availability. *Tree Physiology* 30, 1221–1234.

Krol, M., Gray, G.R., Hurry, V.M., Őkvist, G., Malek, L. and Huner, N.P.A. (1995) Low temperature stress and photoperiod affect an increased tolerance to photoinhibition in *Pinus banksiana* seedlings. *Canadian Journal of Botany* 73, 1119–1127.

Kvaalen, H. and Johnsen, Ø. (2008) Timing of bud set in *Picea abies* is regulated by a memory of temperature during zygotic and somatic embryogenesis. *New Phytologist* 177, 49–59.

Kyriacou, C.P., Peixoto, A.A., Sandrelli, F., Costa, R. and Tauber, E. (2007) Clines in clock genes: fine-tuning circadian rhythms to the environment. *Trends in Genetics* 24, 124–132.

Langley, J.A. and Megonigal, P. (2010) Ecosystem response to elevated CO_2 levels limited by nitrogen-induced plant species shift. *Nature* 466, 96–99.

Litton, C.M. and Giardina, C.P. (2008) Below-ground carbon flux and partitioning: global patterns and response to temperature. *Functional Ecology* 22, 941–954.

Lukac, M., Calfapietra, C., Lagomarsino, A. and Loreto, F. (2010) Global climate change and tree nutrition: effects of elevated CO_2 and temperature. *Tree Physiology* 30, 1209–1220.

Malcolm, J.R., Markham, A., Neilson R.P. and Garaci, M. (2002) Estimated migration rates under scenarios of global climatic change. *Journal of Biogeography* 29, 835–849.

McBean, G., Alekseev, G., Chen, D., Førland, E., Fyfe, J., Groisman, P.Y., King, R., *et al.* (2005) Arctic climate: Past and present. In: Symon, C., Arris, L. and Heal, B. (eds) *Arctic Climate Impact Assessment: Scientific Report*. Cambridge University Press, Cambridge, pp. 21–60.

McDonald, A.J.S., Stadenberg, I. and Sands, R. (1992) Diurnal variation in extension growth of leaves of Salix viminalis. *Tree Physiology* 11, 123–132.

McLachlan, J.S. and Clark, J.S. (2004) Reconstructing historical ranges with fossile data at continental scales. *Forest Ecology and Management* 197, 139–147.

Mebrathu, T., Hannover, J.W., Layne, D.R., and Flore, J.A. (1991) Leaf temperature effects on net photosynthesis, dark respiration, and photorespiration of seedlings of black locust families with contrasting growth rates. *Canadian Journal of Forest Research* 21, 1616–1621.

Medlyn, B.E., Loustau. D. and Delzon, S. (2002) Temperature response of parameters of a biochemically based model of photosynthesis. I. Seasonal changes in mature maritime pine (*Pinus pinaster*). *Plant Cell and Environment* 25, 1155–1165.

Menzel, A., Sparks, T.H., Estrella, N., Koch, E., Aasa, A., Ahass, R., Alm-Kübler, K., *et al.* (2006) European phenological response to climate change matches the warming pattern. *Global Change Biology* 12, 1969–1976.

Millard, P. and Grelet, G.-A. (2010) Nitrogen storage and remobilization by trees: ecophysiological relevance in a changing world. *Tree Physiology* 30, 1083–1095.

Millard, P., Sommerkorn, M. and Grelet, G.A. (2007) Environmental change and carbon limitation in trees: a biochemical, ecophysiological and ecosystem appraisal. *New Phytologist* 175, 11–28.

Mølmann, J.A., Berhanu, A.T., Stormo, S.K., Ernstsen, A., Junttila, O. and Olsen, J.E. (2003) The metabolism of gibberellin A_{19} is under photoperiodic control in *Populus*, *Salix* and *Betula*, but not in daylength-insensitive *Populus* overexpressing phytochrome A. *Physiologia Plantarum* 119, 278–286.

Mølmann, J.A., Asante, D.K., Jensen, J.B., Krane, M.N., Junttila, O. and Olsen, J.E. (2005) Low night temperature and inhibition of gibberellin biosynthesis override phytochrome action, and induce bud set and cold acclimation, but not dormancy in hybrid aspen. *Plant Cell and Environment* 28, 1579–1588.

Mølmann, J.A., Junttila, O., Johnsen, Ø. and Olsen, J.E. (2006) Effects of red, far-red and blue light in maintaining growth in latitudinal populations of Norway spruce (*Picea abies*). *Plant, Cell and Environment* 29, 166–172.

Myster, J. and Moe, R. (1995) Effects of diurnal temperature alternations on plant morphology on some greenhouse crops: A mini review. *Scientia Horticulturae* 62, 205–215.

Nejad, A.R. and van Meeteren, U. (2005) Stomatal response characteristics of Tradescantia virginiana grown at high relative air humidity. *Physiologia Plantarum* 125, 324–332.

Nejad, A.R. and van Meeteren, U. (2007) The role of abscisic acid in disturbed stomata characteristics of Tradescantia virginiana during growth at high relative air humidity. *Journal of Experimental Botany* 58, 627–636.

Nilsen, J. (1985) Light climate in northern areas. In: Kaurin, Å., Junttila, O. and Nilsen, J. (eds) *Plant Production in the North*. Norwegian University Press, Tromsø, Norway, pp. 62–72.

Nordli, Ø., Wielgolaski, F.E., Bakken A.K., Hjeltnes, S.H., Måge, F., Sivle, A. and Skre, O. (2008) Regional trends for bud burst and flowering of woody plants in Norway as related to climate change. *International Journal of Biometereology* 52, 625–639.

Okamoto, M., Tanaka, Y., Abrahms, S.R., Kamiya, I., Seki, M. and Nambara, E. (2009) High humidity induces abscisic acid 8-`hydoxylase in stomata and vasculature to regulate local and systemic abscisic acid responses in Arabidopsis. *Plant Physiology* 149, 825–834.

Olsen, J.E. (2010) Light and temperature sensing and signaling in induction of bud dormancy in woody plants. *Plant Molecular Biology* 73, 37–47.

Olsen, J.E., Jensen, E., Junttila, O. and Moritz, T. (1995a) Photoperiodic control of endogenous gibberellins in seedlings of *Salix pentandra*. *Physiologia Plantarum* 93, 639–644.

Olsen, J.E., Junttila, O. and Moritz, T. (1995b) A localised decrease of GA_1 in shoot tips of *Salix pentandra* seedlings precedes cessation of shoot elongation under short photoperiod. *Physiologia Plantarum* 95, 627–632.

Olsen, J.E., Junttila, O. and Moritz, T. (1997a) Long day-induced bud break in *Salix pentandra* is associated with transiently elevated levels of GA_1 and gradual increase in IAA. *Plant and Cell Physiology* 38, 536–540.

Olsen, J.E., Junttila, O., Nilsen, J., Eriksson, M.E., Martinussen, I., Olsson, O., Sandberg, G., *et al.* (1997b) Ectopic expression of oat phytochrome A in hybrid aspen changes critical daylength for growth and prevents cold acclimatization. *Plant Journal* 12, 1339–1350.

Őquist, G. and Huner, N.P.A. (2003) Photosynthesis of overwintering evergreen plants. *Annual Review of Plant Biology* 54, 329–355.

Pauley, S.S. and Perrey, T.O. (1954) Ecotypic variation of the photoperiodic response in *Populus*. *Journal of the Arnold Arboretum, Harvard University* 35, 167–188.

Ramos, A., Perez-Solis, E., Ibanez, C., Casado, R., Collada, C., Gomez, L., Aragoncillo, C., *et al.* (2005) Winter disruption of the circadian clock in chestnut. *Proceedings of the National Academy of Sciences USA* 102, 7037–7042.

Rehfeldt, G.E., Ying, C.C., Spittlehouse, D.L. and Hamilton, D.A. (1999) Genetic responses to climate in *Pinus contorta*: Niche breadth, climate change and reforestration. *Ecological Monographs* 69, 375–407.

Ren, J., Yao, Y., Yang, Y., Korpelainen, H., Junttila, O. and Li, C. (2006) Growth and physiological responses to supplemental UVB-radiation of two contracting poplar species. *Tree Physiology* 26, 665–672.

Ruonala, R., Rinne, P.L.H., Baghour, M., Moritz, T., Tuominen, H. and Kangasjärvi, J. (2006) Transitions of the functioning of the shoot apical meristem in birch (*Betula pendula*) involve ethylene. *Plant Journal* 46, 628–640.

Ruttink, T., Arend, M., Morreell, K., Storme, V., Rombauts, S., Fromm, J., Bhalerao, R., *et al.* (2007) A molecular timetable for apical bud formation and dormancy induction in Poplar. *Plant Cell* 19, 2370–2390.

Samuelson, L.J. and Teskey, R.O. (1991) Net photosynthesis and leaf conductance of loblolly pine seedlings in 2 and 21% oxygen as influenced by irradiance, temperature and provenance. *Tree Physiology* 8, 205–211.

Saxe, H., Cannell, M.G.R. and Johnsen, Ø. (2001) Tree and forest functioning in response to global warming. *New Phytologist* 149, 369–400.

Schäfer, E. and Schmidt, W. (1974) Temperature dependence of phytochrome dark reversions. *Planta* 116, 257–266.

Sidaway-Lee, K., Josse, E.M., Brown, A., Gan, Y., Halliday, K.J., Graham, I.A. and Penfield, S. (2010) SPATULA links daytime temperature and plant growth rate. *Current Biology* 20, 1493–1497.

Søgaard, G., Johnsen, Ø., Nilsen, J. and Junttila, O. (2008) Climatic control of bud burst in young seedlings of nine provenances of Norway spruce. *Tree Physiology* 28, 311–320.

Spielhagen, R.F., Werner, K., Sørensen, S.A., Zamelczyk, K., Kandiano, E., Budeus, G., Husum, K., *et al.* (2011) Enhanced modern heat transfer to the Arctic by warm Atlantic water. *Science* 331, 450–453.

Stavang, J.A., Lindgaard, B., Ernstsen, A., Lid, S.E., Moe, R. and Olsen, J.E. (2005) Thermoperiodic regulation of shoot elongation is mediated by transcriptional regulation of GA inactivation in pea. *Plant Physiology* 138, 2344–2353.

Stavang, J.A., Junttila, O., Moe, R. and Olsen, J.E. (2007) Differential temperature regulation of GA metabolism in light and darkness in pea. *Journal of Experimental Botany* 58, 3061–3069.

Stavang, J., Gallego-Bartolome, J., Yoshida, S., Asami, T., Olsen, J.E., Garcia-Martinez, J.L., Alabadi, D. *et al.* (2009) Hormonal regulation of temperature-induced growth in *Arabidopsis*. *Plant Journal* 60, 589–601.

Stavang, J.A., Pettersen, R.I., Wendell, M., Solhaug, K.A., Junttila, O., Moe, R. and Olsen, J.E. (2010) Thermoperiodic growth control by gibberellin does not involve changes in photosynthetic or respiratory capacities in pea. *Journal of Experimental Botany* 61, 1015–1029.

Sturm, M., Racine, C. and Tape, K. (2001) Increasing shrub abundance in the Arctic. *Nature* 411, 546–547.

Svendsen, E., Wilen, R., Liu, R. and Tanino, K. (2007) A molecular marker associated with dormancy induction in northern and southern ecotypes of red-osier dogwood. *Tree Physiology* 27, 385–397.

Tanino, K., Kalcsits, L. Silim, S., Kendall, E. and Gray, G.R. (2010) Temperature-driven plasticity in growth cessation and dormancy development in deciduous woody plants: a working hypothesis suggesting how molecular and cellular function is affected by temperature during dormancy induction. *Plant Molecular Biology* 73, 49–75.

Taulavuori, K., Sarala, M., Karhu, J., Taulavuori, E., Kubin, E., Laine, K., Poikolainen, J., *et al.* (2005) Elongation of Scots pine seedlings under blue light depletion. *Silva Fennica* 39, 131–136.

Taulavuori, K., Sarala, M. and Taulavuori, E. (2010) Growth responses of trees to arctic light environment. In: Lüttge, U., Beyschlag, W., Büdel, B. and Francis, D. (eds) *Progress in Botany. 71*. Springer-Verlag, Berlin, Germany, pp. 156–168.

Thingnæs, E., Torre, S., Ernstsen, A. and Moe R. (2003) Day and night temperature responses in *Arabidopsis*: Effect on gibberellin and auxin content, cell size, morphology and flowering time. *Annals of Botany* 92, 601–612.

Thingnæs, E., Torre, S. and Moe, R. (2008) The role of phytochrome B, D and E in thermoperiodic responses of *Arabidopsis thaliana*. *Plant Growth Regulation* 56, 53–59.

Torre, S. and Fjeld, T. (2001) Water loss and postharvest characteristics of cut roses grown at high or moderate relative air humidity. *Scientia Horticulturae* 89, 217–226.

Tsegay, B.A., Lund, L., Nilsen, J., Olsen, J.E., Mølmann, J.A., Ernsten, A., Junttila, O. (2005) Growth responses of *Betula pendula* ecotypes to red and far red light. *Electronic Journal of Biotechnology* 8, 17–23, www.ejbiotechnology.info/index.php/ejbiotechnology/article/viewFile/v8n1-10/425, accessed 17 June 2011.

Vince-Prue, D., Clapham, D.H., Ekberg, I. and Norell, L. (2001) Circadian timekeeping for the photoperiodic control of budset in *Picea abies* (Norway spruce) seedlings. *Biological Rhythm Research* 32, 479–487.

Warren, C.R. (2008) Does growth temperature affect the temperature response of photosynthesis and internal conductance to CO_2? A test with *Eucalyptus regnans*. *Tree Physiology* 28, 11–19.

Way, D. and Sage, R. (2008) Thermal acclimation of photosynthesis in black spruce [*Picea mariana* (Mill.) B.S.]. *Plant Cell and Environment* 31, 1250–1262.

Way, D.A. and Oren, R. (2010) Differential responses to changes in growth temperature between trees from different functional groups and biomes: a review and synthesis of data. *Tree Physiology* 30, 669–688.

Weiser, C.J. (1970) Cold resistance and injury in woody plants. *Science* 169, 1269–1278.

Welling, A. and Palva, E.T. (2006) Molecular control of cold acclimation in trees. *Physiologia Plantarum* 127, 167–181.

Went, F. (1944) Plant growth under controlled conditions. II. Thermoperiodicity in growth and fruiting of tomato. *American Journal of Botany* 31, 135–150.

Westergaard, L. and Eriksen, E.N. (1997) Autumn temperature affects the induction of dormancy in first-year seedlings of *Acer platanoides* L. *Scandinavian Journal of Forest Research* 12, 11–16.

Xu, X., Zhao, H., Zhang, X., Hänninen, H., Korpelainen, H. and Li, C. (2010) Different growth sensitivity to enhanced UVB-radiation between male and female *Populus cathayana*. *Tree Physiology* 30, 1489–1498.

Yakovlev, I., Fossdal C.G. and Johnsen, Ø. (2010a) MicroRNAs, the epigenetic memory and climatic adaptation in Norway spruce. *New Phytologist* 187, 1154–1169.

Yakovlev, I., Asante, D.K.A., Fossdal, C.G., Junttila, O. and Johnsen, Ø. (2010b) Differential gene expression related to an epigenetic memory affecting climate adaptation in Norway spruce. *Plant Science* 180, 132–139.

Zhang, X., Friedl, M.A., Schaaf, C.B. and Strahler, A.H. (2004) Climate control on vegetation phonological patterns in northern mid- and high latitudes inferred from MODIS data. *Global Change Biology* 10, 1133–1145.

12 The Paradoxical Increase in Freezing Injury in a Warming Climate: Frost as a Driver of Change in Cold Climate Vegetation

Marilyn C. Ball, Daniel Harris-Pascal, J.J.G. Egerton and Thomas Lenné

12.1 Introduction

How will vegetation respond to changing atmospheric and climatic conditions? Atmospheric concentrations of greenhouse gases such as CO_2 have been increasing since the industrial revolution. This rise has been coupled with climate warming, leading to uncertainty about future climatic conditions. Measurable increases in minimum temperatures have occurred over the past 30 years, but the pattern is complex. Nadir temperatures have increased relatively more than temperature maxima, with warming being greatest during winter and at higher latitudes. IPCC scenarios predict that atmospheric CO_2 will double and mean global temperatures will increase between 1.1°C–6.4°C by 2100 (IPCC, 2007). The most responsive species to these changes are likely to occur in the cool to cold climates at high latitudes and high altitudes, where seasonal temperatures and the length of the frost-free period are important determinants of the growing season (Chen *et al.*, 1995). Responses of cold climate vegetation to increasing temperature and CO_2 are enormously complex and varied, but many general predictions can be made. These are based on the interplay of temperature with the natural seasonal progression in phenological events and acclimation to temperature extremes. Paradoxically, greater freezing stress is predicted to occur in cold climate vegetation with increasing temperature, with the effects being amplified by increasing CO_2.

12.2 Seasonal Acclimation to Temperature Extremes in Cold Climate Vegetation

Acclimation is the process by which tolerance of abiotic stress is increased in response to environmental cues (Rohde and Bhalerao, 2007). Rates of seasonal change in acclimation and deacclimation to freezing temperatures in cold climate species are a function of temperature, day length, and the prevailing phase of the annual ontogenetic cycle (Hänninen, 2006; Rohde and Bhalerao, 2007). Cold acclimation is activated by exposure to low temperatures and an accompanying reduction in photoperiod. Cold acclimation induces a state of minimum growth or dormancy and generates a level of freezing tolerance that enhances survival through winter. However, the breaking of dormancy requires prolonged exposure to chilling. Once the chilling requirement has been achieved, dormancy is broken, but plants remain quiescent while cold winter temperatures inhibit growth. Bud break and the flushing of flowers and leaves occurs once temperatures become favourable, with warmer temperatures also inducing deacclimation of tolerance to freezing

© CAB International 2012. *Temperature Adaptation in a Changing Climate: Nature at Risk* (eds K.B. Storey and K.K. Tanino)

temperatures. Thus, warmer temperatures can lead to lower levels of freeze tolerance, delay cold acclimation in autumn, and accelerate deacclimation in late winter or early spring, making plants more vulnerable to freezing injury at different times of the year. Indeed, recent studies are showing that climate warming is increasing stress from freezing temperatures in cold climate vegetation. Consequently, frost is emerging as a major driver of change in cold climate vegetation in response to climate warming, as discussed below.

12.3 Vulnerability to Frost in a Warming Climate

Unprecedented frost damage is occurring in warming climates. As expected, numerous studies have reported advances in phenology in association with increasing temperatures (Menzel *et al.*, 2006). The hypothesis that earlier bud break would increase risk of frost damage to developing leaves and flowers in a warming climate (Cannell and Smith, 1986) was tested spectacularly by the 2007 spring frost that caused widespread damage to rapidly growing plants over a vast area in North America (Gu *et al.*, 2008). This extreme frost event occurred during an ongoing phenological study of 20 woody deciduous species which revealed interspecific differences in both vulnerability to frost damage and capacity for recovery through refoliation from dormant buds (Augspurger, 2009). Such differences over time could affect species abundances, particularly if the predicted increase in temperature variability includes the incidence of warm spring temperatures followed by a sudden freezing event (Augspurger, 2009).

12.4 Loss in Snow Cover Can Increase Freezing Injury in a Warming Climate

Since 1974, Inouye (2008) has been studying the phenology of flowering in alpine vegetation at a site 2890 m above sea level in the Rocky Mountains of Colorado, USA. His studies have shown that recent advances in the date of snowmelt attributable to climate change have had major repercussions for herbaceous species. The flower buds are especially sensitive to frost; a single severe frost in early summer can greatly reduce the number of species that flower later in the season (Inouye, 2000). However, earlier onset of the growing season following snowmelt has exposed them to more frequent frosts in early summer. In one common perennial species, *Helianthella quinquenervis*, Inouye (2008) found that the percentage of buds killed by frost averaged 36.1% from 1992 to 1998 and 73.9% from 1999 to 2006. The loss of flowers, and hence also seeds, to frost can reduce recruitment and adversely affect the pollinators, herbivores, and seed predators that depend on them. Although the magnitude and direction of responses to warming differ among species (Lambrecht *et al.*, 2007), disproportionate effects of frost on early flowering species could make them more vulnerable to climate warming, leading to rapid change in the species composition of alpine communities (Inouye, 2000, 2008). Further, species-specific shifts in reproductive phenology in response to earlier snowmelt have altered assemblages of co-flowering species that share the same pollinator; these results raise the prospect that climate warming may disrupt plant–pollinator relationships with unknown consequences for alpine communities (Forrest *et al.*, 2010).

Lack of adequate snow cover can also make highly freeze-tolerant trees more vulnerable to low winter temperatures. For example, Schaberg *et al.* (2008) have linked the decline in yellow cedar (*Chamaecyparis nootkatensis*) in south-western Canada with reduction in the depth and duration of insulating snow cover due to warmer winter temperatures, accompanied by an increasing dominance of precipitation by rainfall. The foliage of yellow cedar is sufficiently cold tolerant to avoid injury from ambient winter temperatures, whereas the roots are much less tolerant and sustain injury at temperatures below −5°C. This difference in temperature tolerance is possible because snow cover usually maintains soil temperatures well above injurious levels during winter. However, the loss of an insulating layer of snow exposed roots to lethal freezing temperatures despite warmer winter temperatures. In a field-based study, Schaberg *et al.* (2008) showed that winter freezing injury to roots of seedlings grown without insulating snow was followed by foliar browning and mortality after the onset of warm spring temperatures. Their studies with seedlings were consistent with the development of symptoms of yellow cedar decline in forests covering over 200,000 hectares. This study shows how major change in vegetation can arise from a relatively small change in ambient temperature that paradoxically increased exposure of roots to injurious freezing conditions in a warming climate.

12.5 Extreme Winter Warming Events

Extreme winter warming events are also increasing in frequency and intensity with devastating consequences for cold climate vegetation (Kreyling, 2010). These events can be damaging for at least two reasons:

1. Temperatures above freezing can melt snow cover, thereby reducing insulation. Upon return of winter conditions, exposed plants and soil can be subject to a greater range of temperatures and more extreme freezing temperatures that could exceed the tolerance limits of plants formerly sheltered by insulating snow.
2. The initiation of growth is inhibited by low winter temperatures once the chilling requirement for the breaking of dormancy has been satisfied. Consequently, extreme warming during winter, even for just a few days, can induce deacclimation and other physiological changes associated with the initiation or resumption of growth, again making plants vulnerable to severe frost injury upon rapid return of winter temperatures (Bokhorst *et al.*, 2010).

An example of a natural extreme winter warming event occurred in the sub-Arctic at Abisko Research Station in northern Sweden on 1 January 2002, when air temperatures rose rapidly from −18°C to 4°C, and remained above freezing for a week before plummeting back to cold winter conditions (Bokhorst *et al.*, 2008). The plants showed no visible signs of frost damage immediately after the event. However, frost damage became apparent when function was impaired in subsequent seasons, as shown by delayed bud break, reduction in shoot and canopy growth, death of stems, and reduction in flowering and fruit production (Bokhorst *et al.*, 2009). The relationship between the extreme winter warming event and performance during the following spring and summer was experimentally established through the use of infra-red heat lamps to impose a simulated extreme winter warming event (Bokhorst *et al.*, 2009). Using this system, Bokhorst *et al.* (2009) also showed that extreme winter warming events in consecutive winters led to greater damage and loss of function than a single event, showing that carry-over effects of extreme events could accelerate change in vegetation as the incidence of extreme winter warming events increases with climate warming.

12.6 Delay in the Breaking of Dormancy

Warming winter temperatures could slow fulfilment of chilling requirements for the breaking of dormancy, thereby delaying spring phenology. Cannell and Smith (1986) indicated an increase in thermal time to bud burst with decreased chilling. Climate warming may particularly delay bud burst in those species such as *Fagus sylvatica* which have inherently high values of thermal time (Murray *et al.*, 1989). Yu *et al.* (2010) tested this hypothesis by using the Normalised Difference Vegetation Index ratio method to determine the beginning, end, and length of the growing season for meadow and steppe vegetation of the Tibetan Plateau between 1982 and 2006. Correlating the observed phenological dates with monthly temperatures showed that spring phenology initially advanced in both vegetation types, but began retreating in the mid-1990s despite continued warming. This led to a shortening of the growing season because the end date of the growth period also advanced. Their analysis showed that warm winters caused a delay in spring phenology, which was attributed to later fulfilment of chilling requirements. They concluded that continued warming may strengthen the effect, and slow or even reverse the advancing trend in spring phenology that has dominated recent responses of cold climate vegetation to warming (Yu *et al.*, 2010). Dormancy-inducing temperatures may also have subsequent impact on depth of dormancy, and therefore on bud break (Kalcsits *et al.*, 2009).

12.7 Elevated CO_2 Enhances Frost Damage

Recent studies have shown that plants can be more vulnerable to frost injury under elevated CO_2 (Repo *et al.*, 1996), with losses in leaf area (Lutze *et al.*, 1998; Barker *et al.*, 2005) or reduction in photosynthetic capacity decreasing the capacity for growth in subsequent seasons (Barker *et al.*, 2005). Elevated CO_2 appears to affect freezing tolerance through at least two processes, the nucleation of ice and the acclimation state of the plant.

Elevated CO_2 increases the temperature at which ice nucleation occurs in plant tissues (Lutze *et al.*, 1998, Beerling *et al.*, 2001, 2002; Royer *et al.*, 2002). These studies show considerable interspecific variation in ice nucleation temperatures, but slopes of the relationship describing ice nucleation temperature as a function of CO_2 were consistent between studies, suggesting a common physical mechanism (Woldendorp *et al.*, 2008). Whatever the mechanism, the implication is that plants growing under present atmospheric conditions may be subject to greater freezing stress today than prior to the Industrial Revolution (Woldendorp *et al.*, 2008). A further doubling of the CO_2 from 350 to 700 ppm can be expected to increase ice nucleation temperatures by as much as 1°C–2°C. Consequently, the incidence of frost damage will not necessarily decrease with a climate warming of 2°C in areas where frosts persist (Lutze *et al.*, 1998).

Whether or not a freezing event injures freeze-tolerant plant species depends on the acclimation state of the plant which is also affected by elevated CO_2 (Loveys *et al.*, 2006). Stomatal conductance is typically lower in plants grown under elevated than ambient CO_2, with lower transpiration rates resulting in higher leaf temperatures due to lower evaporative cooling (Siebke *et al.*, 2002). Barker *et al.* (2005) suggested that warmer daytime leaf temperatures under elevated CO_2 could adversely affect freeze tolerance. Loveys *et al.* (2006) tested this hypothesis in a field-based study using open top chambers (OTC) and infrared lamps for free air temperature increase (FATI) to subject snow gum seedlings (*Eucalyptus pauciflora*) to two experimental regimes: elevated/ambient CO_2 or elevated/ambient day time leaf temperatures. Daytime leaf temperatures in the FATI treatment were raised by as much as 3°C, consistent with the average elevation of daytime leaf temperatures (Barker *et al.*, 2005) due to stomatal closure in snow gum seedlings grown under elevated CO_2 (Roden *et al.*, 1999). Loveys *et al.* (2006) showed that acclimation to freezing was delayed by at least 3 weeks in leaves of seedlings subject to either the artificial daytime warming or elevated CO_2 treatments. Importantly, there were no differences between treatments in the level of freeze tolerance achieved by the end of the study, showing that the treatments affected the timing but not the extent of acclimation. The similarity in effects of these treatments on acclimation to freezing temperatures was interpreted as an indication of a common cause, namely higher daytime leaf temperatures. Both treatments affected leaf temperature only during daytime and had no effect on minimum leaf temperatures at night. These results showed that it was the diurnal range in leaf temperature, not just the temperature minima, that affected acclimation to freezing temperatures. Hence, Loveys *et al.* (2006) concluded that increase in leaf temperature due to stomatal closure under elevated CO_2 could contribute to effects of elevated CO_2 on acclimation to freezing temperatures, delaying acclimation in autumn and accelerating deacclimation in spring.

Woldendorp *et al.* (2008) used a modelling approach to explore whether the increased susceptibility to frost damage under elevated CO_2 would be counteracted by climate warming. Their model focused on the incidence and severity of frost damage to *Eucalyptus pauciflora* in a subalpine region of Australia for current and future conditions using the A2 IPCC elevated CO_2 and climate change scenario. They added effects of elevated CO_2 on acclimation and deacclimation to freezing temperatures to a model (King and Ball, 1998) for predicting frost sensitivity of *E. pauciflora* seedlings. They assumed that stomatal closure in response to increasing CO_2 would increase daytime leaf temperature; in effect, elevated CO_2 would amplify temporal effects of climate warming on acclimation states because the leaves would sense warmer conditions. Despite fewer days with freezing temperatures in the future, with consequently fewer damaging frosts with lower average levels of impact, the model predicted individual weather sequences that still resulted in widespread plant mortality due to severe frosts. The model showed that delayed acclimation due to either warming or rising CO_2 combined with an early severe frost could lead to more frost damage and higher mortality than would occur in current conditions. Importantly, the model showed that effects of elevated CO_2 on frost damage were greater in autumn, while climate warming induced more frost damage in spring. Thus, frost damage will continue to be a management issue for plantation and forest management in regions where frosts persist (Woldendorp *et al.*, 2008).

Field-based studies employing long term *in situ* CO_2 enrichment at tree line in the Swiss Alps have confirmed that both the frequency of freezing events during the early growing season and the vulnerability to freezing of high altitude plants could increase under future atmospheric and climatic conditions (Martin *et al.*, 2010). They studied effects of CO_2 enrichment (566 versus 370 ppm) on the phenology and freezing sensitivity of ten treeline species. Importantly, they used a free air CO_2 enrichment (FACE) system, enabling manipulation of the CO_2 around naturally occurring plants subject to the vagaries of the weather. The results revealed highly divergent responses of the species. Nevertheless, long-term exposure to elevated CO_2 increased the sensitivity to freezing in many species, but did not influence phenology (Martin *et al.*, 2010). These results are consistent with previous reports that spring phenology of conifers enclosed in climate-controlled chambers was affected by temperature but not elevated CO_2 (Slaney *et al.*, 2007; Hall *et al.*, 2009). Thus, warmer temperatures both advance the timing of bud break and reduce acclimation to freezing temperatures, with the latter effect amplified by growth under elevated CO_2.

12.8 Conclusion

The rate of vegetation change is occurring more rapidly than expected and the nature of the changes is more complex than initially hypothesized. The effects, for example, of elevated CO_2 on both the incidence of freezing and the tolerance of freezing temperatures were unexpected, but have major consequences for cold climate vegetation. Nevertheless, the observed changes in vegetation are consistent with predictions based on the role of temperature in mediating acclimation and the annual ontogenetic cycle. Treelines are shifting poleward and to higher elevations as woody species encroach into tundra and alpine vegetation (Kullman 2002; Hallinger *et al.*, 2010). Alpine species are also on the move, as recent studies have reported shifts in their distributions along elevational gradients in apparent response to warming temperatures (Cannone *et al.*, 2007). However, different species do not face the same risks due to climate change, because they differ both in responses to climatic variables and in capacity for dispersal (Morin *et al.*, 2008). The individualistic responses of species to changes in climate and atmospheric composition are likely to produce novel species assemblages or 'ecological surprises' (Williams and Jackson, 2007). Further, concerns have been raised about the vulnerability to extinction of individual species, and even whole ecosystems, particularly where topographic features limit migration to areas where future climatic conditions may enable survival (Dirnböck *et al.*, 2011). Understanding how vegetation will change in response to global warming in combination with increasing atmospheric CO_2 is one of the greatest challenges of the 21st century.

12.9 Acknowledgements

We are grateful to Karen Tanino for encouragement and patience in the writing of this chapter. This review was supported by funding from Australian Research Council Discovery Project Grants DP0881009 and DP110105380 to MC Ball.

References

Augspurger, C.K. (2009) Spring 2007 warmth and frost: phenology, damage and refoliation in a temperate deciduous forest. *Functional Ecology* 23, 1031–1039.

Barker, D.H., Loveys, B.R., Egerton, J.J.G., Gorton, H., Williams, W.E. and Ball, M.C. (2005) CO_2 enrichment predisposes foliage of a eucalypt to freezing injury and reduces spring growth. *Plant, Cell & Environment* 28, 1506–1515.

Beerling, D., Terry, A., Mitchell, P., Callaghan, T., Gwynn-Jones, D. and Lee, J. (2001) Time to chill: effects of simulated global change on leaf ice nucleation temperatures of subarctic vegetation. *American Journal of Botany* 4, 628–633.

Beerling, D., Terry, A., Hopwood, C. and Osborne, C. (2002) Feeling the cold: atmospheric CO_2 enrichment and the frost sensitivity of terrestrial plant foliage. *Palaeogeography, Palaeoclimatology, Palaeoecology* 182, 3–13.

Bokhorst, S., Bjerke, J.W., Bowles, F.W., Melillo, J., Callaghan, T.V. and Phoenix, G.K. (2008) Impacts of extreme winter warming in the sub-Arctic: growing season responses of dwarf shrub heathland. *Global Change Biology* 14, 2603–2612.

Bokhorst, S., Bjerke, J.W., Tømmervik, H., Callaghan, T.V. and Phoenix, G.K. (2009) Winter warming events damage sub-Artic vegetation: consistent evidence from an experimental manipulation and a natural event. *Journal of Ecology* 97, 1408–1415.

Bokhorst, S., Bjerke, J.W., Davey, M.P., Taulavuori, K., Taulavuori, E., Laine, K., Callaghan, T.V., *et al.* (2010) Impacts of extreme winter warming events on plant physiology in sub-Arctic heath community. *Physiologia Plantarum* 140, 128–140.

Cannell, M.G.R. and Smith, R.I. (1986) Climatic warming, spring budburst and forest damage on trees. *Journal of Applied Ecology* 23, 177–191.

Cannone, N., Sgorbati, S. and Guglielmin, M. (2007) Unexpected impacts of climate change on alpine vegetation. *Frontiers in Ecology and the Environment* 5, 360–364.

Chen, T., Burke, M. and Gusta, L. (1995) Freezing tolerance in plants: an overview. In: Lee, R.E., Warren, G.J. and Gusta, L.V. (eds) *Biological Ice Nucleation and its Applications*. APS Press, St Paul, Minnesota, pp. 115–135.

Dirnböck, T., Essl, F. and Rabitsch, W. (2011) Disproportional risk for habitat loss of high-altitude endemic species under climate change. *Global Change Biology* 17, 990–996.

Forrest, J., Inouye, D.W. and Thomson, J.D. (2010) Flowering phenology in subalpine meadows: does climate variation influence community of co-flowering patterns? *Ecology* 91, 431–440.

Gu, L., Hanson, P.J., Mac Post, W., Kaiser, D.P., Yang, B., Nemani, R., Pallardy, S.G., *et al.* (2008) The 2007 Eastern US spring freeze: increased cold damage in a warming world? *BioScience* 58, 253–262.

Hall, M., Räntfors, M., Slaney, M., Linder, S. and Wallin, G. (2009) Carbon dioxide exchange of buds and developing shoots of boreal Norway spruce exposed to elevated or ambient CO_2 concentration and temperature in whole-tree chambers. *Tree Physiology* 29, 467–481.

Hallinger, M., Manthey, M. and Wilmking, M. (2010) Establishing a missing link: warm summers and winter snow cover promote shrub expansion into alpine tundra in Scandinavia. *New Phytologist* 186, 890–899.

Hänninen, H. (2006) Climate warming and the risk of frost damage to boreal forest trees: identification of critical ecophysiological traits. *Tree Physiology* 26, 889–898.

Inouye, D.W. (2000) The ecological and evolutionary significance of frost in the context of climate change. *Ecology Letters* 3, 457–463.

Inouye, D.W. (2008) Effects of climate change on phenology, frost damage and floral abundance of montane wildflowers. *Ecology* 89, 353–362.

IPCC (2007) *Climate Change 2007: Synthesis Report, Summary for Policymakers.* IPCC Plenary XXVII, Valencia, Spain.

Kalcsits, L.A., Silim, S., Tanino, K. (2009) Warm temperature accelerates short photoperiod-induced growth cessation and dormancy induction in hybrid poplar (*Populus* x spp.). *Trees* 23, 971–979.

King, D. and Ball, M.C. (1998) A model of frost impacts on seasonal photosynthesis of *Eucalyptus pauciflora. Australian Journal of Plant Physiology* 25, 27–37.

Kreyling, J. (2010) Winter climate change: a critical factor for temperate vegetation performance. *Ecology* 91, 1939–1948.

Kullman, L. (2002) Rapid recent range-margin rise of tree and shrub species in the Swedish Scandes. *Journal of Ecology* 90, 68–77.

Lambrecht, S.C., Loik, M.E., Inouye, D.W. and Harte, J. (2007) Reproductive and physiological responses to simulated climate warming for four subalpine species. *New Phytologist* 173, 121–134.

Loveys, B.R., Egerton, J.J.G. and Ball, M.C. (2006) Higher daytime leaf temperatures contribute to lower freeze tolerance under elevated CO_2. *Plant, Cell & Environment* 29, 1077–1086.

Lutze, J.L., Roden, J.S., Holly, C., Wolfe, J., Egerton, J.J.G. and Ball, M.C. (1998) Elevated atmospheric CO_2 promotes frost damage in evergreen tree seedlings. *Plant, Cell and Environment* 21, 631–635.

Martin, M., Gavazov, K., Körner, C., Hättenschwiler, S. and Rixen, C. (2010) Reduced early growing season freezing resistance in alpine treeline plants under elevated atmospheric CO_2. *Global Change Biology* 16, 1057–1070.

Menzel, A., Sparks, T.H., Estrella, N., Koch, E., Aasa, A., Ahas, R., Alm-Kübler, K., *et al.* (2006) European phenological response to climate change matches the warming pattern. *Global Change Biology* 12, 1969–1976.

Morin, X., Viner, D. and Chuine, I. (2008) Tree species range shifts at a continental scale: new predictive insights from a process-based model. *Journal of Ecology* 96, 784–794.

Murray, M.B., Cannell, M.G.R. and Smith, R.I. (1989) Date of budburst of fifteen tree species in Britain following climatic warming. *Journal of Applied Ecology* 26, 693–700.

Repo, T., Hänninen, H. and Kellomäki, S. (1996) The effects of long term elevation of air temperature and CO_2 on the frost hardiness of Scots pine. *Plant, Cell & Environment* 19, 209–216.

Roden, J.S., Egerton, J.J.G. and Ball, M.C. (1999) Effect of elevated CO_2 on photosynthesis and growth of snow gum (*Eucalyptus pauciflora*) seedlings during winter and spring. *Australian Journal of Plant Physiology* 26, 37–46.

Rohde, A. and Bhalerao, R.P. (2007) Plant dormancy in the perennial context. *Trends in Plant Science* 12, 217–223.

Royer, D.L., Osborne, C.P. and Beerling, D.J. (2002) High CO2 increases the freezing sensitivity of plants: Implications for paleoclimatic reconstructions from fossil floras. *Geology* 30, 963–966.

Schaberg, P.G., Hennon, P.E., D'amore, D.V. and Hawley, G.J. (2008) Influence of simulated snow cover on the cold tolerance and freezing injury of yellow-cedar seedlings. *Global Change Biology* 14, 1–12.

Siebke, K., Ghannoum, O., Conroy, J.P. and von Caemmerer, S. (2002) Elevated CO_2 increases the leaf temperature of two glasshouse-grown C-4 grasses. *Functional Plant Biology* 29, 1377–1385.

Slaney, M., Wallin, G., Medhurst, J. and Linder, S. (2007) Impact of elevated carbon dioxide concentration and temperature on bud burst and shoot growth of boreal Norway spruce. *Tree Physiology* 27, 301–312.

Williams, J.W. and Jackson, S.T. (2007) Novel climates, no-analog communities, and ecological surprises. *Frontiers in Ecology and Environment* 5, 475–482.

Woldendorp, G., Hill, M.J., Doran, R. and Ball, M.C. (2008) Frost in a future climate: modelling interactive effects of warmer temperatures and rising atmospheric [CO_2] on the incidence and severity of frost damage in a temperate evergreen (*Eucalyptus pauciflora*). *Global Change Biology* 14, 294–308.

Yu, H., Luedeling, E. and Xu, J. (2010) Winter and spring warming result in delayed spring phenology on the Tibetan Plateau. *Proceedings of the National Academy of Sciences* 51, 22151–22156.

13 Annual Field Crops

Klára Kosová and Ilja Tom Prášil

13.1 Introduction

Annuals are those plants whose whole life cycle from seed to seed spans a period of only one year. These plants are also sometimes called 'true annuals'. In higher latitudes, large differences in air and soil temperatures occur regularly during a year. Seasons with relatively high, warm temperatures (summer) are followed by seasons with low, below-zero temperatures (winter). Temperatures below zero (frost) lead to the formation of ice crystal nuclei in living plant cells which can cause serious damage to the fine network of intracellular compartments resulting in cell death. Gusta *et al.* (2009) outlined seven types of freezing processes in plants. Those affecting annual plants include freezing patterns of non-acclimated, partially acclimated, fully cold-hardened, and supercooling mechanisms of non-acclimated annual plants. Plants inhabiting higher latitudes with regular cold seasons (winter) have evolved mechanisms to minimize the adverse effects of low temperatures in living cells, tissues, organs, and at the whole-organism level. However, some perennial plants of tropical and subtropical origin which are grown in temperate climates are not able to survive freezing temperatures during winter; they are therefore grown as annuals in temperate climates. This group includes tomato, *Petunia* and others.

All plants, annual as well as perennial, are poikilothermic organisms: they cannot efficiently regulate the temperature of their own bodies. Thus, an increase as well as a decrease in ambient air and soil temperatures leads to an increase or a decrease of the temperature of a plant. Basically, two major strategies of how annuals can cope with adverse environmental conditions during cold seasons can be distinguished. Annuals which belong to spring ecotypes avoid the adverse effects of cold seasons and survive these periods only in the form of mature seeds which contain only low amounts of water. This is a *stress avoidance* strategy (Levitt, 1980). However, in some regions with hot dry summers and relatively mild wet winters, annuals try to use the relatively wet winter period for their vegetative growth, and produce seeds in the dry summer period. These annuals germinate in autumn, use the winter period for their vegetative growth and produce seeds in the next summer period. Therefore, they are also called winter annuals. Since they survive the cold period of a year in the actively growing vegetative stage, they have to cope with adverse effects of low temperatures on well-hydrated plant tissues. They had to evolve several mechanisms of *stress tolerance*, and these lie in minimizing the adverse effects of low-temperature stress. These mechanisms include short-term reversible changes (generally termed cold acclimation) which include complex physiological processes such as osmotic adjustment of cells which prevents large water losses caused by extracellular freezing; an increase in the content of unsaturated fatty acids in membrane lipids in order to maintain a sufficient level of membrane fluidity; and complex changes in photosynthetic processes. On a long-term scale, cold adaptation mechanisms also include a complex regulation of plant individual development which prevents the plants making a premature transition from a generally more tolerant vegetative stage into a generally less tolerant reproductive stage. These regulatory mechanisms which require a long-term

© CAB International 2012. *Temperature Adaptation in a Changing Climate: Nature at Risk* (eds K.B. Storey and K.K. Tanino)

(several weeks to months) cold period prior to the transition to flowering evolved independently in several groups of angiosperms inhabiting higher latitudes and are generally known as vernalization (Chouard, 1960). More recently, molecular progress on vernalization mechanisms has been made (Horvath *et al.*, 2003; Böhlenius *et al.*, 2006; Horvath, 2009). In the following section, a variety of cold adaptation mechanisms including reversible mechanisms of cold acclimation as well as developmental adaptations (vernalization) will be described. A special focus on the regulatory relationships between vernalization and acquisition of enhanced frost tolerance in the model species *Arabidopsis thaliana* and cereals from the tribe *Triticeae* is given.

13.2 Cold Adaptation Mechanisms in Annuals

13.2.1. Reversible mechanisms – cold acclimation

The term cold acclimation is used for complex, short-term reversible changes which are launched after plant exposure to low, above-zero temperatures (chilling) which are aimed at the enhancement of plant tolerance to the adverse effects of low temperatures, that is, chilling and frost (Guy, 1990; Thomashow, 1999; Xin and Browse, 2000; Ruelland *et al.*, 2009). These changes are complex, involving practically all cellular compartments and physiological processes. Generally, basic mechanisms of cold acclimation processes are shared by all higher plants, winter annuals as well as perennials. According to the ability to survive (acclimate to) low temperatures, plants can basically be divided into three major groups:

1. Chilling-sensitive plants, which can acclimate to low above-zero temperatures (usually 0°C–12°C, but in some cases even 0°C–18°C) only to a limited extent and are seriously damaged by these temperatures. This group includes predominantly spring annual crops of tropical and sub-tropical origin (chickpea, cowpea, cucumber, maize, pepper, rice, sugar cane, soybean, tomato and others) which are affected by cold predominantly in the early vegetative stage during spring or at the end of their growth period in autumn;
2. Chilling-tolerant plants, which are not damaged by chilling, but are susceptible to frost; this group includes some plants of sub-tropical origin such as potato or sugar beet which could be affected by harmful effects of short periods of frost during their vegetative stage in early spring;
3. Frost (freezing)-tolerant plants, which can acclimate not only to low above-zero temperatures, but also to low below-zero temperatures (frost). To this group, some frost-tolerant spring genotypes of *Triticeae* cereals belong; however, this group includes mainly winter annual crops such as winter cereals from the tribe *Triticeae* or winter genotypes of oat or oilseed rape which are genetically adapted to regular periods of low temperatures (winter) in temperate climates during their vegetative stage via vernalization mechanisms.

General cold acclimation mechanisms

No plants, annual as well as perennial, can efficiently regulate the temperature of their own bodies. From a physical point of view, a temperature decrease of a given object is associated with a decrease in kinetic energy of the molecules (also called thermal energy) of the object. The decrease in kinetic energy of the biomolecules which form the plant cell results in a decrease in the rate of all enzyme-catalysed chemical reactions as well as a decrease in the rate of lateral movement of phospholipid molecules forming the membrane bilayer (Stitt and Hurry, 2002). Cell membranes, especially plasmalemma, are the primary sites sensing a temperature decrease. A decrease in ambient temperature results in the decrease of membrane fluidity, which is suggested to serve as a primary signal of cold (Browse and Xin, 2001). A decrease in membrane fluidity strongly affects many physiological processes since it has a large impact on many transmembrane complexes. Conformational changes in transmembrane two-component histidine kinases in a cyanobacterium *Synechocystis* strain PCC6803 have been proposed a primary signal of cold sensing (Murata and Los, 1997; Suzuki *et al*., 2000). A decrease in membrane fluidity has a negative effect on many transmembrane processes; therefore, plants tend to maintain a sufficient level of membrane fluidity via an increase in the content of unsaturated fatty acids in membrane lipids.

The cold signal probably arising in plasmalemma is then transduced via the cytoplasm into the nucleus. Ca^{2+} influx and rearrangement of actin microfilaments may be involved in the cold signal transduction, since cold leads to enhanced Ca^{2+} levels in the cell cytoplasm (Plieth *et al*., 1999; Browse and Xin, 2001). Örvar *et al*. (2000) have shown in alfalfa (*Medicago sativa*) that not only Ca^{2+} influx, but also expression of the specific nuclear-encoded cold-inducible gene *CAS30* can be induced even at 25°C if microfilament rearrangement and a decrease in membrane rigidification are prevented. In the nucleus, many cold-inducible signalling pathways leading to profound changes in gene expression are initiated. Nuclear cold-inducible signalling pathways can generally be divided into two major groups: ABA-dependent and ABA-independent. ABA-dependent signalling pathways involve bZIP AREB/ABF factors which bind to ABRE elements in promoters of their target genes. Other ABA-dependent transcription factors (TF) are CBF4/DREB1D, MYB, and MYC TF, which bind to CRT/DRE, MYBR, and MYCR promoter elements, respectively (for a review, see Yamaguchi-Shinozaki and Shinozaki, 2005, 2006, for example). The major components of ABA-independent cold-inducible signalling pathways are *CBF/DREB1* transcription activators, which bind to CRT/DRE elements in the promoters of several *COR* (*cold-regulated*) genes via the AP2 binding domain (Gilmour *et al*., 1998; Jaglo-Ottosen *et al*., 1998). The expression of *CBF* genes reveals a significant circadian rhythmicity. In *A. thaliana*, the expression of *CBF* genes is regulated by a complex regulatory network. Expression of *A. thaliana CBF* genes or *COR* genes can – directly or indirectly – be positively regulated by a phosphorylated form of *ICE1* (Chinnusamy *et al*., 2003), nuclear-located enolase (encoded by *LOS2* locus) whose cytoplasmic form functions as a key glycolytic enzyme (Lee *et al*., 2002), and a DEAD-box RNA helicase at the *LOS4* locus (Gong *et al*., 2002, 2005). In contrast, *CBF* expression has been shown to be repressed by *HOS1* (Lee *et al*., 2001), *STZ/ZAT10* and *ZAT12* (Lee *et al*., 2002; Vogel *et al*., 2005).

Cold-inducible changes in gene expression are mirrored at the levels of transcript and protein, as well as metabolite. Recently, the burst of high-throughput techniques such as specific cDNA microarray, two-dimensional differential gel electrophoresis (2D-DIGE), tandem mass spectrometry (MS/MS), quantitative MS techniques such as iTRAQ and ICAT, as well as chromatography techniques coupled with tandem MS, has enabled researchers to study complex changes in the plant transcriptome (Seki *et al*., 2001, 2002; Kreps *et al*., 2002), proteome (Kawamura and Uemura, 2003; Amme *et al*., 2006) and metabolome (Cook *et al*., 2004; Kaplan *et al*., 2004, 2007; Gray and Heath, 2005) under cold. The early studies were conducted mostly on *A. thaliana*, but later studies on *Triticeae* (Monroy *et al*., 2007; Kocsy *et al*., 2010) and rice (Rabbani *et al*., 2003), were also published. The studies have revealed profound changes in carbohydrate metabolism leading to the enhancement of catabolic processes such as glycolysis and starch degradation; and conversely, to the decrease in those processes that lead to the synthesis of UDP-glucose, an activated form of glucose necessary for its incorporation into many organic compounds (Kaplan *et al*., 2007; Kocsy *et al*., 2010).

Therefore, cold leads to the increase of low-molecular, highly hydrophilic sugars: the monosaccharides glucose, fructose and disacharide sucrose; and the oligosaccharides raffinose, stachyose and verbascose, which accumulate in the cell cytoplasm and play a significant role in osmotic adjustment. Osmotic adjustment lies in the decrease in intracellular osmotic potential, which results in a decrease of intracellular water potential; this leads to a minimization of water loss. Osmotic adjustment is mediated not only by sugar accumulation, but also by accumulation of other low-molecular organic compounds such as sugar alcohols including mannitol, pinitol and sorbitol; polyamines such as spermine, spermidine and putrescine; the amino acid proline; quaternary ammonium compounds also called betaines (glycine betaine, alanine betaine); as well as inorganic salt ions such as K^+, Na^+, Cl^-, Ca^{2+}, and phosphates. Moreover, retention of intracellular water is also mediated by a synthesis of high-molecular, highly hydrophilic proteins from a large COR/LEA protein family.

The expression of *COR/LEA* genes can be activated by an enhanced expression of cold-inducible *CBF/DREB1* transcription activators (Jaglo-Ottosen *et al.*, 1998) which bind to CRT/DRE *cis*-regulatory elements in their promoters. COR/LEA proteins can accumulate to relatively large amounts (up to 1%–5% of total soluble proteins) (Houde *et al.*, 1995; Close, 1997) in the cell cytoplasm as well as various intracellular compartments (nucleus and semi-autonomous organelles such as mitochondria and plastids, endoplasmic reticulum, etc.) under cold. The accumulation of large amounts of hydrophilic proteins in the cell cytoplasm prevents the cell from excessive water loss during dehydration. Since both cold and frost stress factors have a dehydrative component, the protective features of these proteins are extremely important under these conditions. Some COR/LEA proteins, such as LEA-D11 proteins known as dehydrins, undergo complex structural changes under cellular dehydration which result in the formation of amphipathic α-helical domains capable of binding to partially dehydrated hydrophobic membrane surfaces. The binding of a hydrophilic protein via its amphipathic domain to a partially dehydrated membrane surface prevents the membrane from further dehydration and denaturation. The protective features of COR/LEA proteins, especially dehydrins, under several dehydrative stresses including low-temperature stresses have recently been reviewed in Allagulova *et al.* (2003), Rorat (2006), Kosová *et al.* (2007), Tunnacliffe and Wise (2007), Battaglia *et al.* (2008) and others.

Recently, it has been shown by Achard *et al.* (2008) that the activation of *A. thaliana CBF1/DREB1b* gene also leads to the accumulation of DELLA proteins, which are nuclear growth-repressing regulatory proteins, via stimulation of the expression of gibberellin-inactivating GA *2-oxidase* genes. Increased levels of GA 2-oxidases contribute to an enhanced inactivation of gibberellins which thus cannot stimulate the degradation of DELLA proteins. *CBF1* can thus induce changes in gibberellin metabolism resulting in dwarfism, that is, a repression of plant growth.

Low temperatures also have a powerful impact on energy metabolism, especially primary and secondary photosynthetic processes. Since low temperatures generally lead to a significant decrease in the rate of enzymatic reactions, the rate of carbon assimilation reactions as well as ATP synthesis both in chloroplasts and mitochondria decreases. However, primary electron transport processes are not enzyme-catalysed, thus their rate is not significantly affected by low-temperature stress. Therefore, a combination of low temperatures with a sufficient irradiance leads to an imbalance between the rate of primary electron transport reactions and carbon assimilation processes in chloroplasts, a phenomenon known as chilling-enhanced photoinhibition. These processes can easily be monitored *in vivo* by measurement of changes in chlorophyll *a* (chl *a*) fluorescence in photosystem II (PSII). An increase in thermal energy dissipation in PSII (generally termed non-photochemical fluorescence quenching) leads to an increase in violaxanthin de-epoxidation (the first step in the xanthophyll cycle) (Demmig-Adams and Adams, 1996) and to the decrease in the maximum quantum yield efficiency, defined as the ratio of variable to maximum chl *a* fluorescence Fv/Fm. In chilling-sensitive spring annual crops such as maize (*Zea mays*), tomato (*Lycopersicon esculentum*), cucumber (*Cucumis sativus*), sweet pepper (*Capsicum annuum*) or rice (*Oryza sativa*), chilling leads to a *chilling-enhanced photoinhibition* associated with a decrease in photosynthesis light saturation point, a relative decrease in the content of chl *a* and chl *b* to the content of carotenoids, especially components of the xanthophyll cycle, a decrease in Fv/Fm, an inhibition of activity of CO_2-assimilating enzymes ribuloso-1,5-bisphosphate carboxylase/oxygenase (Rubisco: both in C_3 and C_4 plants), and phosphoenolpyruvate carboxylase (PEPcase: in C_4 plants). In contrast, chilling leads to an increase in non-photochemical quenching and content of xanthophyll-cycle components, especially violaxanthin, in these plants (Brüggemann *et al.*, 1992; Sonoike and Terashima, 1994; Haldimann *et al.*, 1996; Jung and Steffen, 1997; Kingston-Smith *et al.*, 1997; Aguilera *et al.*, 1999; Fracheboud *et al.*, 1999; Xu *et al.*, 2000; Liu *et al.*, 2001). In chilling-sensitive cucumber, photoinhibition at the acceptor side of PSI is associated with damage to Fe-S electron acceptors (Sonoike and Terashima, 1994). In contrast to chilling-sensitive spring annuals, chilling does not lead to long-term photoinhibition in cold-tolerant winter annuals. In fully cold-acclimated winter annuals such as winter wheat or oats which maintain active growth under cold periods, relatively high photosynthetic rates and Fv/Fm values are observed due to development of non-photochemical fluorescence quenching and violaxanthin de-epoxidation (Hurry and Huner, 1991; Rizza *et al.*, 2001; Savitch *et al.*, 2002). For a review of chl *a* fluorescence and fluorescence-quenching mechanisms, see, for example, Lichtenthaler and Rinderle (1988) or Krause and Weis (1991).

The lack of oxidized forms of terminal electron acceptors ($NADP^+$, ferredoxin and thioredoxin) in the photosynthetic electron transport chain leads to an enhanced formation of reactive oxygen species (ROS). Electrons from a reduced ferredoxin could be directly transferred to oxygen resulting in formation of superoxide (a so-called Mehler reaction). Two superoxide anion radicals can form hydrogen peroxide in a reaction catalysed by thylakoid or stroma superoxide dismutase. Hydrogen peroxide could then react with a ferrous ion (Fe^{2+}) resulting in formation of a hydroxide anion (HO^-), hydroxyl radical ($HO^·$), and a ferric ion (Fe^{3+}) (a so-called Fenton reaction). Activation of enzymatic as well as non-enzymatic ROS scavenging mechanisms represents an important part of the plant cold-acclimation response. Among ROS-scavenging enzymes, superoxide dismutases (SOD), ascorbate peroxidases (APX), and catalases (CAT) should be mentioned. Superoxide dismutases use several metal ions as cofactors, namely Cu^{2+}/Zn^{2+} (Cu/Zn-SOD), Mn^{2+} (Mn-SOD) and Fe^{2+} (Fe-SOD). The different forms of SOD are found in different intracellular compartments: Cu/Zn-SOD are found in cytoplasm and chloroplasts, Fe-SOD are found predominantly in chloroplasts, while Mn-SOD are specifically located in mitochondria. Ascorbate peroxidases (APX) use ascorbate as a cofactor for ROS scavenging. They can convert hydrogen peroxide into water while reducing ascorbate into the inactive reduced form monodehydroascorbate. The active ascorbate is regenerated from monodehydroascorbate via a so-called ascorbate–glutathione cycle. Several compartment-specific forms of APX, similar to superoxide dismutases, can be distinguished: stromal APX (sAPX), thylakoid APX (tAPX), cytosolic APX (cAPX), and glyoxysomal APX (gmAPX). Non-enzymatic ROS-scavenging mechanisms include reactive low-molecular compounds which can directly accept electrons from ROS. The most important non-enzymatic ROS-scavenging systems in plant cells include α- and γ-tocopherols, ascorbate, glutathione, and carotenoids (Wise, 1995; Apel and Hirt, 2004).

At the level of plant cell ultrastructure, several changes were observed in chilling-sensitive plants under cold. The main changes include chromatin condensation in the nucleus followed in the later stages by chromatin and nucleoli fragmentation, dilatation of rough endoplasmic reticulum and Golgi vesicles, mitochondria and chloroplast swelling, and vacuolarization, often associated with thylakoid dilatation, development of chloroplast peripheral reticulum, and accumulation of lipid droplets in the chloroplast stroma. These ultrastructural changes indicate enhanced starch degradation (chloroplast swelling), chilling-enhanced photo-oxidation (thylakoid dilatation and grana unstacking), reduced transport capacity of chloroplast inner membrane (development of peripheral reticulum), and other complex metabolic changes (reviewed in Kratsch and Wise, 2000).

Cold acclimation mechanisms in winter annuals

Various aspects of the cold-acclimation mechanisms given above are aimed at the enhancement of plant survival under adverse temperature conditions. Winter annuals had adapted evolutionarily to cold seasons (winter) occurring regularly during the vegetative stage of development. In nature, seasons with prevalent low temperatures both above and below the freezing point are usually characterized not only by low temperatures, but also by several other factors which affect adversely plant vitality and crop yield (Bertrand and Castonguay, 2003); in other words, which act as stress factors (Fig. 13.1). In temperate climate habitats, these adverse factors usually occur irregularly during winter, depending on temperature fluctuations and precipitation form and amount. High precipitation in combination with rapid temperature changes during the winter can have predominantly negative effects on plant winter survival, since these factors often lead to repeated freeze–thaw cycles associated with soil heaving and formation of ice crusts during the freezing phase and waterlogging stress during the thawing phase. Formation of ice crusts can result in a stress of ice encasement which is accompanied by hypoxia (lack of the oxygen necessary for aerobic metabolism). After thawing, plants suffering from hypoxia usually reveal enhanced levels of ROS and phytotoxic products of anaerobic metabolism, resulting in irreversible plant tissue damage and death (Gudleifsson, 2009). Last, but not least, the spreading of various snow moulds in exhausted plant communities under thick snow cover represents an example of biological stresses threatening overwintering crops. Therefore, the ability of winter annuals to survive adverse cool seasons is complex and is usually termed winter hardiness. However, a plant's ability to cold-acclimate and to induce a sufficient level of frost tolerance often represents a crucial component of plant winter hardiness. Moreover, all the above-mentioned winter stresses negatively affect the level of plant frost tolerance (Gusta and Fowler, 1979; Prášil and Zámečník, 1991). Therefore, the determination of acquired frost tolerance

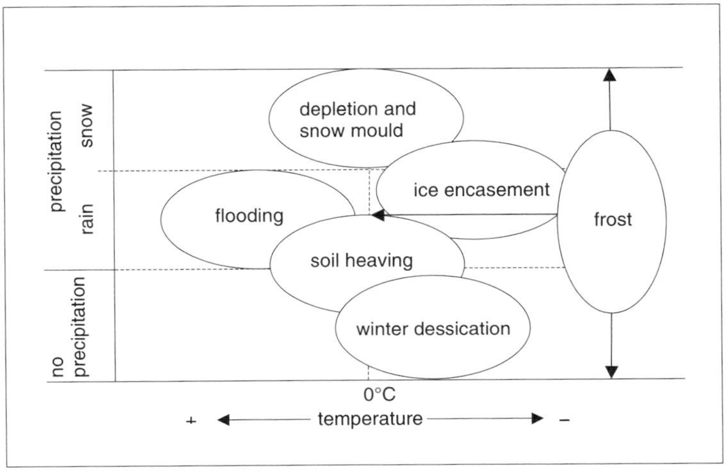

Fig. 13.1 Main stress factors affecting the winter survival of field annuals (modified after Prášil and Zámečník, 1991).

represents an important aspect of the evaluation of plant ability to survive winter in the field.

Acquired frost tolerance of winter annuals, quantitatively expressed as LT50 values (a temperature when 50% of samples are killed during a frost test), varies during the winter period. The level of maximum acquired plant frost tolerance, as well as the time when the maximum frost tolerance is reached, varies during the individual winter periods and is affected by several variety-specific factors as well as environmental factors, especially growth temperature. Since crown tissues are crucial for survival of winter cereals, soil temperature at the depth of crown tissues has a crucial impact on winter survival of whole plants (Fig. 13.2). Despite significant variations in dynamics of plant frost tolerance in different geographical areas and growth seasons, three basic stages shared by various winter cereals can be distinguished (Gusta and Fowler, 1979; Gusta and Chen, 1987; Prášil and Zámečník, 1991; Fig. 13.3):

1. Hardening, when frost tolerance rapidly increases (LT50 values decrease) in the autumn. Cold hardening is an active, energy-demanding process that can be efficiently induced by low, non-freezing temperatures only if a plant has a sufficient pool of organic reserves. In the field, air and soil temperatures decrease continuously during autumn from relatively high temperatures at the time of seeding to low, around-zero temperatures at the end of autumn. The incidence of moderate frost (−2° to −6°C) at the end of the hardening stage is crucial for achievement of maximum acquired frost tolerance (Tumanov, 1979; Gusta and Chen, 1987; Sakai and Larcher, 1987; Fowler *et al.*, 1999). Depending on climatic conditions at a given geographical location and weather conditions in a given season, maximum hardiness is usually achieved in the second half of November in Saskatoon, Canada (Gusta and Fowler, 1979; Fowler, 2002), in the first half of December in St Petersburg, Russia (Tumanov, 1979) and in Norway (Bergjord *et al.*, 2008), or in the second half of December in Prague, Czech Republic (Prášil and Zámečník, 1991; Fig. 13.2).

2. Maintenance, when plants have already achieved a certain level of frost tolerance. However, the acquired frost tolerance level is not constant, but changes during the winter. Once maximum cold hardiness is reached, winter cereals can maintain the enhanced level of frost tolerance for several weeks to months when temperatures stay below the freezing point and plants have sufficient energy sources. In regions with repeated temperature fluctuations around the freezing point, freeze–thaw cycles occur, and these can adversely affect the maintenance of enhanced cold hardiness. Nevertheless, partial de-hardening is a reversible process in this stage, and cold hardiness increases during the freezing phase after thawing. However, plant ability to re-harden after a partial de-hardening generally decreases along with increasing day length and the progression of winter. Thus, the maintenance phase proceeds slowly into the de-hardening phase.

3. De-hardening, when the frost tolerance level decreases irreversibly in the spring. A similar dynamics curve of acquired frost tolerance has also been described in winter oilseed rape (Kacperska-Palacz, 1987) and other winter field crops (Sakai and Larcher, 1987). The rise of temperature above the freezing point and subsequently above severe chilling temperatures (0°–9°C) during the end of winter and/or the beginning of spring initiates plant de-hardening. The increase in air and soil temperatures induce plant active growth which is accompanied by an irreversible decline in the ability to induce cold hardiness. The rate of de-hardening is temperature dependent; that is, the higher the temperature, the faster the rate of plant de-hardening (Fowler *et al.*, 1999; Fowler, 2002).

Each stage of the hardening process is affected by the plant itself (plant genotype, developmental stage, vitality and health) as well as by several environmental factors which determine the final shape of the frost-tolerance curve. The three major stages of plant cold hardiness development have repeatedly been reported not only in field studies, but also under constant cold

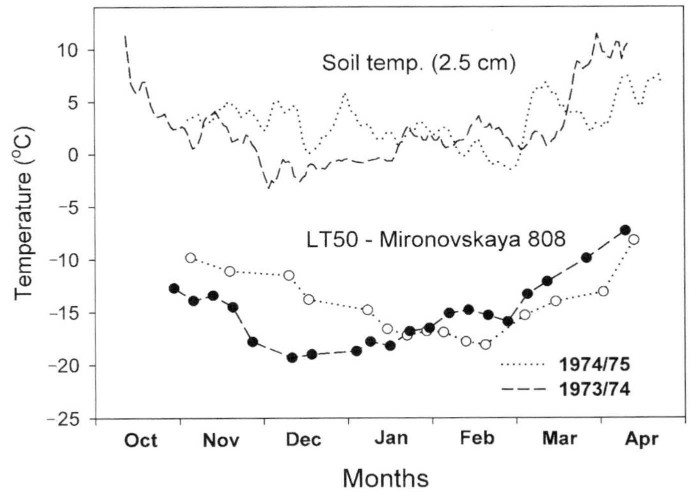

Fig. 13.2 Time course of average daily soil temperature (2.5 cm depth) and frost tolerance (LT50) of the 'Mironovskaya 808' winter wheat in Prague (Czech Republic) during the winter seasons of 1973/74 and 1974/75 (compiled from unpublished data of Vladimír Seget'a).

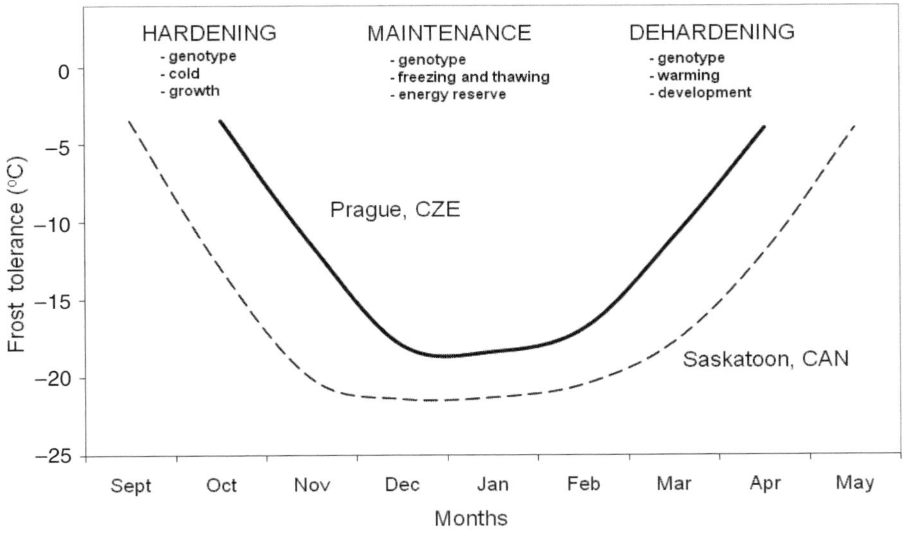

Fig. 13.3 Generalized dynamics of frost tolerance of winter wheat under field conditions in two different places. CZE, Czech Republic; CAN, Canada. The main stages and some factors responsible for the dynamics are shown (modified after Gusta and Fowler, 1979; Gusta and Chen, 1987; Prášil and Zámečník, 1991).

temperatures in controlled environments (growth chambers) (Fowler *et al.*, 1996, 2001; Prášil *et al.*, 2004; Vítámvás and Prášil, 2008). Growth chamber studies have enabled researchers to dissect several factors affecting plant frost tolerance level and to focus on the genetic basis of plant frost tolerance.

The plant genotype is the most important factor affecting dynamics of plant frost tolerance. Frost tolerance is a multigenic trait which is genetically determined by several quantitative trait loci (QTL). In *Triticeae*, two major frost resistance QTL affecting the expression of *COR* genes have been determined. In relation to the major cultivated *Triticeae* species, the values of maximum acquired frost tolerance decline from rye, which generally has the highest values, followed by wheat with a broad range of acquired frost tolerance, to barley with the lowest levels of hardiness. Naturally, there is a great genotypic variability within each *Triticeae* species. The largest genotypic variability in acquired frost tolerance has been reported in wheat, with winter genotypes varying in LT50 values from −24°C to −12°C and spring genotypes approaching LT50 values around −10°C or higher (Gusta *et al.*, 2001; Fowler, 2002; Prášil, unpublished).

Plant developmental stage is also a very important factor affecting plant ability to acquire frost tolerance. Transition from a vegetative into a reproductive stage, which is regulated by several vernalization (*Vrn*) and photoperiodically (*Ppd*) regulated loci, is accompanied by a significant decrease in the ability to acquire enhanced frost tolerance upon inducing environmental conditions. Following the study of relationships between plant developmental stage and its ability to induce frost tolerance in *Triticeae*, Fowler *et al.* (1996) proposed a close association between vernalization saturation and a decrease in the expression of *COR* genes, indicating a decrease in plant ability to induce frost tolerance upon cold-inducing conditions. This hypothesis has been confirmed by further papers from this group (Fowler *et al.*, 2001; Danyluk *et al.*, 2003; Kane *et al.*, 2005). In *Triticeae*, expression of the major vernalization gene *VRN1* indicates the transition into the reproductive stage and precedes morphological changes in the plant shoot apex – a so-called formation of double ridges, indicating formation of spike instead of leaves. Formation of double ridges indicates an irreversible transition into the reproductive stage (Prášil *et al.*, 2004). Parallel investigations of acquired frost tolerance level and morphological changes in shoot apex, as well as changes in *COR* gene expression, have confirmed Fowler's hypothesis that transition into the reproductive stage results in the decrease in *COR* gene expression. A faster dynamic of development in spring genotypes without vernalization requirement prevents them from being able to acquire high frost tolerance under inducing conditions; however, some facultative genotypes without vernalization requirement can also induce high frost tolerance because they remain in the vegetative stage under short-day (SD) photoperiods (Akar *et al.*, 2009).

As already stated above, plant ability to induce frost tolerance is also affected by several environmental factors. The growth temperature itself has a huge impact on frost tolerance induction. Low temperatures lead to inhibition of active growth and slow down several metabolic processes; by contrast, they induce cold acclimation processes resulting in enhancement of acquired frost tolerance. The lower the growth temperature, the faster the induction of frost tolerance. However, as shown by Fowler (2008), differences in the potential to induce frost tolerance between the highly frost-tolerant *Triticeae* genotypes and the less tolerant ones can be observed even at temperatures as high as 15°C. Fowler (2008) and Galiba *et al.* (2009) have formulated a hypothesis about threshold induction temperatures in *Triticeae*: the highly frost-tolerant genotypes are able to induce enhanced frost tolerance at higher growth temperatures than the less tolerant ones. More tolerant genotypes also increase the acquired frost tolerance more rapidly than the less tolerant ones. Moreover, a positive relationship between wheat or barley capacity to cold-acclimate and the level of *COR* transcript or protein accumulation has recently been found not only in fully cold-acclimated plants (Houde *et al.*, 1992; Zhu *et al.*, 2000; Vítámvás *et al.*, 2007; Kosová *et al.*, 2008a, 2010a), but also in plants grown at a relatively broad range of mild temperatures (10°C–18°C) (Crosatti *et al.*, 1995; Vágújfalvi *et al.*, 2000, 2003; Holková *et al.*, 2009; Vítámvás *et al.*, 2010). These studies demonstrate that the highly frost-tolerant *Triticeae* genotypes are genetically adapted to slow temperature decrease during the autumn in such a manner that they start increasing *COR* transcript and protein levels at much higher temperatures than the frost-susceptible genotypes. Several *COR* transcripts or proteins have thus been proposed as reliable indirect markers of plant frost tolerance that could be employed in breeding programmes aimed at improvement of plant frost tolerance (Prášil *et al.*, 2007; Kosová *et al.*, 2010b, 2011).

The other major factor affecting plant ability to cold-acclimate is light. Both the induction of enhanced frost tolerance as well as the maintenance of enhanced frost tolerance place increased demands on energy which are covered by photoassimilates from seed endosperm in early developmental stages and by photosynthetic processes in later developmental stages. Plant cold hardiness is significantly affected not only by the quantity of photosynthetically active radiation (irradiance), but also by the day length (photoperiod). In *Triticeae*, the long day (LD) photoperiod generally enhances plant transition into the less tolerant reproductive stage (Fowler *et al.*, 2001). However, Limin *et al.* (2007) have shown that if the plants are still in the vegetative stage, a LD (20 h) photoperiod leads to a greater decrease in LT50 values than an SD (8 h) photoperiod. One possible explanation of the difference in LT50 values under LD and SD conditions can lie in the fact that under LD conditions, plants are able to produce more assimilates than under SD conditions. As shown by Savitch *et al.* (2000), the plant cold-acclimation process has increased demands for assimilates.

Other environmental factors that strongly affect overwintering of winter annual crops are precipitation and soil humidity. High soil humidity reveals predominantly adverse effects on induction of enhanced frost tolerance and increases plant susceptibility to freezing temperatures (Olien and Smith, 1981). Mild drought stress in the autumn reveals positive effects on the hardening of winter cereals. However, lack of precipitation in the form of snow during the winter usually has negative effects on plant survival. Sufficiently thick snow cover has a protective effect on crown tissues of winter cereals, since it protects the soil from rapid cooling during harsh frosts, and thus prevents rapid temperature decline below the minimum temperature that cold-acclimated crowns can survive (Sakai and Larcher, 1987). In winter wheat, the site of low-temperature stress injury is the lower portion of the crown from which the roots regenerate (Tanino and McKersie, 1982; 1985; Chen *et al.*, 1983). Thus, protecting the crown in fact means protecting this region.

13.2.2 Developmental mechanisms – vernalization

Vernalization – the requirement for a long-term (several weeks to months) cold period prior to the transition to flowering – evolved independently in several groups of dicots as well as monocots inhabiting higher latitudes. It is a major genetic adaptation preventing the plants from a premature transition from a relatively cold-tolerant vegetative stage to a relatively cold-susceptible reproductive stage. Once the plants proceed into a more susceptible reproductive stage, they cannot return to a more tolerant vegetative stage. Therefore, regulation of the vegetative/reproductive transition is extremely important. The organisms in which vernalization has been studied most thoroughly are *Arabidopsis thaliana* (an LD dicot from the *Brassicaceae* family); and cereals from the tribe *Triticeae*, especially a diploid einkorn wheat, a hexaploid common wheat, and a diploid barley, which are long-day monocots from the *Poaceae* family. It has been postulated by several researchers (Jung and Müller, 2009 – see references therein) that vernalization evolved independently in dicots and monocots after their evolutionary split in the middle Jurassic–early Cretaceous era. Palaeographic studies have shown that the climate in that period was significantly warmer than the current climate, and revealed low seasonality, so mechanisms preventing a premature developmental transition into a cold-susceptible reproductive stage were not necessary.

In *Arabidopsis thaliana*, vernalization

results in a down-regulation of the major flowering repressor *FLC* (Michaels and Amasino, 1999), which belongs to a special group within a large family of MADS-box transcription factors and whose expression is positively regulated by *FRI* (Koornneef *et al*., 1994; Johansson *et al*., 2000; Gazzani *et al*., 2003; Shindo *et al*., 2005). Down-regulation of *FLC* results in the up-regulation of several morphogenetic MADS-box genes (e.g. *LFY, AP1* and *SOC1*) which regulate the transition of the shoot apical meristem (SAM) from the vegetative into the reproductive stage. The down-regulation of *FLC* is also accompanied by a decrease in *A. thaliana* low-temperature tolerance (Seo *et al*., 2009). Unlike *Arabidopsis*, vernalization in *Triticeae* results in an up-regulation of the major vernalization gene *VRN1*. This gene is sequentially related to *A. thaliana AP1*; that is, it belongs to the large family of MADS-box transcription factors. The expression of *VRN1* has been proved necessary for the transition to flowering (Shitsukawa *et al*., 2007a). In both cases, the regulation of the expression of *FLC* and *VRN1* genes, respectively, is associated with chromatin modifications. *FLC* expression is down-regulated by the histone methylation complex involving *VRN1* and *VRN2* genes (polycomb factors, no homologues of *VRN1* and *VRN2* genes in *Triticeae*) and *VIN3* gene (PHD finger) (Bastow *et al*., 2004; Sung and Amasino, 2005), and the histone deacetylase complex involving *FVE* gene (Kim *et al*., 2004). Chromatin modifications which lie in histone 3 lysine 9 and lysine 27 (H3K9 and H3K27) methylation and histone 3 deacetylation lead to *FLC* repression. For stable *FLC* repression, binding of *VRN2* protein (a homologue of *Suppressor of Zeste 12* (*Su(z)12*), a PcG component from *Drosophila*) is required. In *Triticeae*, up-regulation of *VRN1* gene expression is also accompanied by changes in histone methylation: the level of H3K27me3 (histone 3 lysine 27 trimethylation) decreases and the level of H3K4me3 increases (Oliver *et al*., 2009). Shitsukawa *et al*. (2007a) have proved that in einkorn wheat (*Triticum monococcum*) the expression of functional *VRN-A^{m}1* gene is essential for flowering, since the *T. monococcum* recessive mutant *maintained vegetative phase 1* (*mvp1*), which encodes a non-functional *VRN-A^{m}1* allele with deletions in promoter and coding regions of *VRN-A^{m}1*, is not able to flower. Vernalization in winter growth habit of *Triticeae* is associated with a down-regulation of the major flowering repressor *VRN2*, which encodes a ZCCT transcription factor (transcription factor with zinc-finger and CCT domains) acting as a major repressor of the *VRN1* gene. *FLC* in *A. thaliana* and *VRN2* in *Triticeae* thus reveal analogous functions (major flowering repressor down-regulated by vernalization), although they belong to different structural classes of transcription factors (*FLC* – a special sub-group of MADS-box TFs, *VRN2* in *Triticeae* – a ZCCT gene having no direct homologues in *A. thaliana*). Up-regulation of *VRN1* gene expression in *Triticeae* is enhanced by an LD signal mediated by *A. thaliana CO, PRR* and *FT/VRN3* homologues (Turner *et al*., 2005; Yan *et al*., 2006).

In both *A. thaliana* and *Triticeae*, considerable experimental evidence has shown that a transition from a vegetative stage into a reproductive stage (vegetative/reproductive stage transition) is associated with a significant decrease in their ability to acquire a high level of frost tolerance upon inducing conditions (for review, see Kosová *et al*., 2008b; Galiba *et al*., 2009, for example). In *Triticeae*, the major vernalization locus *VRN1* which encodes the *VRN1* gene is tightly associated with the major *Frost-resistance* QTL named *Fr1*. Some recent studies (Kim *et al*., 2004; Stockinger *et al*., 2007; Seo *et al*., 2009) have shed more light on the regulatory relationships between the vegetative/reproductive transition and the acquisition of enhanced frost-tolerance level. Mutual relationships between cold acclimation and vernalization regulatory pathways in *A. thaliana* and cereals from the tribe *Triticeae* will be discussed in the following paragraphs. They are also schematically illustrated in Fig. 13.4.

In *A. thaliana*, a mutant *altered cold-responsive gene expression 1* (*acg1*) was described by Kim *et al*. (2004). The mutant had increased levels of *CBF* genes and

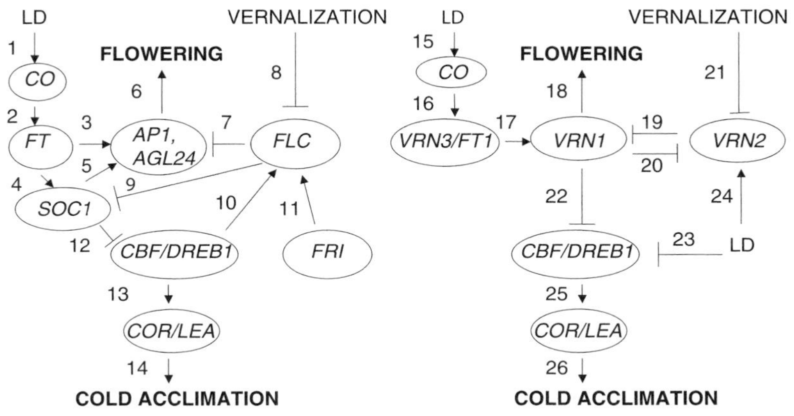

Arabidopsis thaliana *Triticeae*

Fig. 13.4 Schematic relationships between key vernalization genes and cold-acclimation pathways in *Arabidopsis thaliana* and *Triticeae*. LD – a long-day photoperiod.
Arabidopsis thaliana: 1 – Putterill *et al.* (1995) – expression of *CO* is promoted by a long-day photoperiod in *Arabidopsis*; 2 – Kardailsky *et al.* (1999) – characterization of *FT* gene; 3 – Jaeger and Wigge (2007); Valverde *et al.* (2004); Wigge *et al.* (2005) – *FT* activates *AP1* and other flowering meristem identity genes; 4 – Yoo *et al.* (2005) - *CO* and *FT* activate *SOC1*; 5 – Moon *et al.* (2003) – *SOC1* activates flowering-inducing genes; 6 - Bowman *et al.* (1993) – *AP1* and other interacting genes promote flowering; 7 – Sheldon *et al.* (2000) – *FLC* represses transition to flowering; 8 – Michaels and Amasino (1999) – *FLC* is down-regulated by vernalization; 9 – Seo *et al.* (2009) – *FLC* represses *SOC1* prior to vernalization; 10 – Seo *et al.* (2009) – *CBFs* positively regulate *FLC* expression prior to vernalization; 11 – Koornneef *et al.* (1994) – *FRI* and *FLC* are associated with vernalization; 12 – Seo *et al.* (2009) – *SOC1* represses *CBF/DREB1* expression; 13 – Jaglo-Ottosen *et al.* (1998) – *CBF/DREB1* activate *COR/LEA* genes; 14 – Thomashow (1999) – expression of *COR/LEA* genes is associated with cold acclimation.
Triticeae: 15 – Griffiths *et al.* (2003) – *CO*-like gene families in barley, rice, and *Arabidopsis*; 16 – Yan *et al.* (2006) – expression of *VRN3* (*FT1*) is induced under LD conditions; 17 – Yan *et al.* (2006) – a positive effect of *VRN3* on *VRN1* expression; 18 – Shitsukawa *et al.* (2007a) – *mvp* mutant of einkorn wheat reveals that the expression of *VRN1* gene is necessary for flowering; 19 – Von Zitzewitz *et al.* (2005) – a repressive effect of dominant *Vrn2* allele on recessive *vrn1* allele; 20 – Loukoianov *et al.* (2005) – a repressive effect of dominant *Vrn1* allele on the expression of *Vrn2* allele; 21 – Yan *et al.* (2004) – *VRN2* gene is down-regulated by vernalization; 22 – Danyluk *et al.* (2003); Kane *et al.* (2005); Dhillon *et al.* (2010) – expression of *VRN1* gene leads to down-regulation of *CBF* expression; 23 – Dhillon *et al.* (2010) – LD signal is necessary for down-regulation of *CBF/DREB1* gene expression in *Triticeae*; 24 – Dubcovsky *et al.* (2006) – 'short-day vernalization' – short-day photoperiods induce down-regulation of *VRN2* gene expression while long-day photoperiods induce *VRN2* gene expression; 25 – Vágújfalvi *et al.* (2000); Choi *et al.* (2002); Skinner *et al.* (2005) – *CBFs* activate expression of *COR/LEA* genes in *Triticeae*; 26 – Thomashow (1999) – expression of *COR/LEA* genes is associated with cold acclimation.

COR15a and *COR47* genes upon cold exposure in comparison with wild-type plants. In addition, the *acg1* mutant revealed delayed flowering due to an increased level of FLC. Kim *et al.* (2004) have shown that *acg1* mutation is a null allele of the autonomous pathway gene *FVE*, which encodes a component of a histone deacetylase complex (HDAC), involved in the repression of *FLC*. A positive regulatory feedback loop between the expression of the three *CBF/DREB1* genes (*CBF1/DREB1B*, *CBF2/DREB1C*, *CBF3/DREB1A*) and the expression of *FLC* gene before vernalization was described by Seo *et al.* (2009). Overexpression of *CBF* leads to the enhanced expression of *FLC* and vice versa. However, vernalization overrrides these relationships: overexpression of *CBF* in vernalized plants does not result in the up-regulation of *FLC* gene expression. It has also been found by Seo *et al.* (2009) that the expression of all three *CBF* can be repressed directly by the flowering activator *SOC1*, a MADS-box gene involved in the gibberellin flowering pathway (Moon *et al.*, 2003), which can bind to the CArG box regions present in both proximal and distal parts of the promoters of the three *CBF* genes.

Analogously to *A. thaliana*, a negative effect of the expression of the major *Triticeae* flowering-inducing gene *VRN1* on the expression of several cold-inducible *CBF* genes has been reported by several researchers (Danyluk *et al.*, 2003; Kane *et al.*, 2005; Stockinger *et al.*, 2007). The family of *CBF/DREB1* transcription activators is much more diverse in *Triticeae* than in *A. thaliana*. It contains many paralogues which can be divided into ten different sub-classes, six of which are *Pooideae*-specific (Badawi *et al.*, 2007). Thirteen *CBF* genes have been proposed in einkorn wheat (Knox *et al.*, 2008), at least 17 *CBF* genes in barley (Skinner *et al.*, 2005; Francia *et al.*, 2007), and at least 25 *CBF* genes in common wheat (Badawi *et al.*, 2007). This large diversity of *CBF* genes in *Triticeae* seems to represent an evolutionary adaptation to temperate climate habitats with a wide range of temperature changes. The majority of *CBF* genes in *Triticeae* is located at the *Frost-resistance 2* (*Fr2*) locus on the long arm of group 5 homoeologous chromosomes (Vágújfalvi *et al.*, 2003; Francia *et al.*, 2004). Studies on two einkorn wheat lines with contrasting levels of frost tolerance have shown that the *CBF* gene cluster at *Fr-A^{m}2* locus consists of three sub-clusters and that the peak of frost tolerance QTL at *Fr-A^{m}2* coincides with the position of the central sub-cluster which encodes three *TmCBF* genes (*TmCBF12*, *TmCBF14* and *TmCBF15*) with contrasting expression levels in freezing-tolerant and freezing-susceptible varieties of einkorn wheat (Knox *et al.*, 2008). Moreover, in the case of the *TmCBF12* gene, two different alleles were described by Knox *et al.* (2008). The allele of *TmCBF12* gene encoded by a freezing-tolerant winter line G3116 encodes a fully functional *CBF* transcription activator capable of binding to the CRT/DRE element in the promoter regions of several *COR* genes. In contrast, the *TmCBF12* allele encoded by a freezing-susceptible spring line DV92 has a deletion corresponding to a deletion of five amino acids in the AP2 domain which mediates the interaction between the *CBF* factor and CRT/DRE element in the promoter of *COR* genes. Gel-shift assays have proved that the *TmCBF12* allele encoded by a freezing-sensitive spring line DV92 cannot bind to CRT/DRE elements in the promoters of *COR* genes and thus cannot activate *COR* gene expression. Studies carried out on winter and spring barley genotypes have also revealed that the tolerant winter genotypes contain more paralogous *CBF* genes (two paralogues of *HvCBF2* and *HvCBF4* genes) than the sensitive spring genotypes (only one copy of *HvCBF2* and *HvCBF4* genes at the *Fr-H2* locus) (Knox *et al.*, 2010). These differences in *CBF* alleles and the structure of the *Fr2* locus between freezing-tolerant and freezing-susceptible *Triticeae* genotypes can underlie the differences in acquired frost tolerance levels between these genotypes. The data about the regulation of *CBF* gene expression in *Triticeae* are still very scant, although some homologues of *A. thaliana* regulatory genes (*ICE1*, an activator of *CBF* gene expression; *ZAT12*, a repressor of *CBF* gene expression) have already been found (Skinner *et al.*, 2006). Preliminary studies on barley Nure × Tremois mapping population by Stockinger *et al.* (2007) have also suggested that unlike the situation in *A. thaliana*, at least some cold-inducible *HvCBF* genes (*HvCBF2A* and *HvCBF4B*) do not contain CArG elements (MADS-box binding motifs) in their promoter regions. Thus, a direct interaction between the *VRN1* gene product and these cold-inducible *CBF* cannot be proposed. However, since the MADS-box binding CArG elements have been found in all three *A. thaliana* cold-inducible *CBF* genes (*CBF1*, *CBF2*, *CBF3*) (Seo *et al.*, 2009), it can be expected that the CArG elements could be found in the promoters of at least some cold-inducible *CBF* homologues in *Triticeae*. In *A. thaliana*, a direct regulatory interaction between *CBF* genes and the *SOC1* flowering activator has been described (Seo *et al.*, 2009); in *Triticeae*, an *A. thaliana SOC1* homologue named *WSOC1* has already been cloned from common wheat (Shitsukawa *et al.*, 2007b). Similarly to *A. thaliana SOC1*, *WSOC1* also encodes a MADS-box flowering activator whose expression is positively regulated by gibberellins. Taking *A. thaliana* as a model, interactions between *WSOC1* and *CBF* genes could be proposed in *Triticeae*. However, possible regulatory relationships between *CBF* genes and *VRN1* or *WSOC1* MADS-box flowering activators remain to be explored experimentally.

Recently, Dhillon *et al.* (2010) have studied *T. monococcum* plants carrying *Vrn1* alleles with deletions in the CArG box region in the promoter which confer a spring growth habit and a high *VRN1* expression under both LD and SD photoperiods. The expression of *VRN1* gene under LD has led to the decrease in *COR14b* expression; however, under SD conditions, the expression of the *VRN1* gene has not led to decrease in *COR14b* expression. The authors thus conclude that expression of the *VRN1* gene itself is necessary but not sufficient to down-regulate the expression of *COR* genes. In addition to *VRN1*, some LD-inducible regulatory factors may play an important role in the regulation of *COR* gene expression. Unravelling the complex, finely balanced regulatory network between the *VRN1* gene, photoperiodically regulated regulatory factors, and cold-inducible *CBF* and *COR* genes represents a great challenge for researchers dealing with the improvement of low-temperature tolerance in winter annuals, especially in economically important winter *Triticeae* species.

13.3 Possible Impacts of Global Climate Change on Annuals

Global temperature changes, that is, warming of some regions on the Earth's surface or cooling of some others, will cause disturbances in fine-tuned plant physiological responses to cold. Longer periods of sufficiently warm temperatures may be advantageous for chilling-sensitive spring annuals, but prolonged periods of extremely high temperatures during summer may cause severe water deficits which would, in turn, have adverse effects on annuals. In winter periods, changes in geographical and temporal distribution of several stress factors may occur (Bélanger *et al.*, 2002; Gudleifsson, 2009). Harsh freezing temperatures associated with formation of ice crusts leading to ice encasement will probably move to higher latitudes and altitudes; instead of ice, spread of various snow moulds may severely damage the overwintering crops (Bertrand and Castonguay, 2003). Large areas in central and northeastern Europe will probably experience a shift from a continental to a maritime type of climate (Crawford, 2000), which may increase the risk of killing the overwintering crops by sudden frosts due to insufficient snow cover. In 2002–2003, warm and humid weather in autumn did not induce sufficient cold hardiness in overwintering crops, which were then severely damaged by frost accompanied by the absence of snow cover during the winter. In the eastern part of the Czech Republic, up to 90% of winter barley and 70%–80% of winter wheat areas were severely damaged by frost (Prášil, unpublished). Similar effects have also been predicted for climate change in eastern Canada (Bélanger *et al.*, 2002). However, overwintering crops can be severely damaged by several stress factors even under thick snow cover. In 2004–2005, thick snow layers covering unfrozen soil led to development of various snow moulds in several Central European regions. In late winter and early spring, rapid temperature changes may cause disturbances in genetically driven acclimation mechanisms thus increasing the risk of damage in partially de-hardened winter annuals which could be suddenly exposed to freezing temperatures. Therefore, any change in climatic conditions may have adverse effects on overwintering of winter annual crops adapted to certain regions with specific combinations of stress factors acting not only during the winter period, but throughout the whole year.

A huge burst of molecular genetics at the dawn of a new millennium has brought new insights into the study of the complex mechanisms underlying plant response to low temperatures. It should always be borne in mind that complex cold adaptation processes are usually regulated not only by temperature, but also by other environmental cues, including photoperiod and water availability. For example, transition to flowering is regulated not only by vernalization, but also by photoperiod. In *Triticeae*, low temperatures combined with short-day photoperiods result in down-regulation of the major flowering repressor *VRN2* gene (a so-called short-day vernalization); however, these conditions cannot induce up-regulation of the *VRN1* gene and flowering. In order to induce the expression of the *VRN1* gene, a cold-induced signal leading to changes in chromatin methylation (vernalization) combined with an LD-induced photoperiodic signal mediated by *CO* and *FT* (*VRN3* gene) homologues is essential. The differences in cold-inducible vernalization mechanisms among different angiosperm groups of winter annuals stand in contrast to evolutionarily conserved photoperiodic pathways, such as the *CO*-mediated signal transduction pathway which is conserved from *Chlamydomonas reinhardtii* to angiosperms (Serrano *et al.*, 2009). However, a few recent works have shown some functional analogies between heterologous expression of *TaVRN1* (Adam *et al.*, 2007) and *TaVRN2* (Diallo *et al.*, 2010) genes in *A. thaliana* genetic background and their 'natural' roles in wheat with respect to vernalization and frost-tolerance induction. The studies analysing the processes of cold acclimation and vernalization in various plants at the molecular level thus reflect the fact that, analogously to the present, temperature conditions changed while photoperiodic conditions remained constant during different geological eras. Recent work on *T. monococcum* mutants with non-functional *VRN1* gene *mvp1* and *mvp2* by Professor Dubcovsky's group (Dhillon *et al.*, 2010) has shown that besides *VRN1* gene expression, LD-induced cues are necessary for down-regulation of *COR* genes. Therefore, alterations in temperature conditions under unchanged photoperiodic cues may not result in qualitatively different plant responses to a given season throughout the year; however, large temperature changes will surely have important quantitative effects on the acquired level of plant frost tolerance, increasing the risk of plant damage. Changes in temperature that are too large or too rapid would probably cause significant disturbances in the finely tuned regulatory networks between the processes of cold acclimation and vernalization in winter annuals. Coping successfully with these disturbances will surely influence the ability of winter annuals to survive in these new habitats. Recent advances in molecular genetics, together with high-throughput techniques aimed at the study of complex changes in plant transcriptome, proteome, and metabolome under cold conditions, have enabled us to uncover some key players in the complex phenomena of the major annual plant cold adaptation mechanisms – cold acclimation and vernalization.

13.4 Acknowledgements

This manuscript is dedicated to the memory of the late Dr Vladimír Segeťa, who dedicated his life to the study of cold hardening in cereals. The work was supported by grant from the Czech Ministry of Agriculture (MZe 0002700604) and by the Grant Agency of the Czech Republic (GA CR P501/11/P637).

References

Achard, P., Gong, F., Cheminant, S., Alioua, M., Hedden, P. and Genschik, P. (2008) The cold-inducible CBF1 factor-dependent signaling pathway modulates the accumulation of the growth-repressing DELLA proteins via its effect on gibberellin metabolism. *The Plant Cell* 20, 2117–2129.

Adam, H., Ouellet, F., Kane, N.A., Agharbaoui, Z., Major, G., Tominaga, Y. and Sarhan, F. (2007) Overexpression of *TaVRN1* in *Arabidopsis* promotes early flowering and alter development. *Plant and Cell Physiology* 48, 1192–1206.

Aguilera, C., Stirling, C.M. and Long, S.P. (1999) Genotypic variation within *Zea mays* for susceptibility to and rate of recovery from chill-induced photoinhibition of photosynthesis. *Physiologia Plantarum* 106, 429–436.

Akar, T., Francia, E., Tondelli, A., Rizza, F., Stanca, A.M. and Pecchioni, N. (2009) Marker-assisted characterization of frost tolerance in barley (*Hordeum vulgare* L.). *Plant Breeding* 128, 381–386.

Allagulova, Ch., Gimalov, F.R., Shakirova, F.M. and Vakhitov, V.A. (2003) The plant dehydrins: structure and putative functions. *Biochemistry* 68, 945–951.

Amme, S., Matros, A., Schlesier, B. and Mock, H.-P. (2006) Proteome analysis of cold stress response in *Arabidopsis thaliana* using DIGE-technology. *Journal of Experimental Botany* 57, 1537–1546.

Apel, K. and Hirt, H. (2004) Reactive oxygen species: metabolism, oxidative stress, and signal transduction. *Annual Review of Plant Biology* 55, 373–399.

Badawi, M., Danyluk, J., Boucho, B., Houde, M. and Sarhan, F. (2007) The *CBF* gene family in hexaploid wheat and its relationship to the phylogenetic complexity of cereal *CBFs*. *Molecular Genetics and Genomics* 277, 533–554.

Bastow, R., Mylne, J.S., Lister, C., Lippman, Z., Martienssen, R.A. and Dean, C. (2004) Vernalization requires epigenetic silencing of FLC by histone methylation. *Nature* 427, 164–167.

Battaglia, M., Olvera-Carrillo, Y., Garciarrubio, A., Campos, F. and Covarrubias, A.A. (2008) The enigmatic LEA proteins and other hydrophilins. *Plant Physiology* 148, 6–24.

Bélanger, G., Rochette, P., Castonguay, Y., Bootsma, A. Mongrain, D. and Ryan, D.A.J. (2002) Climate change and winter survival of perennial crops in eastern Canada. *Agronomy Journal* 94, 1120–1130.

Bergjord, A.K., Bonesmo, H. and Skjelvåg, A.O. (2008) Modelling the course of frost tolerance in winter wheat. I. Model development. *European Journal of Agronomy* 28, 321–330.

Bertrand, A. and Castonguay, Y. (2003) Plant adaptations to overwintering stresses and implications of climate change. *Canadian Journal of Botany* 81, 1145–1152.

Böhlenius, H., Huang, T., Charbonnel-Campaa, L., Brunner, A.H., Jansson, S., Strauss, S.H. and Nilsson, O. (2006) CO/FT regulatory module controls timing of flowering and seasonal growth cessation in trees. *Science* 312, 1040–1043.

Bowman, J.L., Alvarez, J., Weigel, D., Meyerowitz, E.M. and Smyth, D.R. (1993) Control of flower development in *Arabidopsis thaliana* by *APETALA1* and interacting genes. *Development* 119, 721–743.

Browse, J. and Xin, Z. (2001) Temperature sensing and cold acclimation. *Current Opinion in Plant Biology* 4, 241–246.

Brüggemann, W., van der Kooij, T.A.W. and van Hasselt, P.R. (1992) Long-term chilling of young tomato plants under low light and subsequent recovery. *Planta* 186, 172–178.

Chen, T.H.H., Gusta, L.V. and Fowler, D.B. (1983) Freezing injury and root development in winter cereals. *Plant Physiology* 73, 773–777.

Chinnusamy, V., Ohta, M., Kanrar, S., Lee, B., Hong, X., Agarwal, M. and Zhu, J.-K. (2003) ICE1: a regulator of cold-induced transcriptome and freezing tolerance in *Arabidopsis*. *Genes and Development* 17, 1043–1054.

Choi, D.-W., Rodriguez, E.M. and Close, T.J. (2002) Barley *Cbf3* gene identification, expression pattern, and map location. *Plant Physiology* 129, 1781–1787.

Chouard, P. (1960) Vernalization and its relations to dormancy. *Annual Review of Plant Physiology* 11, 191–238.

Close, T.J. (1997) Dehydrins: a commonalty in the response of plants to dehydration and low temperature. *Physiologia Plantarum* 100, 291–296.

Cook, D., Fowler, S., Fiehn, O. and Thomashow, M.F. (2004) A prominent role for the *CBF* cold response pathway in configuring the low-temperature metabolome of *Arabidopsis*. *Proceedings of the National Academy of Sciences USA* 101, 15243–15248.

Crawford, R.M.M. (2000) Ecological hazards of oceanic environments. *New Phytologist* 147, 257–281.

Crosatti, C., Soncini, C., Stanca, A.M. and Cattivelli, L. (1995) The accumulation of a cold-regulated chloroplastic protein is light-dependent. *Planta* 196, 458–463.

Danyluk, J., Kane, N.A., Breton, G., Limin, A.E., Fowler, D.B. and Sarhan, F. (2003) TaVRT-1, a putative transcription factor associated with vegetative to reproductive transition in cereals. *Plant Physiology* 132, 1849–1860.

Demmig-Adams, B. and Adams, W.W. III (1996) The role of xanthophyll cycle carotenoids in the protection of photosynthesis. *Trends in Plant Science* 1, 21–26.

Dhillon, T., Pearce, S.P., Stockinger, E.J., Distelfeld, A., Li, C., Knox, A.K., Vashegyi, I., *et al.* (2010) Regulation of freezing tolerance and flowering in cereals: the *VRN-1* connection. *Plant Physiology* 153, 1846–1858.

Diallo, A., Kane, N., Agharbaoui, Z., Badawi, M. and Sarhan, F. (2010) Heterologous expression of wheat VERNALIZATION2 (TaVRN2) gene in Arabidopsis delays flowering and enhances freezing tolerance. *PLoS One* 5, e8690, doi:10.1371/journal.pone.0008690.

Dubcovsky, J., Loukoianov, A., Fu, D., Valarik, M., Sanchez, A. and Yan, L. (2006) Effect of photoperiod on the regulation of wheat vernalization genes *VRN1* and *VRN2*. *Plant Molecular Biology* 60, 469–480.

Fowler, D.B. (2002) Winter survival. In: Fowler, D.B. (ed.) *Winter Wheat Production Manual.* University of Saskatchewan, Saskatoon, Canada, www.usask.ca/agriculture/plantsci/winter_cereals/Winter_wheat/CHAPT12/cvchpt12.php, accessed 12 July 2010.

Fowler, D.B. (2008) Cold acclimation threshold induction temperatures in cereals. *Crop Science* 48, 1147–1154.

Fowler, D.B., Chauvin, L.P., Limin, A.E. and Sarhan, F. (1996) The regulatory role of vernalization in the expression of low-temperature-induced genes in wheat and rye. *Theoretical and Applied Genetics* 93, 554–559.

Fowler, D.B., Limin, A.E. and Ritchie, J.T. (1999) Low-temperature tolerance in cereals: model and genetic interpretation. Interpretive paper. *Crop Science* 39, 626–633.

Fowler, D.B., Breton, G., Limin, A.E., Mahfoozi, S. and Sarhan, F. (2001) Photoperiod and temperature interactions regulate low-temperature-induced gene expression in barley. *Plant Physiology* 127, 1676–1681.

Fracheboud, Y., Haldimann, P., Leipner, J. and Stamp, P. (1999) Chlorophyll fluorescence as a selection tool for cold tolerance of photosynthesis in maize (*Zea mays* L.). *Journal of Experimental Botany* 50, 1533–1540.

Francia, E., Rizza, F., Cattivelli, L., Stanca, A.M., Galiba, G., Tóth, B., Hayes, P.M., *et al.* (2004) Two loci on chromosome 5H determine low-temperature tolerance in a 'Nure' (winter) x 'Tremois' (spring) barley map. *Theoretical and Applied Genetics* 108, 670–680.

Francia, E., Barabaschi, D., Tondelli, A., Laidò, G., Rizza, F., Stanca, A.M., Busconi, M., *et al.* (2007) Fine mapping of a Hv*CBF* gene cluster at the frost resistance locus *Fr-H2* in barley. *Theoretical and Applied Genetics* 115, 1083–1091.

Galiba, G., Vágújfalvi, A., Li, C., Soltész, A. and Dubcovsky, J. (2009) Regulatory genes involved in the determination of frost tolerance in temperate cereals. *Plant Science* 176, 12–19.

Gazzani, S., Gendall, A.R., Lister, C. and Dean, C. (2003) Analysis of the molecular basis of flowering time variation in *Arabidopsis* accessions. *Plant Physiology* 132, 1107–1114.

Gilmour, S.J., Zarka, D.G., Stockinger, E.J., Salazar, M.P., Houghton, J.M. and Thomashow, M.F. (1998) Low temperature regulation of the *Arabidopsis CBF* family of AP2 transcriptional activators as an early step in cold-induced *COR* gene expression. *The Plant Journal* 16, 433–442.

Gong, Z., Lee, H., Xiong, L., Jagendorf, A., Stevenson, B. and Zhu, J.-K. (2002) RNA helicase-like protein as an early regulator of transcription factors for plant chilling and freezing tolerance. *Proceedings of the National Academy of Sciences USA* 99, 11507–11512.

Gong, Z., Dong, C.-H., Lee, H., Zhu, J., Xiong, L., Gong, D., Stevenson, B., *et al.* (2005) A DEAD box RNA helicase is essential for mRNA export and important for development and stress responses in *Arabidopsis. The Plant Cell* 17, 256–267.

Gray, G.R. and Heath, D. (2005) A global reorganization of the metabolome in *Arabidopsis* during cold acclimation is revealed by metabolic fingerprinting. *Physiologia Plantarum* 124, 236–248.

Griffiths, S., Dunford, R.P., Coupland, G. and Laurie, D.A. (2003) The evolution of *CONSTANS*-like gene families in barley, rice, and *Arabidopsis. Plant Physiology* 131, 1855–1867.

Gudleifsson, B.E. (2009) Ice encasement damage on grass crops and alpine plants in Iceland – impact of climate change. In: Gusta, L.V., Wisniewski, M.E. and Tanino, K.K. (eds) *Plant Cold Hardiness. From the Laboratory to the Field.* CAB International, Wallingford, pp. 163–172.

Gusta, L.V. and Chen, T.H.H. (1987) The physiology of water and temperature stress. In: Heyne, E.G. (ed.) *Wheat and Wheat Improvement.* 2nd ed., ASA, CSSA, SSSA, Madison, Wisconsin, pp. 115–150.

Gusta, L.V. and Fowler, D.B. (1979) Cold resistance and injury in winter cereals. In: Mussell, H. and Staples, R.C. (eds) *Stress Physiology in Crop Plants.* John Wiley and Sons Inc., New York, pp. 159–178.

Gusta, L.V., O´Connor, B.J., Gao, Y.-P. and Jana, S. (2001) A re-evaluation of controlled freeze-tests and controlled environment hardening conditions to estimate the winter survival potential of hardy winter wheats. *Canadian Journal of Plant Science* 81, 241–246.

Gusta, L.V., Wisniewski, M.E. and Trischuk, R.G. (2009) Patterns of freezing in plants: the influence of species, environment and experiential procedures. In: Gusta, L.V., Wisniewski, M.E. and Tanino, K.K. (eds) *Plant Cold Hardiness. From the Laboratory to the Field.* CAB International, Wallingford, pp. 214–225.

Guy, C.L. (1990) Cold acclimation and freezing stress tolerance: role of protein metabolism. *Annual Review of Plant Physiology and Plant Molecular Biology* 41, 187–223.

Haldimann, P., Fracheboud, P. and Stamp, P. (1996) Photosynthetic performance and resistance to photoinhibition of *Zea mays* L. leaves grown at sub-optimal temperature. *Plant, Cell and Environment* 19, 85–92.

Holková, L., Prášil, I.T., Bradáčová, M., Vítámvás, P. and Chloupek, O. (2009) Screening for frost tolerance in wheat using the expression of dehydrin genes *Wcs120* and *Wdhn13* at 17°C. *Plant Breeding* 128, 420–422.

Horvath, D.P. (2009) Common mechanisms regulate flowering and dormancy. *Plant Science* 177, 523–531.

Horvath, D.P., Anderson, J.V., Chao, W.S. and Foley, M.E. (2003) Knowing when to grow: signals regulating bud dormancy. *Trends in Plant Science* 8, 534–540.

Houde, M., Dhindsa, R.S. and Sarhan, F. (1992) A molecular marker to select for freezing tolerance in *Gramineae. Molecular and General Genetics* 234, 43–48.

Houde, M., Daniel, C., Lachapelle, M., Allard, F., Laliberté, S. and Sarhan, F. (1995) Immunolocalization of freezing-tolerance-associated proteins in the cytoplasm and nucleoplasm of wheat crown tissues. *The Plant Journal* 8, 583–593.

Hurry, V.M. and Huner, N.P.A. (1991) Low growth temperature effects a differential inhibition of photosynthesis in spring and winter wheat. *Plant Physiology* 96, 491–497.

Jaeger, K.E. and Wigge, P.A. (2007) FT protein acts as a long-range signal in *Arabidopsis. Current Biology* 17, 1050–1054.

Jaglo-Ottosen, K.R., Gilmour, S.J., Zarka, D.G., Schabenberger, O. and Thomashow, M.F. (1998) *Arabidopsis CBF1* overexpression induces *COR* genes and enhances freezing tolerance. *Science* 280, 104–106.

Johansson, U., West, J., Lister, C., Michaels, S., Amasino, R. and Dean, C. (2000) Molecular analysis of *FRIGIDA*, a major determinant of natural variation in *Arabidopsis* flowering time. *Science* 290, 344–347.

Jung, C. and Müller, A.E. (2009) Flowering time control and applications in plant breeding. *Trends in Plant Science* 14, 563–573.

Jung, S. and Steffen, K.L. (1997) Influence of photosynthetic photon flux densities before and during long-term chilling on xanthophyll cycle and chlorophyll fluorescence quenching in leaves of tomato (*Lycopersicon hirsutum*). *Physiologia Plantarum* 100, 958–966.

Kacperska-Palacz, A. (1987) Mechanism of cold acclimation in herbaceous plants. In: Kaurin A., Juntila, O. and Nilsen, J. (eds) *Plant Production in the North.* Norwegian University Press, Tromso, Norway, pp. 99–115.

Kane, N.A., Danyluk, J., Tardif, G., Ouellet, F., Laliberté, J.-F., Limin, A.E., Fowler, D.B., *et al.* (2005) *TaVRT-2*, a member of the *St*-MADS-11 clade of flowering repressors, is regulated by vernalization and photoperiod in wheat. *Plant Physiology* 138, 2354–2363.

Kaplan, F., Kopka, J., Haskell, D.W., Zhao, W., Schiller, C., Gatzke, N., Sung, D.Y., *et al.* (2004) Exploring the temperature-stress metabolome of *Arabidopsis. Plant Physiology* 136, 4159–4168.

Kaplan, F., Kopka, J., Sung, D.Y., Zhao, W., Popp, M., Porat, R. and Guy, C.L. (2007) Transcript and metabolite profiling during cold acclimation of *Arabidopsis* reveals an intricate relationship of cold-regulated gene expression with modifications in metabolite content. *The Plant Journal* 50, 967–981.

Kardailsky, I., Shukla, V.K., Ahn, J.H., Dagenais, N., Christensen, S.K., Nguyen, J.T., Chory, J., *et al.* (1999) Activation tagging of the floral inducer *FT. Science* 286, 1962–1965.

Kawamura, Y. and Uemura, M. (2003) Mass spectrometric approach for identifying putative plasma membrane proteins of *Arabidopsis* leaves asociated with cold acclimation. *The Plant Journal* 36, 141–154.

Kim, H.-J., Hyun, Y., Park, J.-Y., Park, M.-J., Park, M.-K., Kim, M.D., Kim, H.-J., *et al.* (2004) A genetic link between cold responses and flowering time through *FVE* in *Arabidopsis thaliana. Nature Genetics* 36, 167–171.

Kingston-Smith, A.H., Harbinson, J., Williams, J. and Foyer, C.H. (1997) Effect of chilling on carbon assimilation, enzyme activation, and photosynthetic electron transport in the absence of photoinhibition in maize leaves. *Plant Physiology* 114, 1039–1046.

Knox, A.K., Li, C., Vágújfalvi, A., Galiba, G., Stockinger, E.J. and Dubcovsky, J. (2008) Identification of candidate *CBF* genes for the frost tolerance locus *Fr-A^{m}2* in *Triticum monococcum. Plant Molecular Biology* 67, 257–270.

Knox, A.K., Dhillon, T., Cheng, H., Tondelli, A., Pecchioni, N. and Stockinger, E.J. (2010) *CBF* gene number variation at *Frost Resistance-2* is associated with levels of freezing tolerance in temperate-climate cereals. *Theoretical and Applied Genetics* 121, 21–35.

Kocsy, G., Athmer, B., Perovic, D., Himmelbach, A., Szücs, A., Vashegyi, I., Schweizer, P., *et al.* (2010) Regulation of gene expression by chromosome 5A during cold hardening in wheat. *Molecular Genetics and Genomics* 283, 351–363.

Koornneef, M., Blankestijn-de Vries, H., Hanhart, C., Soppe, W. and Peeters, T. (1994) The phenotype of some late-flowering mutants is enhanced by a locus on chromosome 5 that is not effective in the Landsberg *erecta* wild-type. *The Plant Journal* 6, 911–919.

Kosová, K., Vítámvás, P. and Prášil, I.T. (2007) The role of dehydrins in plant response to cold. *Biologia Plantarum* 51, 601–617.

Kosová, K., Holková, L., Prášil, I.T., Prášilová, P., Bradáčová, M., Vítámvás, P. and Čapková, V. (2008a) Expression of dehydrin 5 during the development of frost tolerance in barley (*Hordeum vulgare*). *Journal of Plant Physiology* 165, 1142–1151.

Kosová, K., Prášil, I.T. and Vítámvás, P. (2008b) The relationship between vernalization- and photoperiodically-regulated genes and the development of frost tolerance in wheat and barley. *Biologia Plantarum* 52, 601–615.

Kosová, K., Prášil, I.T., Prášilová, P., Vítámvás, P. and Chrpová, J. (2010a) The development of frost tolerance and DHN5 protein accumulation in barley (*Hordeum vulgare*) doubled haploid lines derived from Atlas 68 x Igri cross during cold acclimation. *Journal of Plant Physiology* 167, 343–350.

Kosová, K., Prášil, I.T. and Vítámvás, P. (2010b) Role of dehydrins in plant stress response. In: Pessarakli, M. (ed.) *Handbook of Plant and Crop Stress*. 3rd ed., CRC Press Inc., Boca Raton, Florida, pp. 239–285.

Kosová, K., Vítámvás, P. and Prášil, I.T. (2011) Expression of dehydrins in wheat and barley under different temperatures. *Plant Science* 180, 46–52.

Kratsch, H.A. and Wise, R.R. (2000) The ultrastructure of chilling stress. *Plant, Cell and Environment* 23, 337–350.

Krause, G.H. and Weis, E. (1991) Chlorophyll fluorescence and photosynthesis: the basics. *Annual Review of Plant Physiology and Plant Molecular Biology* 42, 313–349.

Kreps, J.A., Wu, Y., Chang, H.-S., Zhu, T., Wang, X. and Harper, J.F. (2002) Transcriptome changes for *Arabidopsis* in response to salt, osmotic, and cold stress. *Plant Physiology* 130, 2129–2141.

Lee, H., Xiong, L., Gong, Z., Ishitani, M., Stevenson, B. and Zhu, J.K. (2001) The *Arabidopsis HOS1* gene negatively regulates cold signal transduction and encodes a RING finger protein that displays cold-regulated nucleo-cytoplasmic partitioning. *Genes and Development* 15, 912–924.

Lee, H., Guo, Y., Ohta, M., Xiong, L., Stevenson, B. and Zhu, J.-K. (2002) *LOS2*, a genetic locus required for cold-responsive gene transcription encodes a bi-functional enolase. *The EMBO Journal* 21, 2692–2702.

Levitt, J. (1980) *Responses of Plants to Environmental Stress. Chilling, Freezing and High Temperature Stresses*. 2nd ed., Academic Press, New York.

Lichtenthaler, H.K. and Rinderle, U. (1988) The role of chlorophyll fluorescence in the detection of stress conditions in plants. *Critical Reviews in Analytical Chemistry* 19, s29–s85.

Limin, A., Corey, A., Hayes, P. and Fowler, D.B. (2007) Low-temperature acclimation of barley cultivars used as parents in mapping populations: response to photoperiod, vernalization and phenological development. *Planta* 226, 139–146.

Liu, P., Meng, Q.-W., Zou, Q., Zhao, S.-J. and Liu, Q.-Z. (2001) Effects of cold-hardening on chilling-induced photoinhibition of photosynthesis and on xanthophyll cycle pigments in sweet pepper. *Photosynthetica* 39, 467–472.

Loukoianov, A., Yan, L., Blechl, A., Sanchez, A. and Dubcovsky, J (2005) Regulation of *VRN-1* vernalization genes in normal and transgenic polyploid wheat. *Plant Physiology* 138, 2364–2373.

Michaels, S.D. and Amasino, R.M. (1999) *FLOWERING LOCUS C* encodes a novel MADS domain protein that acts as a repressor of flowering. *The Plant Cell* 11, 949–956.

Monroy, A.F., Dryanova, A., Malette, B., Oren, D.H., Farajalla, M.R., Liu, W., Danyluk, J., *et al.* (2007) Regulatory gene candidates and gene expression analysis of cold acclimation in winter and spring wheat. *Plant Molecular Biology* 64, 409–423.

Moon, J., Suh, S.-S., Lee, H., Choi, K.-R., Hong, C.-B., Paek, N.-C., Kim, S.-G., *et al.* (2003) The *SOC1* MADS-box gene integrates vernalization and gibberellin signals for flowering in *Arabidopsis*. *The Plant Journal* 35, 613–623.

Murata, N. and Los, D.A. (1997) Membrane fluidity and temperature perception. *Plant Physiology* 115, 875–879.

Olien, C.R. and Smith, M.N. (1981) *Analysis and Improvement of Plant Cold Hardiness*. CRC Press Inc., Boca Raton, Florida.

Oliver, S.N., Finnegan, E.J., Dennis, E.S., Peacock, W.J. and Trevaskis, B. (2009) Vernalization-induced flowering in cereals is associated with changes in histone methylation at the *VERNALIZATION1* gene. *Proceedings of the National Academy of Sciences USA* 106, 8386–8391.

Örvar, B.L., Sangwan, V., Omann, F. and Dhindsa, R.S. (2000) Early steps in cold sensing by plant cells: the role of actin cytoskeleton and membrane fluidity. *The Plant Journal* 23, 785–794.

Plieth, C., Hansen, U.-P., Knight, H. and Knight, M.R. (1999) Temperature sensing by plants: the primary characteristics of signal perception and calcium response. *The Plant Journal* 18, 491–497.

Prášil, I. and Zámečník, J. (1991) The effect of weather on the overwintering of winter crops. In: Petr, J. (ed.) *Weather and Yield. Developments in Crop Science 20*. Elsevier Science Publisher, Amsterdam, the Netherlands, pp. 141–151.

Prášil, I.T., Prášilová, P. and Pánková, K. (2004) Relationships among vernalization, shoot apex development and frost tolerance in wheat. *Annals of Botany* 94, 413–418.

Prášil, I.T., Prášilová, P. and Mařík, P. (2007) Comparative study of direct and indirect evaluations of frost tolerance in barley. *Field Crops Research* 102, 1–8.

Putterill, J., Robson, F., Lee, K., Simon, R. and Coupland, G. (1995) The *CONSTANS* gene of *Arabidopsis* promotes flowering and encodes a protein showing similarities to zinc finger transcription factors. *Cell* 80, 847–857.

Rabbani, M.A., Maruyama, K., Abe, H., Khan, M.A., Katsura, K., Ito, Y., Yoshiwara, K., *et al.* (2003) Monitoring expression profiles of rice genes under cold, drought, and high-salinity stresses and abscisic acid application using cDNA microarray and RNA gel-blot analyses. *Plant Physiology* 133, 1755–1767.

Rizza, F., Pagani, D., Stanca, A.M. and Cattivelli, L. (2001) Use of chlorophyll fluorescence to evaluate the cold acclimation and freezing tolerance of winter and spring oats. *Plant Breeding* 120, 389–396.

Rorat, T. (2006) Plant dehydrins – tissue location, structure and functions. *Cellular and Molecular Biology Letters* 11, 536–556.

Ruelland, E., Vaultier, M.-N., Zachowski, A. and Hurry, V. (2009) Cold signalling and cold acclimation in plants. In: Kader, J.-C. and Delseny, M. (eds) *Advances in Botanical Research 49*. Elsevier Ltd, Amsterdam, the Netherlands, pp. 35–150.

Sakai, A. and Larcher, W. (1987) *Frost Survival of Plants. Responses and Adaptation to Freezing Stress*. Springer-Verlag, New York.

Savitch, L.V., Harney, T. and Huner, N.P.A. (2000) Sucrose metabolism in spring and winter wheat in response to high irradiance, cold stress and cold acclimation. *Physiologia Plantarum* 108, 270–278.

Savitch, L.V., Leonardos, E.D., Krol, M., Jansson, S., Grodzinski, B., Huner, N.P.A. and Öquist, G. (2002) Two different strategies for light utilization in photosynthesis in relation to growth and cold acclimation. *Plant, Cell and Environment* 25, 761–771.

Seki, M., Narusaka, M., Abe, H., Kasuga, M., Yamaguchi-Shinozaki, K., Carninci, P., Hayashizaki, Y., *et al.* (2001) Monitoring the expression pattern of 1,300 *Arabidopsis* genes under drought and cold stresses by using a full-length cDNA microarray. *The Plant Cell* 13, 61–72.

Seki, M., Narusaka, M., Ishida, J., Nanjo, T., Fujita, M., Oono, Y., Kamiya, A., *et al.* (2002) Monitoring the expression profiles of 7,000 *Arabidopsis* genes under drought, cold and high-salinity stresses using a full-length cDNA microarray. *The Plant Journal* 31, 279–292.

Seo, E., Lee, H., Jeon, J., Park, H., Kim, J., Noh, Y.-S. and Lee, I. (2009) Crosstalk between cold response and flowering in *Arabidopsis* is mediated through the flowering-time gene *SOC1* and its upstream negative regulator *FLC*. *The Plant Cell* 21, 3185–3197.

Serrano, G., Herrera-Palau, R., Romero, J.M., Serrano, A., Coupland, G. and Valverde, F. (2009) *Chlamydomonas CONSTANS* and the evolution of plant photoperiodic signaling. *Current Biology* 19, 359–368.

Sheldon, C., Rouse, D.T., Finnegan, E.J., Peacock, W.J. and Dennis, E.S. (2000) The molecular basis of vernalization: the central role of *FLOWERING LOCUS C* (*FLC*). *Proceedings of the National Academy of Sciences USA* 97, 3753–3758.

Shindo, C., Aranzana, M.J., Lister, C., Baxter, C., Nicholls, C., Nordborg, M. and Dean, C. (2005) Role of *FRIGIDA* and *FLOWERING LOCUS C* in determining variation in flowering time of *Arabidopsis*. *Plant Physiology* 138, 1163–1173.

Shitsukawa, N., Ikari, C., Shimada, S., Kitagawa, S., Sakamoto, K., Saito, H., Ryuto, H., *et al.* (2007a) The einkorn wheat (*Triticum monococcum*) mutant, *maintained vegetative phase*, is caused by a deletion in the *VRN1* gene. *Genes and Genetic Systems* 82, 167–170.

Shitsukawa, N., Ikari, C., Mitsuya, T., Sakiyama, T., Ishikawa, A., Takumi, S. and Murai, K. (2007b) Wheat *SOC1* functions independently of *WAP1/VRN1*, an integrator of vernalization and photoperiod flowering promotion pathways. *Physiologia Plantarum* 130, 627–636.

Skinner, J.S., von Zitzewitz, J., Szücs, P., Marquez-Cedillo, L., Filichkin, T., Amundsen, K., Stockinger, E.J., *et al.* (2005) Structural, functional and phylogenetic characterization of a large *CBF* gene family in barley. *Plant Molecular Biology* 59, 533–551.

Skinner, J.S., Szücs, P., von Zitzewitz, J., Marquez-Cedillo, L., Filichkin, T., Stockinger, E.J., Thomashow, M.F., *et al.* (2006) Mapping of barley homologs to genes that regulate low temperature tolerance in *Arabidopsis*. *Theoretical and Applied Genetics* 112, 832–842.

Sonoike, K. and Terashima, I. (1994) Mechanism of photosystem-I photoinhibition in leaves of *Cucumis sativus* L. *Planta* 194, 287–293.

Stitt, M. and Hurry, V. (2002) A plant for all seasons: alterations in photosynthetic carbon metabolism during cold acclimation in *Arabidopsis*. *Current Opinion in Plant Biology* 5, 199–206.

Stockinger, E.J., Skinner, J.S., Gardner, K.G., Francia, E. and Pecchioni, N. (2007) Expression levels of barley *Cbf* genes at the *Frost resistance-H2* locus are dependent upon alleles at *Fr-H1* and *Fr-H2*. *The Plant Journal* 51, 308–321.

Sung, S. and Amasino, R.M. (2005) Remembering winter: toward a molecular understanding of vernalization. *Annual Review of Plant Biology* 56, 491–508.

Suzuki, I., Los, D.A., Kanesaki, Y., Mikami, K. and Murata, N. (2000) The pathway for perception and transduction of low-temperature signals in *Synechocystis*. *The EMBO Journal* 19, 1327–1334.

Tanino, K.K. and McKersie, B.D. (1982) Injury within the crown of winter wheat seedlings after freezing stress. *Plant Physiology* 69, 108.

Tanino, K.K. and McKersie, B.D. (1985) Injury within the crown of winter wheat seedlings after freezing and icing stress. *Canadian Journal of Botany* 63, 432–436.

Thomashow, M.F. (1999) Plant cold acclimation: freezing tolerance genes and regulatory mechanisms. *Annual Review of Plant Physiology and Plant Molecular Biology* 50, 571–599.

Tumanov, I.I. (1979) *Physiology of Cold Hardening and Frost Tolerance in Plants.* [In Russian]. Izdatelstvo Nauka, Moskva, USSR.

Tunnacliffe, A. and Wise, M.J. (2007) The continuing conundrum of LEA proteins. *Naturwissenschaften* 94, 791–812.

Turner, A., Beales, J., Faure, S., Dunford, R.P. and Laurie, D.A. (2005) The *Pseudo-response regulator Ppd-H1* provides adaptation to photoperiod in barley. *Science* 310, 1031–1034.

Vágújfalvi, A., Crosatti, C., Galiba, G., Dubcovsky, J. and Cattivelli, L. (2000) Two loci on wheat chromosome 5A regulate the differential cold-dependent expression of the *cor14b* gene in frost-tolerant and frost-sensitive genotypes. *Molecular and General Genetics* 263, 194–200.

Vágújfalvi, A., Galiba, G., Cattivelli, L. and Dubcovsky, J. (2003) The cold-regulated transcriptional activator *Cbf3* is linked to the frost-tolerance locus *Fr-A2* on wheat chromosome 5A. *Molecular Genetics and Genomics* 269, 60–67.

Valverde, F., Mouradov, A., Soppe, W., Ravenscroft, D., Samach, A. and Coupland, G. (2004) Photoreceptor regulation of CONSTANS protein in photoperiodic flowering. *Science* 303, 1003–1006.

Vítámvás, P. and Prášil, I.T. (2008) WCS120 protein family and frost tolerance during cold acclimation, deacclimation and reacclimation of winter wheat. *Plant Physiology and Biochemistry* 46, 970–976.

Vítámvás, P., Saalbach, G., Prášil, I.T., Čapková, V., Opatrná, J. and Jahoor, A. (2007) WCS120 protein family and proteins soluble upon boiling in cold-acclimated winter wheat. *Journal of Plant Physiology* 164, 1197–1207.

Vítámvás, P., Kosová, K., Prášilová, P. and Prášil, I.T. (2010) Accumulation of WCS120 protein in wheat cultivars grown at 9°C or 17°C in relation to their winter survival. *Plant Breeding* 129, 611–616.

Vogel, J.T., Zarka, D.G., Van Buskirk, H.A., Fowler, S.G. and Thomashow, M.F. (2005) Roles of the *CBF2* and *ZAT12* transcription factors in configuring the low temperature transcriptome of *Arabidopsis*. *The Plant Journal* 41, 195–211.

Von Zitzewitz, J., Szücs, P., Dubcovsky, J., Yan, L., Francia, E., Pecchioni, N., Casas, A., *et al.* (2005) Molecular and structural characterization of barley vernalization genes. *Plant Molecular Biology* 59, 449–467.

Wigge, P.A., Kim, M.C., Jaeger, K.E., Busch, W., Schmid, M., Lohmann, J.U. and Weigel, D. (2005) Integration of spatial and temporal information during floral induction in *Arabidopsis*. *Science* 309, 1056–1059.

Wise, R.R. (1995) Chilling-enhanced photooxidation: the production, action and study of reactive oxygen species produced during chilling in the light. *Photosynthesis Research* 45, 79–97.

Xin, Z. and Browse, J. (2000) Cold comfort farm:

the acclimation of plants to freezing temperatures. *Plant, Cell and Environment* 23, 893–902.

Xu, C.-C., Lin, R.-C., Li, L.-B. and Kuang, T.-Y. (2000) Increase in resistance to low temperature photoinhibition following ascorbate feeding is attributable to an enhanced xanthophyll cycle activity in rice (*Oryza sativa* L.) leaves. *Photosynthetica* 38, 221–226.

Yamaguchi-Shinozaki, K. and Shinozaki, K. (2005) Organization of *cis*-acting regulatory elements in osmotic- and cold-stress-responsive promoters. *Trends in Plant Science* 10, 88–94.

Yamaguchi-Shinozaki, K. and Shinozaki, K. (2006) Transcriptional regulatory networks in cellular responses and tolerance to dehydration and cold stresses. *Annual Review of Plant Biology* 57, 781–803.

Yan, L., Loukoianov, A., Blechl, A., Tranquilli, G., Ramakrishna, W., SanMiguel, P., Bennetzen, J.L., *et al.* (2004) The wheat *VRN2* gene is a flowering repressor down-regulated by vernalization. *Science* 303, 1640–1644.

Yan, L., Fu, D., Li, C., Blechl, A., Tranquilli, G., Bonafede, M., Sanchez, A., *et al.* (2006) The wheat and barley vernalization gene *VRN3* is an orthologue of *FT*. *Proceedings of the National Academy of Sciences USA* 103, 19581–19586.

Yoo, S.K., Chung, K.S., Kim, J., Lee, J.H., Hong, S.M., Yoo, S.J., Yoo, S.Y., *et al.* (2005) *CONSTANS* activates *SUPPRESSOR OF OVEREXPRESSION OF CONSTANS 1* through *FLOWERING LOCUS T* to promote flowering in *Arabidopsis*. *Plant Physiology* 139, 770–778.

Zhu, B., Choi, D.-W., Fenton, R. and Close, T.J. (2000) Expression of the barley dehydrin multigene family and the development of freezing tolerance. *Molecular and General Genetics* 264, 145–153.

14 Perennial Field Crops

Annick Bertrand

14.1 Introduction

Perennial field crops grown in northern areas are mostly used for feeding cattle, and cover a large proportion of cultivated lands. In Canada, for instance, perennial forage crops and rangelands are grown on over 34 million ha, which represent more than 50% of all agricultural land (Statistics Canada, 2009). As for all field crops, the productivity of perennial crops depends on climatic conditions. For perennial crops, however, both summer and winter climatic conditions are determining factors. Yield is mainly dependent on the conditions prevailing during the growing period, while long-term persistence depends largely on the severity of winter conditions. This review focuses on current and future climatic conditions of eastern Canada, but conclusions could be applied to countries where the winter is characterized by freezing temperatures and snow cover.

The changes in climatic conditions that have been recorded since the beginning of the 20th century are bound to accelerate due to increases in greenhouse gas (GHG) emissions attributable to the combustion of fossil fuel. At the current rate of increase (>2 µmol/mol/year), and without mitigation, the concentration of atmospheric CO_2 could reach between 600 and 800 µmol/mol by the end of the century (Field *et al.*, 2007). This increase in the concentration of atmospheric GHG induces positive radiative forcing, resulting in temperature increases. Temperatures are predicted to increase during the next 75 years by 3.8°C during the summer in eastern Canada and by 7.5°C during the winter (Christensen *et al.*, 2007). Increases in precipitation throughout Canada are projected to be in the range of 10%–20% (Hulme and Sheard, 1999), but their impact could be buffered by an increase in evapotranspiration. Projections also include widespread increases in precipitation extremes which could cause flooding, and drought from greater temporal variability in distribution (Christensen *et al.*, 2007). Such changes are bound to impact perennial field crop growth and development, as well as winter survival.

14.2 Impact of Climate Change During the Growing Season

Higher temperatures and CO_2 concentrations may produce positive effects on perennial crops growing in northern areas through the introduction of new crop species and the expansion of suitable areas for crop cultivation. Furthermore, these conditions should increase perennial crop productivity due to higher net photosynthetic rates, particularly for C3 plant species, if nutrient or water do not limit their growth (Hungate *et al.*, 2003; Morgan *et al.*, 2004). Bloom *et al.* (2010) recently documented a decrease in photosynthesis and an associated decline in nitrogen status in C3 plants acclimated to elevated atmospheric CO_2. This response was attributed to reduced nitrate assimilation into organic compounds. However, perennial legumes such as alfalfa and clovers should benefit from increased atmospheric CO_2, because their symbiotic nitrogen fixation capacity allows them to counter the usual nitrogen limitation to growth (Yu *et al.*, 2002; Rogers *et al.*, 2006). Additional factors such as symbiotic rhizobial strains could modulate the response of legumes to elevated atmospheric CO_2 and, by selecting strains better adapted to future climatic

© CAB International 2012. *Temperature Adaptation in a Changing Climate: Nature at Risk* (eds K.B. Storey and K.K. Tanino)

conditions, legume performance could be further improved (Bertrand *et al.*, 2007a; Bertrand *et al.*, 2007b).

While rising atmospheric CO_2 concentration has the potential to increase forage quantity, it could have negative impacts on forage nutritive value. Increasing CO_2 concentrations generally induce a decline in forage nitrogen and protein concentrations (Cotrufo *et al.*, 1998; Bertrand *et al.*, 2007b) concurrently with a decrease in digestibility (Morgan *et al.*, 2004). Reductions in forage nutritive value could lead to unsustainable livestock and milk production unless animal diets are supplemented with protein, a situation that would raise production costs. For instance, growing shortgrass prairie in large open-top chambers under 720 μmol/mol of CO_2 over four years reduced the protein concentration of forage below critical maintenance levels for livestock, and reduced the digestibility of forage by more than 10% (Milchunas *et al.*, 2005). Reduction in forage nutritive value could thus counter any positive effects of elevated CO_2 on other forage attributes, such as increased foliar carbohydrate content (Rogers and Ainsworth, 2006). Increasing the concentration of carbohydrates in alfalfa has been shown to increase nitrogen use efficiency of dairy cows (Miller *et al.*, 2001; Moorby *et al.*, 2006; Brito *et al.*, 2009), as well as forage intake and production of ruminants (Brito *et al.*, 2008).

For cool-season grasses such as timothy, the most prevalent forage grass species in eastern Canada and the Scandinavian countries, the optimum air temperature for growth and yield is between 17° and 21°C (Smith, 1972; Bertrand *et al.*, 2008), and any significant increase of the air temperature over this threshold could result in lower productivity. Nutritive value of cool-season grasses is affected by air temperature, and higher nutritive value is expected when forage crops are grown under cool conditions (Buxton and Fales, 1994). Consequently, predicted higher temperature during summer could negatively impact forage yield and nutritive value in northern areas. Bertrand *et al.* (2008) observed that cell wall degradability decreased while lignin concentration increased in timothy grown under a day/night temperature regime of a few degrees above the current conditions in eastern Canada. The digestibility of neutral detergent fibre (Thorvaldsson *et al.*, 2007) and the concentration of non-structural carbohydrates (Chatterton *et al.*, 1989) have also been shown to decrease with increasing growth temperature for many temperate grass species.

The length of the growing season is defined as the number of days between the last frost in the spring and the first frost in the autumn. With global warming, an extended growing season due to higher temperatures and CO_2 concentrations is expected. In most of eastern Canada, the growing season is expected to increase by approximately 400 degree days by 2050, which is about the interval between forage cuts. Hence, an additional harvest of cultivated forage crop could be possible under the projected future climate (Baron and Bélanger, 2007). An extended growing season could, however, have negative consequences on winter acclimation of forage crops.

14.3 Impact of Climate Change During Cold Acclimation and Overwintering

A trade-off between active growth and growth cessation has to be reached by perennial crops at the end of summer and early autumn in order to cold acclimate and survive winter. Elevated temperatures have been shown to promote autumn growth and delay the acquisition of maximum frost tolerance of winter wheat (Haslin and Mortensen, 2010). Moreover, elevated CO_2 has been shown to stimulate growth at low temperature and to reduce the freezing tolerance of alfalfa (Bertrand *et al.*, 2007a) and of other cold climate species (Ball and Hill, 2009).

Perennial forage crops are exposed to biotic and abiotic stresses during long overwintering periods, resulting in yield reduction and declining stand density.

Abiotic stresses are caused directly or indirectly by low temperatures (frost, ice encasement, soil heaving and drought) (Bertrand and Castonguay, 2003) or can be due to additional factors, such as environmental conditions that prevail during the growth and in the cold acclimation period (soil moisture, crop management and fertilization). In many cases, plant injuries result from the combined action of a mixture of stresses, all depending largely on climatic conditions. One of the expected impacts of climate change in northern regions is a change in the extent and the location of different types of winter injuries. Plants currently adapted to a given environment may be exposed to new types of winter stresses. For instance, if the climate becomes milder, freezing might take over where ice encasement dominates today, whereas ice encasement might take over where snow molds currently prevail, particularly at higher altitudes (Gudleifsson, 2009). This may result in greater winter injury since native plants will be exposed to new stresses to which they are not adapted.

14.3.1 Cold acclimation

At the cellular level, frost damage is caused by freeze-induced membrane disruption. Winter survival is thus determined by the ability of plants to maintain maximum cellular stability during periods of freeze-induced desiccation. The most common form of freezing in herbaceous perennials is caused by ice nucleation in the extracellular spaces at subzero temperatures (Levitt, 1980; Guy, 2003; Volenec and Nelson, 2007). As the temperature decreases, water molecules migrate outside the cell to the surface of ice crystals and cells become severely dehydrated, resulting in an extensive freeze-induced desiccation (Gusta *et al.*, 1975; Pearce, 2001). Freezing injuries result essentially from a breach in structural and functional integrity of cell membranes (Steponkus, 1984; Steponkus *et al.*, 1993; Guy, 2003). The main physiological consequence is the loss of cell compartmentation that leads to leakage of electrolytes and other cell solutes (Pearce, 2001). Plant acclimation to low temperature involves a cascade of molecular events to evoke the adaptive responses. Plants need:

1. To perceive low temperatures;
2. To transduce that signal to activate or repress the expression of cold-regulated (COR) genes; and
3. To translate these transcriptomic changes into biochemical and ultrastructural modifications that will confer superior tolerance to freezing at the cellular level (Kaur and Gupta, 2005).

The numerous physiological and molecular changes that plants undergo during cold acclimation include alteration of membrane lipid composition, accumulation of cryoprotective sugars and free amino acids, and modification of expression of several COR genes that allow plants to tolerate low temperatures, mainly through maintenance of membrane integrity (Thomashow, 2001). Alfalfa, for instance, is killed at −5°C during the summer while it can survive temperatures as low as −20°C following the molecular changes induced during cold acclimation (Castonguay *et al.*, 2009b). Another feature of cold acclimation is the accumulation of carbohydrates and nitrogen reserves, essential for spring regrowth, in perennial organs such as tiller bases, crowns, stolons and rhizomes (Volenec *et al.*, 1996; Dionne *et al.*, 2001; Avice *et al.*, 2003).

Tolerance to low subfreezing temperatures is the single most important factor determining winter survival of perennial crops. To achieve high levels of freezing tolerance, perennial plants undergo the process of cold acclimation, triggered by environmental changes occurring during the autumn. Initial stages of cold acclimation are triggered by lower temperatures and shorter photoperiods that occur in autumn (Fig. 14.1A). When the temperature falls below 5°C, the process of cold acclimation of perennial crops accelerates (Paquin and Pelletier, 1980) and there is a strong correlation between the air temperature and the level of hardening of forage crops (Fig. 14.1A). The maximum freezing tolerance is achieved after the soil has frozen (Castonguay *et al.*, 1993; Dionne *et al.*, 2001;

Bertrand *et al.*, 2001). Based on climate change data generated by the first-generation Canadian Global Coupled General Circulation Model (CGCM1), Bélanger *et al.* (2002) estimated that, as a result of the warming of autumn temperatures under projected climatic conditions, autumn acclimation with temperatures below 5°C would be delayed and that the length of the acclimation period would be 17% shorter in eastern Canada (Fig. 14.1B). Consequently, these changes would result in lower levels of cold tolerance and greater winter damage than under current climatic conditions.

Proper cold acclimation also depends on additional climatic factors such as adequate soil humidity (Suzuki, 1977). Plants exposed to excessively wet soils have a higher crown water content, resulting from passive water absorption by damaged tissues, which impairs the process of acclimation (Paquin and Mehuys, 1980). Excess water also causes a reduction in carbohydrate reserve accumulations due to higher glycolytic or respiration rates. Bélanger *et al.* (2002) predicted only a slight decrease of autumn precipitation in eastern Canada (<7%), which will not be likely to have a significant effect on cold acclimation of perennial crops.

In addition, increased atmospheric CO_2 has been reported to have direct negative effects on the acquisition of freezing tolerance. According to Ball and Hill (2009), elevated CO_2 affects freezing tolerance through at least two processes: it increases ice nucleation temperature and it decreases plant acclimation to subfreezing temperature by inducing stomatal closure. The reduction of stomatal conductance often observed under elevated CO_2 can reduce transpiration and cause an increase in leaf temperature (Barker *et al.*, 2005) which has been observed to delay cold acclimation of snow gum (*E. pauciflora* Sieb. ex Spreng.) by at least three weeks (Loveys *et al.*, 2006).

14.3.2 Overwintering

A high level of freezing tolerance is maintained in fully acclimated plants if soil temperature around perennial organs is maintained close to the freezing point, and if adequate levels of organic reserves are maintained in these perennial organs. Conversely, exposure to temperatures above 0°C in winter can cause plant deacclimation and a loss of freezing tolerance, leading to a higher sensitivity to subsequent exposure to low subfreezing temperatures. In Canada and other Nordic countries, the air temperature often drops below a threshold limit for lethal damage to perennial crops. This threshold can vary from −9° to −16°C for pasture grasses (Gudleifsson *et al.*, 1986), and from −17° to <−27°C for turfgrass ecotypes (Dionne *et al.*, 2010), and could reach −20°C for various alfalfa cultivars (Castonguay *et al.*, 2009b). Therefore, winter survival of perennial crops depends on insulation of perennial organs by snow cover (Fig. 14.1C). A snow cover of 0.1 m is considered sufficient to maintain the temperature at root level around the freezing point (Leep *et al.*, 2001).

Rainfall and snowmelt during winter not only expose plants to very low air temperatures, but also cause the formation of an ice layer over the plants (Andrews, 1996). Ice encasement limits gas exchange at the soil–atmosphere interface, resulting in the development of anaerobic conditions (Bertrand *et al.*, 2003). Castonguay *et al.* (2009a) highlighted the complex interaction between two causes of plant damage under ice, O_2 shortage and CO_2 accumulation, and concluded that plant injury is mainly caused by lack of O_2. Ice-encased plants are further damaged by the accumulation of toxic end-products of anaerobic metabolism, by the lowering of energy metabolism, or by the lack of substrates for respiration (Gudleifsson, 1997). Large interspecific differences in resistance to anoxia under ice cover were reported (Gudleiffson *et al.*, 1986), and the first line of defence in regions of high incidence of ice encasement should be the choice of resistant species. Under simulated anoxic conditions, timothy has been shown to be more resistant than alfalfa, while orchardgrass and red clover were the most sensitive species (Bertrand *et al.*, 2001). A closer look at the metabolism of these species revealed that timothy main-

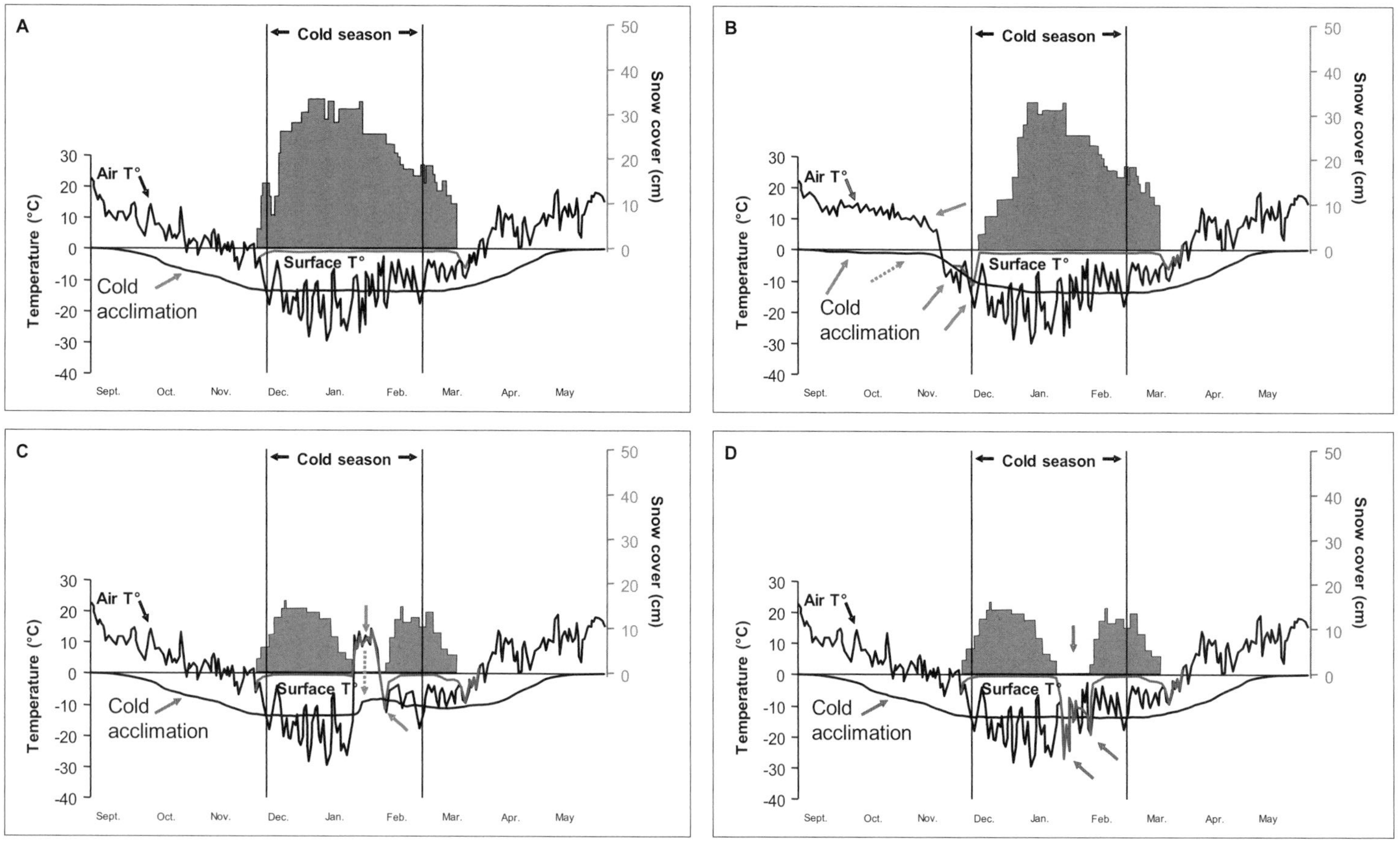

Figure 14.1. Schematic illustration of overwintering scenarios reflecting: A, ideal conditions; B, suboptimal autumn acclimation temperatures; C, deacclimation temperatures in winter; D, lack of snow insulation in midwinter. Adapted from Bélanger *et al.* (2006).

tained higher carbohydrate reserves under anoxia than the three other species, thus favouring its winter survival and spring regrowth (Bertrand *et al.*, 2003).

The greatest warming in North America is projected to occur in winter over northern parts of Alaska and Canada, due to the positive feedback from a reduced period of snow cover. The mean warming might reach up to 7.5°C in winter as compared to the 1980–1999 period, along with 34% more precipitation (Christensen *et al.*, 2007). As a consequence, a greater proportion of the precipitation would be in the form of rain instead of snow. Temperature warming will have a major repercussion on winter survival mainly because of the threat related to subfreezing temperatures in the absence of a protective snow cover. According to Bélanger *et al.* (2002), the number of days with a protective snow cover of at least 0.1 m during the cold period (the period during which potentially low lethal temperatures can occur) will be reduced by 15.6 days in eastern Canada (Fig. 14.1D). Reduced duration of snow cover is due to higher occurrence of high temperatures during the establishment and maintenance of snow cover, to freeze–thaw events, and to rain during the winter. Rainfall or snow melt during winter may directly result in ice encasement of plants. Furthermore, deacclimation could occur rapidly, and even short periods of exposure to warm temperature during winter could affect plant survival (Fig. 14.1C). Altogether, predicted winter climatic conditions are bound to have negative impacts on winter survival of perennial crops. Year to year variations in winter losses will depend on the climatic conditions, as well as their interaction with species, cultivars and crop management. According to Bélanger *et al.* (2002), the expected increase of 20–28% in precipitation including rainfall in December, January and February in eastern Canada (Christensen *et al.*, 2007), will increase the risk of winter damage due to ice, particularly if it is accompanied with a limited snow cover.

In Norway, however, the assessment of winter survival of timothy and perennial ryegrass under future climate conditions (2020–2100) revealed that, even though cold tolerance would be reduced, a general decrease in the risk of winter injury is expected because the plants will remain hardy enough. A slightly increased risk of ice encasement is the only potential cause of increase in winter damage expected for these two species in Norway (Thorsen and Höglind, 2010). This suggests that increased risk of winter damage to perennial field crops in response to climate change will vary from one area to another.

14.3.3 Spring deacclimation and regrowth

The last critical step for winter survival of perennial field crops occurs during springtime (Gu *et al.*, 2008). Deacclimation caused by increasing temperatures and photoperiod occurs much more rapidly than cold acclimation (Kalberer *et al.*, 2006). As a result, frost spells during spring can kill unprotected plants. Shallow snow cover or early melting of snow under a future climate will increase the exposure of perennial plants to potential damage from frost spells.

The capacity of perennial crops to resume their growth in spring relies on the availability of carbon and nitrogen reserves stored in vegetative organs during the cold acclimation period and remaining at the end of the overwintering period. Autumn accumulation of carbohydrates in roots is considered a determining factor for cold tolerance and winter survival of many plant species including alfalfa (Dhont *et al.*, 2002) and turfgrasses (Dionne *et al.*, 2001; Shahba *et al.*, 2003). Nitrogen reserves of vegetative organs have also been found to be important for winter hardiness of perennials, since a large proportion of nitrogen supporting spring regrowth (such as free amino acids and proteins) is derived from nitrogen reserves (Dhont *et al.*, 2006; Volenec *et al.*, 1996). Any climatic factor affecting the accumulation of organic reserves, such as temperature and precipitation, will impact winter survival of perennial crops, either positively or negatively. Increasing atmospheric CO_2 could have a direct positive effect on spring regrowth by increasing biomass allocation to the root system, as it was shown previously with winter wheat in open-top chamber experiments (Haslin and Mortensen, 2010).

14.4 Conclusions

The future climate will be warmer with an increased concentration of atmospheric CO_2. These conditions are bound to increase the quantity of perennial crops produced in northern areas through improved yield and extended cultivated areas. This climate change-driven increase in forage quantity could, however, be accompanied by a reduction in forage nutritive value. The warmer autumn and winter conditions predicted are likely to increase the risk of winter damage through suboptimal autumn acclimation, dehardening temperatures in winter, and lack of insulating snow cover. The extent of increase in damage due to these future winter conditions in northern areas will probably vary by region.

References

Andrews, C.J. (1996) How do plants survive ice? *Annals of Botany* 78, 529–536.

Avice, J.C., Le Dily, F., Goulas, E., Noquet, C., Meuriot, F., Volenec, J.J., Cunningham, S.M., *et al.* (2003) Vegetative storage proteins in overwintering storage organs of forage legumes: Roles and regulation. *Canadian Journal of Botany* 81, 1198–1212.

Ball, M.C., and Hill, M.J. (2009) Elevated atmospheric CO_2 concentrations enhance vulnerability to frost damage in a warming world. In: Gusta, L.V., Wisniewski, M.E. and Tanino, K.K. (eds) *Plant Cold Hardiness from the Laboratory to the Field.* CAB International, Wallingford, UK, pp. 183–189.

Barker, D.H., Loveys, B.R., Egerton, J.J.G., Gorton, H., Williams, W.E. and Ball, M.C. (2005) CO_2 enrichment predisposes foliage of a eucalypt to freezing injury and reduces spring growth. *Plant, Cell & Environment* 28, 1506–1515.

Baron, V.S. and Bélanger, G. (2007) Climate and forage adaptation. In: Barnes, R.F., Nelson, C.J., Moore, K.J. and Collins, M. (eds) *Forages: The Science of Grassland Agriculture.* Vol. II, 6th ed, Wiley-Blackwell Publishing, Ames, Iowa, pp. 83–104.

Bélanger, G., Rochette, P., Castonguay, Y., Bootsma, A., Mongrain, D. and Ryan, D.A.J. (2002) Climate change and winter survival of perennial forage crops in eastern Canada. *Agronomy Journal* 94, 1120–1130.

Bélanger, G., Castonguay, Y., Bertrand, A., Dhont, C., Rochette, P., Couture, L., Drapeau, R., *et al.* (2006) Winter damage to perennial forage crops in eastern Canada: Causes, mitigation and prediction. *Canadian Journal of Plant Science* 86, 33–47.

Bertrand, A. and Castonguay, Y. (2003) Plant adaptations to overwintering stresses and implication of climate change. *Canadian Journal of Botany* 81, 1145–1152.

Bertrand, A., Castonguay, Y., Nadeau, P., Laberge, S., Rochette, P., Michaud, R., Bélanger, G., *et al.* (2001) Molecular and biochemical responses of perennial forage crops to oxygen deprivation at low temperature. *Plant, Cell and Environment* 24, 1085–1093.

Bertrand, A., Castonguay, Y., Nadeau, P., Laberge, S., Michaud, R., Belanger, G. and Rochette, P. (2003) Oxygen deficiency affects carbohydrate reserves in overwintering forage crops. *Journal of Experimental Botany* 54, 1721–1730.

Bertrand, A., Prévost, D., Bigras, F.J. and Castonguay, Y. (2007a) Elevated atmospheric CO_2 and strain of rhizobium alter freezing tolerance and cold-induced molecular changes in alfalfa (*Medicago sativa*). *Annals of Botany* 99, 275–284.

Bertrand, A., Prévost, D., Bigras, F.J., Lalande, R., Tremblay, G.F., Castonguay Y. and Bélanger, G. (2007b) Alfalfa response to elevated atmospheric CO_2 varies with the symbiotic rhizobial strain. *Plant and Soil* 301, 173–187.

Bertrand, A., Tremblay, G.F., Pelletier, S., Castonguay, Y. and Bélanger, G. (2008) Yield and nutritive value of timothy as affected by temperature, photoperiod and time of harvest. *Grass and Forage Science* 63, 421–432.

Bloom, A.J., Burger, M., Asensio, J.S.R. and Cousins, A.B. (2010) Carbon dioxide enrichment inhibits nitrate assimilation in wheat and *Arabidopsis*. *Science* 328: 899–903.

Brito, A.F., Tremblay, G.F., Bertrand, A., Castonguay, Y., Bélanger, G., Michaud, R., Lapierre, H., *et al.* (2008) Alfalfa cut at sundown and harvested as baleage improves milk yield in late-lactation dairy cows. *Journal of Dairy Science* 91, 3968–3982.

Brito, A.F., Tremblay, G.F., Lapierre, H., Bertrand, A., Castonguay, Y., Bélanger, G., Michaud, R., *et al.* (2009) Alfalfa cut at sundown and harvested as baleage increases bacterial protein synthesis in late-lactation dairy cows. *Journal of Dairy Science* 92, 1092–1107.

Buxton, D.R. and Fales, S.L. (1994) Plant environment and quality. In: Fahey, G.C. (ed.) *Forage Quality, Evaluation, and Utilization.* American Society of Agronomy, Madison, Wisconsin, pp. 155–199.

Castonguay, Y., Nadeau, P. and Laberge, S. (1993) Freezing tolerance and alteration of translatable mRNAs in alfalfa (*Medicago sativa* L.) hardened at subzero temperatures. *Plant and Cell Physiology* 34, 31–38.

Castonguay, Y., Thibault, G., Rochette, P., Bertrand, A., Rochefort, S. and Dionne, J. (2009a) Physiological responses of annual bluegrass and creeping bentgrass to contrasted levels of O_2 and CO_2 at low temperatures. *Crop Science* 49, 671–689.

Castonguay, Y., Michaud, R., Nadeau, P. and Bertrand, A. (2009b) An indoor screening method for improvement of freezing tolerance in alfalfa. *Crop Science* 49, 809–818.

Chatterton, N.J., Harrison, P.A., Bennett, J.H. and Asay, K.H. (1989) Carbohydrate partitioning in 185 accessions of Gramineae grown under warm and cool temperatures. *Journal of Plant Physiology* 134, 169–179.

Christensen, J.H., Hewitson, B., Busuioc, A., Chen, A., Gao, X., Held, I., Jones, R., *et al.* (2007) Regional climate projections. In: Solomon, S., Qin, D., Manning, M., Chen, Z., Marquis, M., Averyt, K., Tignor, M., *et al.* (eds) *Climate Change 2007: The Physical Science Basis. Contribution of Working Group I to the Fourth Assessment Report of the Intergovernmental Panel on Climate Change.* Cambridge University Press, Cambridge, pp. 847–940.

Cotrufo, M.F., Ineson, P. and Scott, A. (1998) Elevated CO_2 reduces the nitrogen concentration of plant tissues. *Global Change Biology* 4, 43–54.

Dhont, C., Castonguay, Y., Nadeau, P., Bélanger, G. and Chalifour, F.-P. (2002) Alfalfa root carbohydrates and regrowth potential in response to fall harvest. *Crop Science* 42, 754–765.

Dhont, C., Castonguay, Y., Nadeau, P., Bélanger, G., Drapeau, R., Laberge, S., Avice, J.-C., *et al.* (2006) Nitrogen reserves, spring regrowth and winter survival of field-grown alfalfa (*Medicago sativa*) defoliated in the autumn. *Annals of Botany* 97, 109–120.

Dionne, J., Castonguay, Y., Nadeau, P. and Desjardins, Y. (2001) Freezing tolerance and carbohydrate changes during cold acclimation of green-type annual bluegrass (*Poa annua* L.) ecotypes. *Crop Science* 41, 443–451.

Dionne, J., Rochefort, S., Huff, D.R., Desjardins, Y., Bertrand, A. and Castonguay, Y. (2010) Variability for freezing tolerance among 42 ecotypes of green-type annual bluegrass. *Crop Science* 50, 321–336.

Field, C.B., Mortsch, L.D., Brklacich, M., Forbes, D.L., Kovacs, P., Patz, J.A., Running, S.W., *et al.* (2007) North America. In: Parry, M.L., Canziani, O.F., Palutikof, P., van der Linden, P.J. and Hanson, C.E. (eds) *Climate Change 2007: Impacts, Adaptation and Vulnerability. Contribution of Working Group II to the Fourth Assessment Report of the Intergovernmental Panel on Climate Change.* Cambridge University Press, Cambridge, UK, pp. 617–652.

Gu, L., Hanson, P.J., Post, W.M., Kaiser, D.P., Yang, B., Nemani, R., Pallardy, S.G., *et al.* (2008) The 2007 eastern US spring freeze: Increased cold damage in a warming world. *Bioscience* 58, 253–262.

Gudleifsson, B.E. (1997) Survival and metabolite accumulation by seedlings and mature plants of timothy grass during ice encasement. *Annals of Botany* 79, 93–96.

Gudleifsson, B.E. (2009) Ice encasement damage on grass crops and alpine plant in Iceland - Impact of climate change. In: Gusta, L.V., Wisniewski, M.E. and Tanino, K.K. (eds) *Plant Cold Hardiness from the Laboratory to the Field*, CAB International, Wallingford, UK, pp. 163–172.

Gudleifsson, B.E., Andrews, C.J. and Bjornsson, H. (1986) Cold hardiness and ice tolerance of pasture grasses grown and tested in controlled environments. *Canadian Journal of Plant Science* 66, 601–608.

Gusta, L.V., Burke, M.J. and Kapoor, A. (1975) Determination of unfrozen water in winter cereals at subfreezing temperatures. *Plant Physiology* 56, 706–709.

Guy, C.L. (2003) Freezing tolerance of plants: current understanding and selected emerging concepts. *Canadian Journal of Botany* 81, 1216–1223.

Haslin, H.M. and Mortensen, L.M. (2010) Autumn growth and cold hardening of winter wheat under simulated climate change. *Acta Agriculturae Scandinavica Section B-Soil and Plant Science* 60, 437–449.

Hulme, M.A. and Sheard, N. (1999) *Climate Change Scenarios for Canada.* Climatic Research Unit, University of East Anglia, Norwich.

Hungate, B.A., Dukes, J.S., Shaw, M.R., Luo, Y.Q. and Field, C.B. (2003) Nitrogen and climate change. *Science* 302, 1512–1513.

Kalberer, S.R., Wisniewski, M. and Arora, R. (2006) Deacclimation and reacclimation of cold-hardy plants: Current understanding and emerging concepts. *Plant Science* 171, 3–16.

Kaur, N. and Gupta, A.K. (2005) Signal transduction pathways under abiotic stresses in plants. *Current Science* 88, 1771–1780.

Leep, R.H., Andresen, J.A. and Jeranyama, P. (2001) Fall dormancy and snow depth effects on winterkill of alfalfa. *Agronomy Journal* 93, 1142–1148.

Levitt, J. (1980) Chilling, freezing, and high temperature stresses. In: Kozlowski, T.T. (ed.) *Responses of Plants to Environmental Stresses*. Vol. 1, Academic Press, New York.

Loveys, B.R., Egerton, J.J.G. and Ball, M.C. (2006) Higher daytime leaf temperatures contribute to lower freeze tolerance under elevated CO_2. *Plant, Cell & Environment* 29, 1077–1086.

Milchunas, D.G., Mosier, A.R., Morgan, J.A., LeCain, D.R., King, J.Y. and Nelson, J.A. (2005) Elevated CO_2 and defoliation effects on a shortgrass steppe: forage quality versus quantity for ruminants. *Agriculture, Ecosystems and Environment* 111, 166–184.

Miller, L.A., Moorby, J.M., Davies, D.R., Humphreys, M.O., Scollan, N.D, MacRae, J.C. and Theodorou, M.K. (2001) Increased concentration of water-soluble carbohydrate in perennial ryegrass (*Lolium perenne* L.): milk production from late-lactation dairy cows. *Grass and Forage Science* 56, 383–394.

Moorby, J.M., Evans, R.T., Scollan, N.D., MacRae, J.C. and Theodorou, M.K. (2006) Increased concentration of water-soluble carbohydrate in perennial ryegrass (*Lolium perenne* L.). Evaluation in dairy cows in early lactation. *Grass and Forage Science* 61, 52–59.

Morgan, J.A., Mosier, A.R., Milchunas, D.G., Lecain, D.R., Nelson, J.A. and Parton, W.J. (2004) CO_2 enhances productivity, alters species composition, and reduces productivity of shortgrass steppe vegetation. *Ecological Applications* 14, 208–219.

Paquin, R. and Mehuys, G.R. (1980) Influence of soil moisture on cold tolerance of alfalfa. *Canadian Journal of Plant Science* 60, 139–147.

Paquin, R. and Pelletier, H. (1980) Influence de l'environnement sur l'endurcissement au froid de la luzerne (*Medicago media* pers.) et sa résistance au gel. *Canadian Journal of Plant Science* 60, 1351–1366.

Pearce, R.S. (2001) Plant freezing and damage. *Annals of Botany* 87, 417–424.

Rogers, A. and Ainsworth, E.A. (2006) The response of foliar carbohydrates to elevated carbon dioxide concentration. In: Nösberger, J., Long, S.P., Norby, R.J., Stitt, M., Hendrey, G.R. and Blum, H. (eds) *Managed Ecosystems and CO_2. Case Studies, Processes and Perspective.* Springer-Verlag, Heidelberg, Berlin, pp. 293–308.

Rogers, A., Gibon, Y., Stitt, M., Morgan, P.B., Bernacchi, C.J., Ort, D.R. and Long, S.P. (2006) Increased C availability at elevated carbon dioxide concentration improves N assimilation in a legume. *Plant, Cell and Environment* 29, 1651–1658.

Shahba, M.A., Qian, Y.L., Hughes, H.G., Koski, A.J. and Christensen, D. (2003) Relationships of soluble carbohydrates and freeze tolerance in saltgrass. *Crop Science* 43, 2148–2153.

Smith, D. (1972) Effect of day-night temperature regimes on growth and morphological development of timothy plants derived from winter and summer tillers. *Journal of the British Grassland Society* 27, 107–110.

Statistics Canada (2009) Agricultural profile of Canada. www40.statcan.gc.ca/l01/cst01/PRIM11A-eng.htm, accessed 27 September 2010.

Steponkus, P.L. (1984) Role of the plasma membrane in freezing injury and cold acclimation. *Annual Review of Plant Physiology* 35, 543–584.

Steponkus, P.L., Uemura, M. and Webb, M.S. (1993) Membrane destabilization during freeze-induced dehydration. In: Close, T.J. and Bray, E.A. (eds) *Plant Responses to Cellular Dehydration During Environmental Stress.* American Society of Plant Physiology, Rockville, Maryland, pp. 37–47.

Suzuki, M. (1977) Effects of soil moisture on cold resistance of alfalfa. *Canadian Journal of Plant Science* 57, 315.

Thomashow, M.F. (2001) So what's new in the field of plant cold acclimation? Lots! *Plant Physiology* 125, 89–93.

Thorsen, S.M. and Höglind, M. (2010) Assessing winter survival of forage grasses in Norway under future climate scenarios by simulating potential frost tolerance in combination with simple agroclimatic indices. *Agricultural and Forest Meteorology* 150, 1272–1282.

Thorvaldsson, G., Tremblay, G.F. and Kunelius, H.T. (2007) The effects of growth temperature on digestibility and fibre concentration of seven temperate grass species. *Acta Agriculturae*

Scandinavica Section B - Soil and Plant Science 57, 322–328.

Volenec, J.J. and Nelson, C.J. (2007) Physiology of forage plants. In: Barnes, R.F., Nelson, C.J., Moore, K.J. and Collins, M. (eds) *The Science of Grassland Agriculture.* Volume II, 6th ed., Wiley-Blackwell Publishing, Ames, Iowa, pp. 37–52.

Volenec, J.J., Ourry, A. and Joern, B.C. (1996) A role for nitrogen reserves in forage regrowth and stress tolerance. *Physiologia Plantarum* 97, 185–193.

Yu, M., Gao, Q. and Shaffer, M.J. (2002) Simulating interactive effects of symbiotic nitrogen fixation, carbon dioxide elevation, and climatic change on legume growth. *Journal of Environmental Quality* 31, 634–641.

15 The Potential Impact of Climate Change on Temperate Zone Woody Perennial Crops

H.A. Quamme and D. Neilsen

15.1 Introduction

In this chapter we discuss the implication of climate change on production of horticultural-grown woody perennials that overwinter in the temperate zone. Freezing injury affects the survival of many temperate zone woody perennials and is a major constraint to the production of fruit and nut crop and many landscape plants. Although these plants have evolved a capacity to avoid or tolerate freezing, they suffer winter damage frequently and this affects survival or productivity. Low-temperature injury can occur in all parts of the plant depending upon tissue susceptibility, exposure, and stage of acclimation. The level of cold hardiness during acclimation varies with species and cultivar (Sakai and Larcher, 1987; Wisniewski *et al.*, 2003).

Often species have characteristic forms of injury. For example, in apple the tissues of the bark and xylem of the trunk are the most prone to injury (Quamme and Hampson, 2004), whereas leaf and flower buds survive even to liquid nitrogen temperatures (Sakai, 1965). On the other hand, the flower bud tissues of apricots, peaches and sweet cherries are the most susceptible to injury, but all tissues succumb to freezing temperatures just below those that injure the flower bud (Quamme *et al.*, 1982). The flower bud hardiness of these species limits commercial production. Freezing injury can occur in roots as well, especially in the absence of snow cover (Quamme, 1990; Wildung *et al.*, 1973). In spring, developing flower buds and flowers of most fruit and landscape plants are prone to frost injury just below freezing.

In a recent review of the cold hardiness of woody plants, Wisniewski *et al.* (2003) pointed out that although the freezing injury to plants has been the subject of much study since the 19th century, the mechanism(s) is still not fully understood. In many woody plants, tolerance and avoidance mechanisms appear to operate, often in the same plant. However, the mechanism of avoidance by deep supercooling appears to be common in tissues critical to the survival of many fruit, nut and landscape species. Although this survival mechanism affords protection to these species, most have a limited capability to withstand winter temperatures.

It has been shown that the maximum northern survival range of many woody plant species is limited by the capacity of the flower buds, xylem parenchyma, and bark tissues to deep supercool (George *et al.*, 1974; Quamme, 1976; Pierquet *et al.*, 1977; Rajashekar and Burke, 1978; Quamme *et al.*, 1982; Quamme, 1989). The lowest deep supercooling point of one or more of the tissues of these plant species often coincides with the average annual minimum temperature isotherm at their extreme boundary. The average annual minimum temperature isotherms are the basis of plant hardiness zone maps first developed at the National Arboretum (1965). Maps are now available for most parts of the world, indicating the regions where different plant species can be grown (Wikipedia, 2011).

As flower buds of these species develop in

© CAB International 2012. *Temperature Adaptation in a Changing Climate: Nature at Risk* (eds K.B. Storey and K.K. Tanino)

late winter and spring, the capacity to deep supercool is lost and survival depends on the tolerance to formation of ice within their tissues, which occurs just below freezing (Quamme, 1989). Roots show less capacity to withstand low temperature than aerial vegetative parts of the plant (Sakai and Larcher, 1987).

15.2 Historical Evidence of the Impact of Climate Change on Survival of Several Horticultural Woody Plants

Access to over 90 years of records in the Okanagan Valley has allowed us the opportunity to study the relationship between fruit production and weather, and to determine the impact of climate change. The method used was the iterative χ^2 statistical technique developed by Caprio (1966) to compare weather with non-continuous variables and generate critical threshold values and measures of significance. The iterative χ^2 technique was used to determine the weather associated with production of apple (Caprio and Quamme, 1999); apricot, peach and sweet cherry (Caprio and Quamme, 2006); grape (Caprio and Quamme, 2002); and pear (Quamme *et al.*, 2010). One of the significant findings was that a strong association of minimum temperature and poor production existed during late autumn and winter in all six fruit crops that was greater than all other weather associations. This association was attributed to the effects of winter injury. During this time period, cardinal temperatures (CT) determined by iterative χ^2 analysis agreed with the killing temperatures reported for these crops.

Caprio and Quamme (1999) gave an in-depth discussion of the application of the iterative χ^2 method to determine the weather that was associated with fruit production, including examples of the interpretation of computer output. To summarize, annual fruit production was first indexed by comparing each year with three previous years and three subsequent years. This was done to minimize the effects of variable tree populations, cultivars, and advancements in horticultural technology. The indexed production data were then ranked and separated into quartiles (lowest quartile, poor; highest quartile, good; and two mid-quartiles, neutral years).

Next, the χ^2 test was used to compare daily weather occurrence of poor and good years with the normal years. The minimum or maximum temperature variable, which was recorded in Imperial units, was broken down into classes of 2°F and tallies were made of the frequency with which the daily measurements fell into each class. The χ^2 test was used to test the difference in frequency of days between each quartile (poor and good) and the two mid-quartiles for each of the classes, independently. The χ^2 test was applied to the total number of days accumulating in each class in succession, ordered with minimum temperature (low to high), or maximum temperature (high to low), generating a χ^2 value for each class in the scan. If the frequency distribution of the poor or good quartile deviated from that of the combined mid-quartiles, the χ^2 values generated in the scan increased and then decreased. When the χ^2 reached its maximum (or turning) point in the scan, the temperature at that point was referred to as the 'cardinal temperature' (CT). A χ^2 value of 7 was taken as the critical value for significant effect of temperature, ($P \leq 0.01$, df = 1). A sliding window of 3 weeks was used in the analysis, with each period in succession adding a new week and dropping the last to give a running average of the χ^2 values. A maximum χ^2 and an associated cardinal value were generated for each week.

Temperatures were converted to metric after the χ^2 scan. In the study of the impact of winter temperature on production, the highest χ^2 value and CT are given for the significant period. Weekly χ^2 and CT values are given in the study of impact of spring frost on production.

15.2.1 Climate change in the Okanagan Valley

Brewer and Taylor (2001) first reported a rise in winter minimum temperature at the Coldstream Ranch, a site in the Okanagan with weather records from 1900 to 1997, and several other sites in southern British

Columbia with weather records from 1970 to 2000. Taylor and Barton (2004) later reported that the average minimum temperature had risen at Summerland CDA/CS, another site in the Okanagan with long-term weather records (1916–2011), and at Coldstream Ranch, and that the trend was greatest in the winter and spring. These changes reflect a larger pattern identified for Western Canada (Zhang *et al.*, 2000).

The iterative χ^2 method can be used to compare weather in recent years with that of previous years, and thus measure climate change (Caprio and Quamme, 2002; Caprio *et al.*, 2009). The output of this comparison indicates the season when climate change has been greatest and the critical thresholds that are associated with this change. Minimum temperature (recorded at Summerland CDA/CS) during the recent 20-year periods of 1970–1989 and 1990–2009, compared to that of 1920–1959 (late summer, autumn and winter of the year preceding harvest, and winter and early spring of the harvest year) is shown in Fig. 15.1A. The greatest warming for this period (deficit days of minimum temperature ≤ than the given CT value) occurred in winter and early spring (17 December–14 April for the period 1990–2009, and 21 January–14 April for the period 1970–1989). The association was stronger in 1990–2009 than in 1970–1989. One period, 26 November–23 December of 1970–1989 (excess days, χ^2 =18, CT ≤-12°C) was actually colder than in earlier times. This cold period was also reported before for the Coldstream site (Quamme *et al.*, 2010) and was correlated with greater frequency of Arctic outbreaks. In the period 1990–2009, the effect was much diminished as the warming trend continued. The 1970–1989 set of years was also cooler in late summer and autumn (23 August (not shown)–21 October, 13 September, χ^2 =21, CT ≤6°C) than that of 1920–1959 (χ^2 not significant).

15.2.2 Winter injury

Here, we report how peach and apricot production in the Okanagan Valley has been affected by climate change during the autumn and winter prior to harvest, for an 88-year period. During 1921–2008, χ^2 values of association (excess days) between poor production of peach and minimum temperature began on 29 October, rose sharply in mid-December, reached a peak on 24 January (χ^2 = 60, CT = ≤-18°C) and declined to 7 March (Fig. 15.1B). The lowest CT, ≤-23°C, occurred on 10 January. Good production was also associated with excess days of low minimum temperature (5 November–7 March), but the χ^2s were lower and CT higher throughout the period of significance. Low temperature above the killing temperatures during this cold weather period cause peach flower buds to harden (Quamme, 1983). Associations between minimum temperature and poor and good production in March and April appear to be related to freezing of the flower buds, which we discuss in Section 15.2.3.

The χ^2 test of maximum temperature and production indicated only weak associations (associations with maximum temperature not shown in figures). Poor production was associated with maximum temperature during 1–27 January with a deficit of warm days (17 January, χ^2 = -12, CT ≤2°C). This coincided with a period when minimum temperature was strongly associated with poor production. Good production was also weakly associated with maximum temperature (deficit days) during 4–17 February (February 14, χ^2 = 8, CT ≤ 6°C).

To determine if production was affected by climate change, we divided the 88 (longest record period that could be divided by 8) into two sets (1920–1964 and 1965–2008) and analysed each set with the χ^2 test (Figs. 15.1C and 15.1D). During 1920–1964, we observed that a strong association between poor production and minimum temperature (24 January, χ^2 = 51, excess days, CT ≤-22°C) from 17 December–3 March that was greater than that of 1965–2008. During 1965–2008 the association between poor production and minimum temperature (excess days) began earlier (5 November) and peaked on 22 November (χ^2 = 24, CT ≤-7°C), again on 27 December (χ^2 = 28, CT ≤-14°C), for a third time on 24 January (χ^2 = 14,

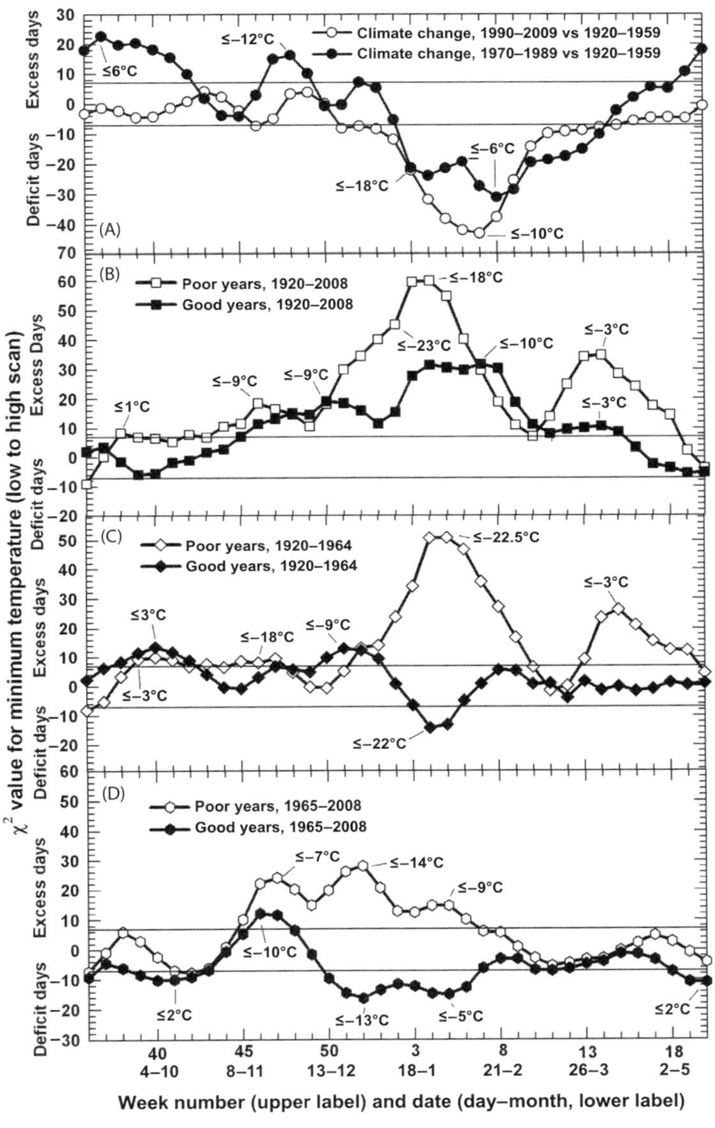

Fig. 15.1. The change in minimum temperature for the years 1920–1959 is compared to 1970–1989 (closed circles) and 1990–2009 (open circles) by the χ^2 test (A). The association of daily minimum temperature with peach production is presented for years 1921–2008 (B), 1920–1964 (C), and 1965–2008 (D). The time period is week 36 of the preceding year to week 20 of the harvest year (6 September–16 May). For the association of minimum temperature with production, the χ^2 values compare the daily weather occurrences of poor years (extreme low quartile of years ranked by production) and good years (extreme high quarter of years) each with that of neutral years (combined mid-quartiles of years ranked by production). For 1921–2008 (B), poor production indicated by open squares and good production by closed squares, for 1920–1964 (C), poor production is indicated by open diamonds, and good production by closed diamonds; and for 1965–2008 (D), poor production is indicated by open hexagons, and good production by closed hexagons. All comparisons are based on a 3-week running average. Cardinal temperatures (the weather value at the maximum χ^2) are indicated on the graph. $\chi^2 = 7$ (indicated by the solid lines) is the critical value for significance ($P \leq 0.01$, 1 df).

CT ≤−9°C), and ended earlier (10 February). We attribute the weaker associations between poor production and minimum temperature during the recent set of years to a warmer climate during winter (Fig. 15.1A). The earlier association of poor production with minimum temperature in the period 1965–2008 (Fig. 15.1D) appeared to be associated with a short period in November and December when temperatures were colder in recent years than previous years (1920–1959 versus 1970–1989, Fig. 15.1A).

From 10 September–18 November of 1920–1964, poor production was also weakly associated with excess days of minimum temperature (4 October, χ^2 = 10, CT ≤3°C, and 15 November χ^2 = 10, CT ≤−18°C) (Fig. 15.1C). The causes are unknown.

Similarly, weak associations between good production and minimum temperatures (excess days) existed during the late summer and autumn of 1920–1964 (23 August–21 October, 4 October, χ^2 = 13, CT ≤3°C; 10 December–6 January, 20 December, χ^2 = 13, CT ≤−9°C) (Fig. 15.1C). During 1965–2008, weak associations between minimum temperature and good production, also of unknown cause, were present (24 September–21 October, χ^2 = 10, deficit days; 4 October, CT ≤2°C; 12–25 November, 22 November, χ^2 = 12, excess days, CT ≤−10°C) (Fig. 15.1D).

During the winters of 1920–1964 (21 January–6 February, 24 January, χ^2 = 14, CT ≤−22°C) (Fig. 15.1C) and 1965–2008 (10 December–10 February, 27 December, χ^2 = 16, CT ≤−13°C; 31 January, χ^2 = 15, CT ≤−8°C), good production was associated with deficit days of minimum temperature (Fig. 15.1D). This association appeared to reflect the influence of minimum temperature on bud survival.

During 1920–1964, a weak association (deficit days) between maximum temperature and poor production (17 December–27 January, χ^2 = 14, CT ≥ 7°C) was present, whereas during the period 1965–2008 a weak association (excess days) (3–23 December, χ^2 = 10, CT ≥ 4°C) and (7 January–10 February, χ^2 = 15, CT ≥ 10°C) was present; the latter relationship indicated a weak detrimental effect of warmer temperatures. No significant association between maximum temperature and good production was found during autumn and winter.

A similar strong association was also found between minimum temperature and apricot production in autumn and winter over the 88-year period, with a reduction in the effects of minimum temperature in the recent 44 years (unpublished data).

15.2.3 Spring frosts

Spring frosts can kill flowers and injure the fruit of many temperate zone perennials. These killing frosts usually result from radiation freezes that occur when night-time temperatures drop below freezing. Skies are usually clear, and humidity and wind speed are low (Westwood, 1991). The effects of spring frosts have mainly been studied in fruit crop species because they cause losses in production. Depending on the stage of development, fruit tree blossoms are killed at temperatures in the range of −15°C (green tip) to −2°C (full bloom) (Proebsting, Jr and Mills, 1978). A number of studies have shown that bud phenology and flower development of many species has advanced in a number of regions of the world as a result of rising temperatures (Fitter and Fitter, 2002; Chemeilewski *et al.*, 2004; Wolf *et al.*, 2005; Lu *et al.*, 2006; Fujisawa and Kobayashi 2010). As frost susceptibility is highly dependent on the stage of development and the stage of development is dependent on temperature, the question is whether the advancing development puts flowers at greater risk of frost injury.

We present evidence that in the Okanagan Valley peach (Fig. 15.2A and B) and apricot production (Figs 15.3A and 15.3B) have benefited from a warmer climate during bloom. During the whole period 1921–2008 (Figs 15.2A and 15.2B), the χ^2 values for peach for both poor and good production declined to the beginning of the flowering period, increased, and then declined again. This period, 14 March–2 May, coincided with the flower development in peach,

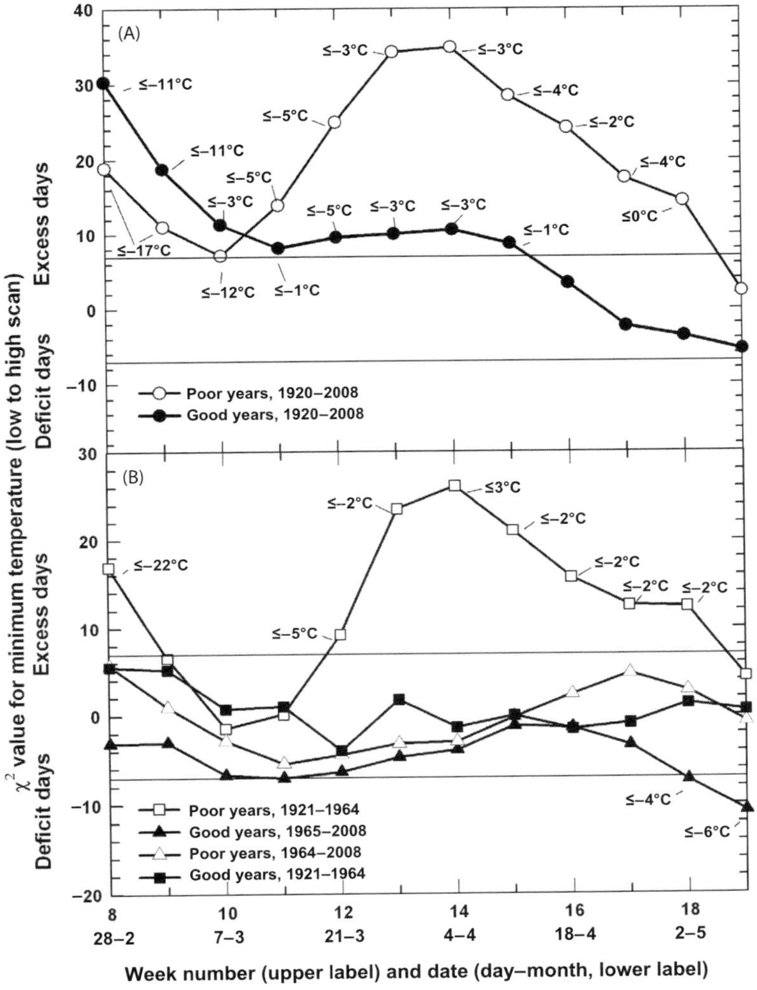

Fig. 15.2. The association of daily minimum temperature with peach production for 1921–2008 (A), for 1920–1964, and for 1965–2008 (B) is presented for weeks 8–19 (2 February–9 May), the period of bud development. For the association of minimum temperature with production, the χ^2 values compare the daily weather occurrences of poor years (extreme low quartile of years ranked by production) and good years (extreme high quarter of years) each with that of neutral years (combined mid-quartiles of years ranked by production). For 1921–2008, poor production is presented by open circles, and good production by closed circles (A); for years 1920–1964, poor production is presented by open squares, and good production by closed squares; and for years 1965–2008, poor production is presented by open triangles, and good production by closed triangles (B). All comparisons are based on a 3-week running average. Cardinal temperatures (the weather value at the maximum χ^2) are indicated on the graph. $\chi^2 = 7$ (indicated by the solid lines) is the critical value for significance ($P \leq 0.01$, 1 df).

exemplified by the range of full bloom dates for Elberta peach (15 April–15 May) (PARC records 1937 to 1964; Denise Neilsen, personal communication). The range of CT values during flowering time for poor years coincided with the range of killing temperatures (–2°C to –5°C). The association of minimum temperature with poor production was greater than that with good production, and CT were lower for the poor

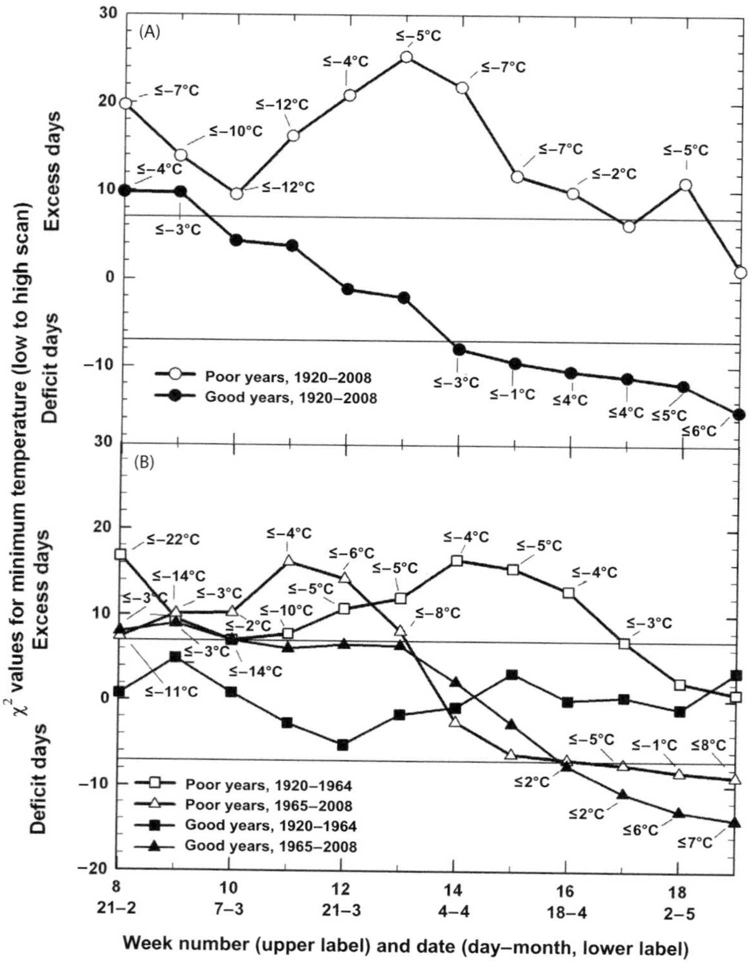

Fig. 15.3. The association of daily minimum temperature with apricot production for 1921–2008 (A), for 1920–1964, and for 1965–2008 (B) is presented for weeks 8–19 (2 February–9 May), the period of bud development. For the association of minimum temperature with production, the χ^2 values compare the daily weather occurrences of poor years (extreme low quartile of years ranked by production) and good years (extreme high quarter of years) each with that of neutral years (combined mid-quartiles of years ranked by production). For 1921–2008, poor production is presented by open circles, and good production closed circles (A); for years 1920–1964, poor production by open squares, and good production by closed squares; and for years 1965–2008 (B), poor production is represented by open triangles, and good production by closed triangles. All comparisons are based on a 3-week running average. Cardinal temperatures (the weather value at the maximum χ^2) are indicated on the graph. χ^2 = 7 (indicated by the solid lines) is the critical value for significance ($P \leq 0.01$, 1 df).

production years than for the good years. The beneficial effect of low temperatures just above the killing point may result from delayed flower development and retarded dehardening.

When we divided the set of 88 years into two sets and ran the χ^2 analysis separately on each set, a strong association was observed for years 1920–1964 with the CT (excess days at ≤−2° to ≤−5°C) in the killing range but not for years 1965–2008 (Fig. 15.2B). When we tested poor years from the two sets of years with a χ^2 test, we found significant differences for all dates from 7 March–18 April. We attribute the lack of the association of minimum temperature with production in years 1965–2008 to a lowered risk of spring frosts associated with the warmer temperatures during this period (Fig. 15.1A). The association with minimum temperature and good peach production was not present in the 44-year comparison (Fig. 15.2B).

The relationship of production with maximum temperatures (excess days) occurred in late April, May and early June (good production, 15 April–19 May, 2 May, χ^2 = 17, CT = ≥22°C; and poor production, 30 April–3 June, 9 May, χ^2 = 14, CT = ≥27°C), and appeared to be associated with the influence of high temperatures on fruit pollination and set.

We observed similar trends in the associations between production and minimum temperature over the 88-year period with apricots as we did with peaches, and the CT were in a similar range (Fig. 15.3A). On the other hand, the 44-year comparisons differed in that the low minimum temperature associated with poor production in 1964–2008 occurred earlier and over a shorter time period than in 1920–1964 (21 February–28 March, 6 weeks versus 7 March–23 April, 7 weeks) (Fig. 15.3B).

15.3 Predicting Impacts of Climate Change on Woody Perennials Using Climatic Models

Global climate models predict 2°C–6°C increase in temperature by 2100 (Pachauri and Reisinger, 2007). The outputs have been downscaled to regional levels and used to predict the impact of climate change on crop production. Rochette *et al.* (2004) assessed the impact of climate change on winter survival of fruit trees in eastern Canada using agro-climatic indices that represent risk of winter injury. Using the Canadian Global General Circulation Model, they found that by 2040–2069 the first autumn frost would be delayed by 16 days, whereas the last spring frost would be advanced by 15 days. The influence of shortening photoperiod at the time of first frost would induce earlier acclimation and greater cold hardiness. The reduction in annual minimum temperatures and accumulation of cold degree days (< −15°C) would reduce the risk of winter injury. However, warmer weather would cause loss of hardiness and thinner snow cover that would result in root damage. Kaukoranta *et al.* (2010) reached similar conclusions with a downscaled model using agro-climatic indices to predict extension of apple production in Finland.

Climate models also predict that the risk of spring frosts becomes lower as the climate gets warmer (Bennie *et al.*, 2009; Eccel *et al.*, 2009). This is in agreement with the historical data presented in this review. However, both Rochette *et al.* (2004) and Kaukoranta *et al.* (2010) predicted that for fruit trees, the change in spring frost risk would not be uniform. In coastal regions, early blooming fruit trees could be at greater risk from frost.

This type of analysis using agro-climatic indices predicts the general response of fruit trees (Rochette *et al.*, 2004), but we suggest that an analysis of the historical response of a crop to weather using the χ^2 analysis could provide a more accurate determination of agro-climatic indices that could be used to determine the impact on specific crops. For example, peach flower bud hardiness is critical to the sites where peaches are produced (Quamme *et al.*, 1982). The flowers are fully formed within the flower bud during summer and early autumn and supercool to temperatures well below freezing (−15°C, Redhaven cultivar, 15 October) (Quamme, 1983). They harden in response to temperatures just below freezing, which causes water to be withdrawn from a barrier below the flower to enable full deep supercooling to occur. Photoperiod may initiate flowering but does not appear to play a role in hardiness once the flower is formed. Freezing temperatures above the deep supercooling point usually increase hardiness. Duration of the freezing temperatures at this stage does not appear to be damaging as long as the temperature does not fall below the deep supercooling point. However, flowers cannot be induced to deep supercool much below a critical temperature, which for Redhaven is –25°C. Thus, over-wintering flowers of most peach cultivars are killed when minimum temperatures reach a range of about –21°C to –25°C, which usually occur in the Okanagan Valley during late December and January (Quamme, 1983). The χ^2 analysis of peach production and temperature records verifies this hypothesis (Fig. 15.1B). The lowest average survival temperatures during late autumn and early winter would be the best agro-climactic index to determine the impacts of climate changes on peach.

We also suggest that synoptic weather patterns be examined to determine the impacts of climate change. For example, during the period 1948–2006, the synoptic weather pattern that was most closely associated with winter freeze events in the five tree fruit crops and grapes was the Arctic outflow (Quamme *et al.*, 2010). This weather pattern was an infrequent weather event but had a great impact on production. A decrease in frequency and increase in minimum temperature of Arctic outflows were associated with the warming trends of the region during winter and early spring, although a slight increase in frequency of Arctic outflows was observed during late autumn. It was concluded that if this pattern in climate change continues, an extension of the northern range of some of these crops in this region (grapes, apples and sweet cherries) that acclimate late might not be as great as anticipated, whereas the northern range of others (pear, peach and apricot) that acclimate early might be extended. Synoptic data may be more readily downscaled from global models and used to predict climatic impact.

Climate change has already occurred, and the changes have been incorporated into the production range of hardiness zone maps that indicate an extension of many horticultural crops, including woody perennials, into regions that were formerly of high risk (McKenney *et al.*, 2001). However, the ultimate impact of climate change on horticultural production is highly unpredictable because it depends on human activity, and perhaps intervention. While some would welcome an extension of horticultural crop production into northern regions as result of global warming, such beneficial effects would probably be trivial compared to other unforeseen detrimental consequences.

References

Bennie, J., Kubin, E., Wiltshire, A., Huntley, B. Baxster, R. (2009) Predicting spatial and temporal patterns of bud-burst and spring frost risk in north-west Europe: the implications of local adaptation to climate. *Global Change Biology* 16,1503–1514.

Brewer, R. and Taylor, B. (2001) Climate change in the Okanagan Basin. In: Cohen, S. and Kulkarni, T. (eds) *Water Management and Climate Change in the Okanagan Basin.* Environment Canada and the University of British Columbia, Vancouver, British Columbia, pp. 14–20.

Caprio, J.M. (1966) A statistical procedure for determining the association between weather and non-measurement biological data. *Agricultural Meteorology* 3, 55–72.

Caprio, J.M. and Quamme, H.A. (1999) Weather conditions associated with apple production in the Okanagan Valley of British Columbia. *Canadian Journal of Plant Science* 79, 129–137.

Caprio, J.M. and Quamme, H.A. (2002) Weather conditions associated with grape production in the Okanagan Valley of British Columbia and potential impact of climate change. *Canadian Journal of Plant Science* 82, 755–763.

Caprio, J.M. and Quamme, H.A. (2006) Influence of weather on apricot peach and sweet cherry production in the Okanagan Valley of British Columbia. *Canadian Journal of Plant Science* 86, 259–267.

Caprio, J.M., Quamme, H.A. and Redmond, K.T. (2009) A statistical procedure to determine recent climate change of extreme daily meteorological data as applied at two locations in Northwestern North America. *Climate Change* 92, 65–82.

Chemeilewski, F.M., Müller, A. and Bruns, E. (2004) Climate changes and trends in phenology of fruit trees and field crops in Germany. *Agriculture and Forest Biology* 121, 69–78.

Eccel, E., Rea, R., Caffarra, A. and Crisci, A.

(2009). Risk of spring frost to apple production under future climate scenarios: the role of phenological acclimation. *International Journal of Biometeorology* 53, 273–286.

Fitter, A.H. and Fitter, R.S.R. (2002) Rapid changes in flowering time in British plants. *Science* 296, 1689–1691.

Fujisawa, M., and Kobayashi, K. (2010) Apple (*Malus pumila* var. *domestica*) phenology is advancing due to rising air temperature in northern Japan. *Global Climate Biology* 16, 2651–2660.

George, M.F., Burke, M.J., Pellet, H.M. and Johnson, A.G. (1974) Low temperature exotherms and woody plant distribution. *Hortscience* 9, 519–522.

Kaukoranta, T., Tahvonen, R. and Ylämäki, A. (2010) Climatic potential and risks for apple growing by 2040. *Agricultural and Food Science* 19, 144–159.

Lu, P., Yu, Q., Liu, and Lee, X. (2006) Advance in tree-flowering dates in response to urban climate change. *Agricultural and Forest Meteorology* 138, 120–131.

McKenny, D.W., Hutchinson, M.F., Kesteven, J.L. and Venier, L.A. (2001) Canada's plant hardiness zones revisited using modern climate interpolation techniques. *Canadian Journal of Plant Science* 81, 129–143.

National Arboretum (1965) *Plant hardiness zone map.* US Dept. Agriculture Miscellaneous Publication 814.

Pachauri, R.K. and Reisinger, A. (eds) (2007) *Climate Change 2007: Synthesis Report.* Contribution of Working Groups I, II and III to the Fourth Assessment Report of the Intergovernmental Panel on Climate Change. IPCC, Geneva, Switzerland, 104 pp.

Pierquet, P., Stushnoff, C. and Burke, M.J. (1977) Low temperature exotherms in stem and bud tissues. *Journal of the American Society for Horticultural Science* 97, 608–613.

Proebsting, E.L. Jr and Mills, H.H. (1978) Low temperature resistance of developing flower buds of six deciduous fruit species. *Journal of American Society Horticulture Science* 103, 192–198.

Quamme, H.A. (1976) Relationship of the low temperature exotherm to apple and pear production in North America. *Canadian Journal of Plant Science* 56, 493–500.

Quamme, H.A. (1983) Relationship of air temperature to water content and supercooling of overwintering peach flower buds. *Journal of the American Society of Horticultural Science* 108, 697–701.

Quamme, H.A. (1989) Deep supercooling in buds of woody plants. In: Lee, R.E. Jr, Warren, G.J. and Gusta, L.V. (eds) *Biological Ice Nucleation and Its Applications.* APS Press St. Paul, Minnesota, 370 pp.

Quamme, H.A. (1990) Cold hardiness of apple rootstocks. *Compact Fruit Tree* 2, 11–12.

Quamme, H.A. and Hampson, C.R. (2004) Winter hardiness measurements on 15 new apple cultivars. *Journal of American Pomological Society* 58, 98–107.

Quamme, H.A., Layne, R.E.C. and Ronald, W.G. (1982) Relationship of supercooling to cold hardiness and the northern distribution of several cultivated and native *Prunus* species and hybrids. *Canadian Journal of Plant Science* 62, 137–148.

Quamme, H.A., Cannon, A.J., Neilsen, D., Caprio, J.M. and Taylor, W.G. (2010) The occurrence of winter freeze events in six fruit crops grown in the Okanagan Valley and the potential impact of climate change. *Canadian Journal of Plant Science* 90, 85–92.

Rajashekar, C. and Burke, M.J. (1978) The occurrence of deep undercooling in the genera, *Pyrus, Prunus and Rosa*: A preliminary report. In: Li, P.H. and Sakai, A. (eds) *Plant Cold Hardiness and Freezing Stress.* Academic Press, New York, pp. 213–225.

Rochette, P., Bélanger, G., Castonguay, Y., Bootsma, A. and Mongrain, D. (2004) Climate change and winter damage to fruit trees in eastern Canada. *Canadian Journal of Plant Science* 84, 1113–1125.

Sakai, A. (1965) Survival at super-low temperatures. III. Relation between effective prefreezing temperatures and degree of frost hardiness. *Plant Physiology* 41, 882–887.

Sakai, A. and Larcher, W. (1987) *Frost Survival of Plants.* Springer, Berlin, 321 pp.

Taylor, B. and Barton, M. (2004) Climate. In: Cohen, S., Neilsen, D. and Welbourne, R. (eds) *Expanding the Dialogue in the Okanagan Basin, British Columbia, Final Report.* Environment Canada and the University of British Columbia, Canada, pp. 25–44.

Westwood, M.N. (1991) *Temperate-Zone Pomology: Physiology and Culture.* Timber Press, Portland, Oregon, 523 pp.

Wikipedia (2011) Hardiness zone. http://en.wikipedia.org, accessed 24 June 2011.

Wildung, D.K., Weiser, C.J. and Pellett, H.M. (1973) Cold hardiness of Malling clonal rootstocks under different conditions of winter soil cover. *Canadian Journal of Plant Science* 53, 323–329.

Wisniewski, M., Bassett, C. and Gusta, L.V. (2003) An overview of cold hardiness in woody plants: seeing the forest through the trees. *HortScience* 38, 952–959.

Wolf, D.W., Shwartz, M.D., Lakso, A.N., Yuka, O., Pool, R.M. and Shaulis, N.J. (2005) Climate change and shifts in spring phenology of three horticultural woody perennials in northeastern USA. *International Journal of Biometeorology* 49, 303–309.

Zhang, X., Vincent, L., Hogg, W.D. and Niitsoo, A. (2000) Temperature and precipitation trends in Canada during the 20th century. *Atmosphere-Ocean* 38, 395–429.

16 Conclusion: Temperature Adaptation Across Organisms

Karen K. Tanino

'The essence of nature is a true harmony of its parts. It is a balance between action and reaction at all levels of organization. Perhaps no place is this balance more evident than in the interaction of an organism with its environment' (Tanino, 1990). Nature at its most fundamental level shows mechanistic similarities between organisms, but we do not often see a book which examines these facets and brings them together under one theme. This volume took a comprehensive, multi-organismal approach and addressed the challenges and impacts of climate change on temperature adaptation from micro-organisms through invertebrates to plants to mammals in terrestrial and marine environments. Ken Storey and I had asked the authors to provide reviews which were both factual as well as futuristic. The result is an exciting seminal volume and we thank the contributors very much for their thoughtful and well-written chapters. Collectively, we hope this book not only adds an important knowledge piece, but will also provide an applied function to, for example, help breed better-adapted plants. My own personal note of thanks to Gord Gray (whose office is right next door to mine) for suggesting the modification which led to our current book title, and for our frequent end-of-the-day discussions, one of which produced the brilliant idea of asking Ken Storey to co-edit this book – and for that I will always be grateful.

Temperature is a primary driver inducing change and thus all organisms have acquired some form of thermal sensor. Mechanisms can be found at all biological levels, but ultimately involve the natural laws of physics. So, are there common temperature-sensing mechanisms which transcend kingdoms? In their important lead chapter, Steve Penfield and colleagues (Chapter 2) have taken:

> a broad view of temperature sensing across organisms [to] discover how evolution has adopted the properties of different molecules to construct biological thermometers. A central theme is that organisms have evolved to exploit the biophysical properties of many molecules to create temperature sensors, and have coupled these to common signal transduction pathways to enable them to elicit responses.

They examine thermo-transient receptor potential (thermoTRP) Ca^{2+} influx channels that are directly activated by temperature, in the absence of other signals (Patapoutian *et al.*, 2003). Penfield *et al.* also report on biophysical responses of membranes, DNA topology, RNA and proteins to temperature. Finally, they discuss chromatin as a thermal sensor in plants and explore heat shock factors as a unifying theme across taxa.

Global carbon fixation is reliant upon 'light-dependent primary production'. The impact of temperature on the photosynthetic ability of psychrophiles ('photopsychrophiles') living in permanently cold ecosystems ($\leq 0°C$) is comprehensively examined in Chapter 3 by Rachel Morgan-Kiss and Jenna Dolhi. Distributional shifts under climate change will in turn depend upon the ability of species to adapt photosynthetically. They report on the photochemical structure/functional changes necessary to maintain photostasis and avoid photooxidative stress, and the molecular adaptations of the carbon fixation enzymes. Their discussion of temperature-modified circadian clock and photosynthesis is both 'timely' and a key

© CAB International 2012. *Temperature Adaptation in a Changing Climate: Nature at Risk* (eds K.B. Storey and K.K. Tanino)

area which has long been under-researched. Morgan-Kiss and Dolhi also indicate that one of the main challenges in photopsychrophiles is to maintain functional membrane fluidity under low temperatures, and that plants have evolved various mechanisms through fatty acid metabolic pathways to deal with this, including increasing specific poly-unsaturated fatty acids (PUFA – possessing more than 1 carbon double bond).

Mammals, unlike plants, are unable to synthesize PUFA. In Chapter 8, Craig Frank presents the results of a 7-year study on the torpor patterns of free-ranging Eastern chipmunks (*Tamias striatus*). He indicates that mammals incorporate PUFA into their cellular storage and membrane lipids (Gunstone, 1996) through the consumption of plants (seeds, nuts). Frank states that:

> The predicted effects of climate warming on the torpor of heterothermic mammals and birds have received little previous consideration. One major mechanism by which mammalian torpor may be affected by climate warming is by influencing the concentration of polyunsaturated fatty acids found in their natural diets.

Temperature is generally well known to have a significant impact on the lipid constituents of plant material during the autumn cold-acclimation period. Hilditch and Williams (1964) have been credited as the first to document the elevation of unsaturation in plant lipids upon exposure to low non-freezing temperatures. This area of plant research was rapidly expanded based on the reports by the Australian group (Lyons, 1972; Raison, 1973) that chilling-tolerant plant species had a greater proportion of linolenic acid in polar lipids than chilling-sensitive species, and that this reduced the injurious membrane phase transition temperature of chilling-tolerant plants. Interest in membranes accelerated following Lovelock's (1955) initial suggestion that the ratio of phospholipids to cholesterols was important in maintaining the integrity of human red blood cell membranes during a thermal shock, in combination with the later ability to extract the plasma membrane (Widell *et al.*, 1982; Uemura and Yoshida, 1983) which was considered the site of freezing injury (Singh *et al.*, 1975; Steponkus, 1984). Craig Frank's long-term study supports the hypothesis of a reduced tendency of free-ranging facultative hibernators to enter torpor as the climate warms in temperate areas. Less clear is whether a reduced PUFA content of the diet is one cause of this response. Nevertheless, it is intriguing to speculate on the potential effect of temperature on plant-based food sources and its subsequent influence on the hibernation/torpor of animals. If correct, then the impact of climate change will be broader than anticipated and will necessarily need to expand beyond the current specific organism-based research to adopt a more integrated approach between organisms across kingdoms.

Steve Chown (Chapter 4) reminds us of the multiple stresses associated with temperature change including moisture stress and oxygen deprivation – all of which have a profound influence on terrestrial and aquatic ectothermal insect populations. He takes an in-depth approach to examine geographic and spatial thermal profiles at the micro- and macro-scales as a driver to the subsequent physiologic adaptations, plasticity and behaviour. Temperature and its influence on insect populations and biodiversity can have profound implications:

> 'Because of their significance as agricultural pests, pollinators, biological control agents, and disease vectors, population-level models of insect responses to the environment, and to ongoing climate change, will continue to form a major component of entomological research (see, for example, Gray, 2010).'

Interestingly, the ability to acclimate thermally appears to be more restricted, with less variation for upper thermal limits than at the lower temperatures. Furthermore,

> 'if periodic warm periods become more common in areas previously characterized by low temperatures throughout winter, such as in far north temperate and Arctic areas, then freezing-tolerant species may be more affected than freeze intolerant ones.'

He also discusses the potential of geoengineering and reminds us that investigating history may tell us a lot about the future. It is certainly a humbling reality that insects have been here at least since the Devonian period and their current 29 Orders have adapted and survived significantly warmer and cooler periods than those predicted in the next 90 years under climate change models.

The Storeys (Ken and Jan) reinforce the message of the many indirect consequences on ectothermic vertebrates including food, shelter and competition, in addition to the direct effects of high and low temperatures. In Chapter 7 they provide an excellent overview of the biochemical and molecular adaptations in ectothermic vertebrates to cope with environmental stress. They address potential stress markers and mechanisms: low molecular-weight metabolites, chaperone proteins, increased antioxidant defence mechanisms, gene and protein expression alterations, differential regulation of the signal transduction pathway, and epigenetic and chromatin remodelling. Importantly, they also indicate potential areas of risk, for example:

> All organisms have a variety of biochemical strategies that they can draw on to accomplish adaptation to new environmental realities, although some species (such as those that are used to highly stable or highly specialized environments) are inherently poorer at this than others. Cold-hardy reptiles and amphibians living at high latitudes already deal with wide variation in seasonal parameters and, overall, may be better able to implement biochemical adaptations to deal with a changing climate than comparable tropical species.

The aquatic ecosystems are thoroughly addressed in the chapters by Tommi Linnansaari and Rick Cunjak (Chapter 6, freshwater fish), Doris Abele (Chapter 5, marine fish and invertebrates) and Michael Castellini (Chapter 9, marine mammals). Shifts in the 'thermal isolines' of freshwater fish are largely dictated by the summer versus winter water temperatures. Stenothermic coldwater species are anticipated to decrease their distribution with warm summer water temperatures exceeding their thermal and hypoxic tolerance limits. Conversely, the thermal isoline of eurythermic cool and warm freshwater fish species is anticipated to expand with warming winter water temperatures. Linnansaari and Cunjak also note that:

> The future fish distribution will also be largely dependent on altered species interactions affecting competition, food availability, predation and parasitism. ... A shorter ice-covered period translates into a longer growth period and shorter winter stratification in many lakes. In fluvial habitats, reduction in ice may lead to changes in predation pressure and increased subsurface ice production that may affect some fish species.

Hypoxia will also be a major stress in marine environments, as it is in freshwaters. Stabilizing the hypoxia-inducing factor (HIF) and thus associated target genes will be an important factor in increasing tissue oxygenation. In the marine fish environment, Doris Abele additionally discusses the impact of the widening thermal window of northern latitudes and narrowing thermal window of tropical regions. The biochemical and molecular consequences are detailed:

> Broader thermal windows will raise costs of stress hardening in cold temperate and polar species, including an increased expression of stress proteins such as antioxidants, HSP, and cellular autophagic and apoptotic mechanisms in marine invertebrates and fish. Costs of cellular and mitochondrial turnover increase as ROS formation exacerbates during warming, reducing the energy availability for growth. For some slow-growing species, a reduction of maximum body size and also of maximum attainable lifespan can already be observed in stressful environments.

She concludes that mechanisms leading to better resistance to broadening thermal windows involve up-regulation of cell maintenance/turnover combined with a down-regulation of growth and reproduction.

As Michael Castellini says, 'The core concepts of the impact of potential climate change on marine mammals are simple in theory, but complex in nature.' Since mammals are endotherms and already have the biochemical and physiological adaptations to buffer environmental stresses, other factors become more dominant – 'bottom up ecosystem changes'. These include shifts in distribution of prey, behavioural changes related to the ability to switch food preference if their original diet is no longer available (ability to alter migration routes), and changes in ice distribution impacting breeding success (moulting, resting areas, predator avoidance and raising young).

Plants are ectotherms but unlike their counterparts above, they lead a stationary existence. Thus, their inherent adaptation and acclimation systems to resist biotic and abiotic stress are central to their survival. The various aspects of plant responses are well covered in the chapters of Denis Gaudet, Anna-Marte Tronsmo and André Laroche (Chapter 10, plant disease); J.E. Olsen and Y.K. Lee (Chapter 11, trees and boreal forests); Marilyn Ball, Daniel Harris-Pascal, J.J.G. Egerton and Thomas Lenné (Chapter 12, native vegetation responses); Klára Kosová and Tom Prášil (Chapter 13, annual field crops); Annick Bertrand (Chapter 14, perennial field crops); and Harvey Quamme and D. Neilsen (Chapter 15, fruit tree crops).

The impact of plant diseases are highest in developing countries where over 1 billion of the world's poorest people live. Pathogens are opportunists and can more quickly adapt/develop pesticide resistance than their host plants can evolve to changing environments. Nevertheless, as Gaudet *et al.* indicate, it is hoped that a combined management approach interfacing breeding, molecular tools, and pest control methods (cultural, chemical and biological) will allow the negative impact of diseases on crops to be mitigated.

Forests are a significant factor in carbon sequestration, representing about 75% of terrestrial biomass. In Chapter 12, Marilyn Ball *et al.* outline that:

> Treelines are shifting poleward and to higher elevations as woody species encroach into tundra and alpine vegetation (Kullman 2002; Hallinger *et al.*, 2010). Alpine species are also on the move, as recent studies have reported shifts in their distributions along elevational gradients in apparent response to warming temperatures (Cannone *et al.*, 2007). However, different species do not face the same risks due to climate change, because they differ in both responses to climatic variables and in capacity for dispersal (Morin *et al.*, 2008).

What was surprising was the unexpected negative consequence of CO_2 on freezing resistance and tolerance on vegetation. This is consistent with a report by Guak *et al.* (1998).

Olsen and Lee (Chapter 11) reiterate the issue of the rapidity of the environmental change compared to plant adaptation, and the additional effects of: 'seasonal and diurnal temperature patterns, precipitation, air humidity, cloud patterns, light intensity, storms, fires, snow cover, and freeze–thaw cycles (Saxe *et al.*, 2001; IPCC, 2007)'. Nevertheless, 'boreal and temperate trees appear to harbour an inherent flexibility in control of growth cessation and bud set...The [epigenetic] memory of temperature under embryogenesis, as best described in Norway spruce, indicates that the plasticity of critical conditions with respect to overwintering processes is larger than thought earlier.' Research focus on temperature effects on plant response is normally placed on that season or developmental stage of interest. However, Olsen and Lee's chapter infers the importance of previous thermal environments. A recent report by Hänninen and Tanino (2011) reinforces this concept and puts forth an integrated conceptual model: in which 'one part of the annual cycle impacts subsequent parts and that individual seasonal responses should not be examined in isolation without taking into account the history of the plant'.

On a longer time scale, Quamme and Nielsen (Chapter 15) examine synoptic weather patterns spanning 70+ years in order to determine the impact of climate change on fruit tree production. They report that through this time scale Arctic outflows were the most critical to production, with a trend of increasing minimum temperature of the outflow event and increasing (autumn) and decreasing (winter) patterns of incidence. Consequently, those trees with an early acclimation ability (pear, peach and apricot) were proposed to be better able to expand their northward ranges compared to those with late acclimation characteristics (grapes, apple and sweet cherries).

In Chapter 13 Klára Kosová and Tom Prášil outline the molecular genetic responses in annual field crops and remind us of the other environmental cues which combine to induce a response including photoperiod and water availability. They also indicate:

> Changes in temperature that are too large or too rapid would probably cause significant disturbances in the finely tuned regulatory networks between the processes of cold acclimation and vernalization in winter annuals. Coping successfully with these disturbances will surely influence the ability of winter annuals to survive in these new habitats.

As Annick Bertrand outlines in Chapter 14, perennial forage crops and rangelands encompass a tremendous land base in Canada, grown on over 34 million ha (more than 50% of all agricultural land (Statistics Canada, 2009)). Due to the perennial habit of forage, the winter environment is also critical to productivity and survival, in addition to summer conditions. Although responses will vary by region, it is anticipated that the yield and distribution of perennial forage crops will increase in northern regions. However, as in the chapter by Ball *et al.*, there are unexpected consequences:

> This climate change-driven increase in forage quantity could, however, be accompanied by a reduction in forage nutritive value. The warmer autumn and winter conditions predicted are likely to increase the risk of winter damage through suboptimal autumn acclimation, dehardening temperatures in winter, and lack of insulating snow cover.

Climate change discussions have mainly focused on large-scale average global temperature shifts and their resultant impacts on organisms. However, the importance of knowing specific geographic and spatial thermal profiles in both micro- and macro-environments (see Chapter 4 by Steve Chown) in order to understand organismal response is becoming clear. Furthermore, temperature patterns and widening/narrowing thermal windows will undoubtedly be more significant to ecosystem adjustments than mean temperatures alone. The resultant shifts are particularly acute when interacting organisms are involved, such as the increasing disruption in synchrony between egg hatching of the moth *Operophtera brumata* and spring bud burst of the oak (*Quercus robur*) on which it feeds (Visser and Holleman, 2001). Reduced insect pollination of horticultural crops due to changing insect populations and lack of synchrony will have significant and negative consequences on food supply.

Of significance is the realization that low temperature stresses will be even more (not less) critical in a warming climate. Some of these aspects are outlined in Chapter 1. In Chapter 12, Ball *et al.* indicate that increased CO_2 levels may in fact reduce freezing resistance due to elevated leaf temperatures, and reduced snow cover may place root zones at greater mid-winter injury risk. Furthermore, lack of sufficient chilling requirement is already negatively impacting temperate fruit and nut crop productivity in borderline regions. Since one part of the annual dormancy cycle may impact subsequent stages, an integrated view is important to appreciate how climate change may shift responses (Hänninen and Tanino, 2011). The chapters in this volume reveal the multiple drivers *across organisms* from the biophysical to the molecular to the behavioural levels which combine to shift the ecosystem. Framing all of this are the 'bottom up ecosystem changes' of Michael Castellini's chapter, which link us all. Finally, Steve Chown (Chapter 4) also challenges us to avoid complacency:

> Research about the biological impacts of climate change is fast becoming one of the largest areas of environmental science. However, with notable exceptions and most frequently in less formal considerations of climate change and its impacts, a sense of urgency is rarely felt in formal papers.

Rather, what is known and how further responses and outcomes can be forecast are usually discussed dispassionately. This measured approach is misleading as to the gravity of the current climate crisis. Changes in the earth system that have already taken place and that are forecast are truly alarming. Only with profound action will further transformation be slowed.

References

Cannone, N., Sgorbati, S. and Guglielmin, M. (2007) Unexpected impacts of climate change on alpine vegetation. *Frontiers in Ecology and the Environment* 5, 360–364.

Gray, D.R. (2010) Hitchhikers on trade routes: a phenology model estimates the probabilities of gypsy moth introduction and establishment. *Ecological Applications* 20, 2300–2309.

Guak, S., Olsyzk, D.M., Fuchigami, L.H. and Tingey, D.T. (1998) The effects of elevated CO2 and temperature on cold hardiness and spring bud burst and growth in Douglas-fir (*Pseudotsuga menziesii*). *Tree Physiology* 18, 671–679.

Gunstone, F.D. (1996) *Fatty Acid and Lipid Chemistry*. Blackie Academic and Professional, Glasgow, Scotland.

Hallinger, M., Manthey, M. and Wilmking, M. (2010) Establishing a missing link: warm summers and winter snow cover promote shrub expansion into alpine tundra in Scandinavia. *New Phytologist* 186, 890–899.

Hänninen, H. and Tanino, K. (2011) Tree seasonality in a warming climate. *Trends in Plant Science*, doi:10.1016/j.tplants.2011.05.001.

Hilditch, T.P. and Williams, P.N. (1964) *Chemical Constitution of Natural Fats*. Chapman and Hall, London.

IPCC (2007) Climate Change 2007: Synthesis report. www.ipcc.ch/pdf/assessment-report/ar4/syr/ar4_syr.pdf, accessed 1 March 2011.

Kullman, L. (2002) Rapid recent range-margin rise of tree and shrub species in the Swedish Scandes. *Journal of Ecology* 90, 68–77.

Lovelock, J.E. (1955) Haemolysis by thermal shock. *British Journal of Haematology* 1, 117–129.

Lyons, J.M. (1972) Phase transitions and control of cellular metabolism at low temperatures. *Cryobiology* 9, 341–350.

Morin, X., Viner, D. and Chuine, I. (2008) Tree species range shifts at a continental scale: new predictive insights from a process-based model. *Journal of Ecology* 96, 784–794.

Patapoutian, A., Peier, A., Story, G. and Viswanath, V. (2003) ThermoTRP channels and beyond: mechanisms of temperature sensation. *Nature Reviews Neuroscience* 4, 529–539.

Raison, J.K. (1973) Temperature-induced phase changes in membrane lipids and their influence on metabolic regulation. *Symposium of the Society for Experimental Biology* 27, 485–512.

Saxe, H., Cannell, M.G.R. and Johnsen, Ø. (2001) Tree and forest functioning in response to global warming. *New Phytologist* 149, 369–400.

Singh, J., De La Roche, I.A. and Siminovitch, D. (1975) Membrane augmentation in freezing tolerance of plant cells. *Nature* 257, 669–670.

Statistics Canada (2009) Agricultural profile of Canada. www40.statcan.gc.ca/l01/cst01/PRIM11A-eng.htm, accessed 27 September 2010.

Steponkus, P.L. (1984) Role of the plasma membrane in freezing injury and cold acclimation. *Annual Review of Plant Physiology* 35, 543–584.

Tanino, K.K. (1990) Abscisic acid induction of environmental stress tolerance in plant cells. PhD thesis, Oregon State University, Corvallis, Oregon.

Uemura, M. and Yoshida, S. (1983) Isolation and identification of plasma membrane from light-grown winter rye seedlings (*Secale cereale* L. cv Puma). *Plant Physiology* 73, 586–597.

Visser, M.E. and Holleman, L.J.M. (2001) Warmer springs disrupt the synchrony of oak and winter moth phenology. *Proceedings of the Royal Society London B* 268, 289–294.

Widell, S., Lundborg, T. and Larsson, C. (1982) Plasma membranes from oats prepared by partition in an aqueous polymer two-phase system: on the use of light-induced cytochrome b reduction as a marker for the plasma membrane. *Plant Physiology* 70, 1429–1435.

Index